THE VOLCANIC EARTH

HG COGGER/AUSTRALIAN MUSEUM

Mamye Jarrett Memorial Library
East Texas Baptist University
Marshall, Texas

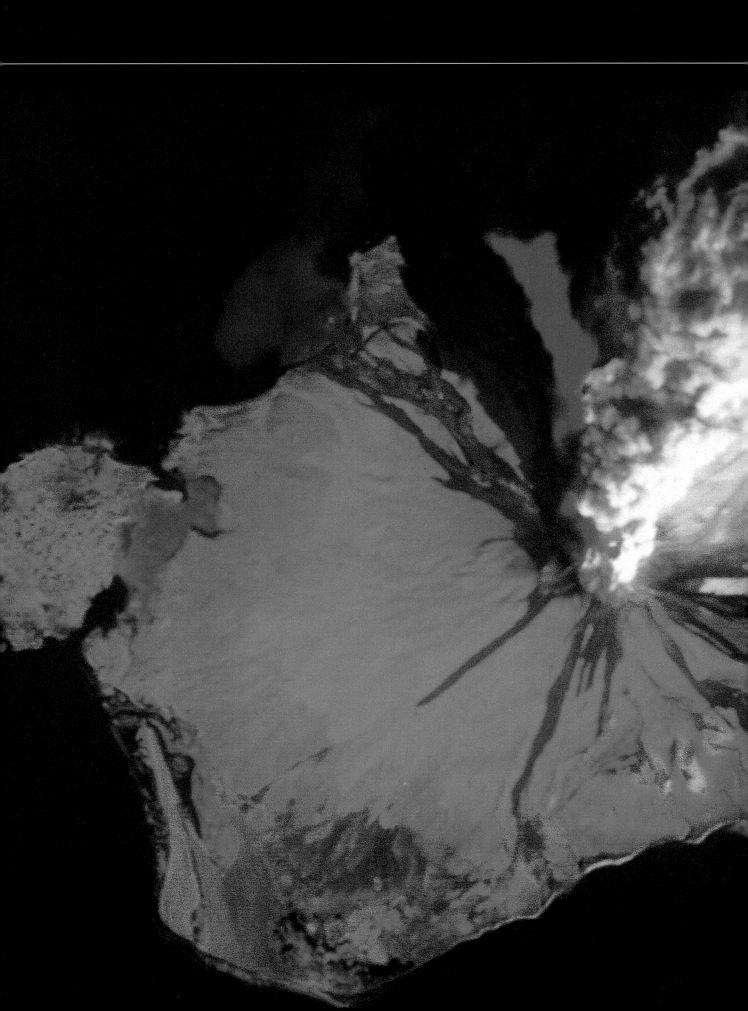

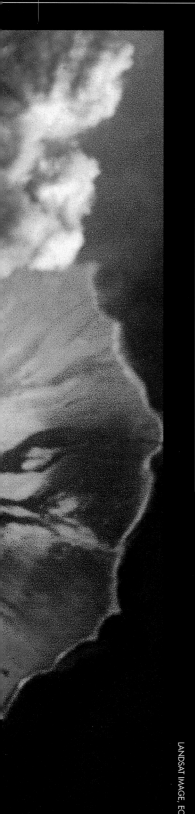

VOLCANOES AND PLATE TECTONICS PAST, PRESENT & FUTURE

LANDSAT IMAGE, EOSAT, USA

THE VOLCANIC EARTH

LIN SUTHERLAND

PRINCIPAL RESEARCH SCIENTIST
DIVISION OF EARTH AND ENVIRONMENTAL SCIENCES
AUSTRALIAN MUSEUM

UNSW PRESS

Published by
UNIVERSITY OF NEW SOUTH WALES PRESS
Sydney 2052 Australia
Telephone (02) 398 8900
Fax (02) 398 3408

© Australian Museum 1995
First printed in 1995

This book is copyright. Apart from any fair dealing for the purpose of private study, research, criticism or review, as permitted under the Copyright Act, no part may be reproduced by any process without written permission from the publisher.

National Library of Australia Cataloguing-in-Publication entry:

Sutherland, F.L. (F. Lin).
   The volcanic earth.

   Bibliography.
   Includes index.
   ISBN 0 86840 071 8
   (UNSW Press)

   1. Volcanoes. 2. Volcanoes – Australia. 3. Plate tectonics. 4. Plate tectonics – Australia. I. Australian Museum. II. Title.

551.21

Designed by Di Quick
Printed by South China Printing, Hong Kong

Photographs, previous pages:
Page i Manam Island, an active basaltic volcano off Papua Niugini, August 1969. The cloud over the island includes steam from the main vent. In 1992 the island's largest recorded explosive and flow eruption took place; the latest lava eruption took place in October 1994.

Page ii Satellite view of Augustine Volcano, Alaska, in eruption 27 March 1986. This EOSAT false colour image shows the surrounding sea as dark blue, and ice on the volcano flanks as lighter blue. A hot pyroclastic flow streaming from the erupting vent shows as red, an earlier cooling pyroclastic flow as pink. Cooled flows show as dark radial streaks. The eruption cloud shows the prevailing wind direction. This volcano formed behind the collision zone where the Pacific plate descends below the North American plate.

CONTENTS

Acknowledgements vi
Preface vii

1
PRELIMINARY ERUPTION
The human side 2
The documentary side 3
Profile of an eruption 4
The geographic side 6
On other worlds 7

2
MAIN ERUPTION: THE VOLCANOES
Explosions and avalanches: blasts from hell 10
Craters: leftover holes 12
Volcanoes: variations on a theme 18
Eventful evolution: volcanic profiles 20
The watery interface: Vulcan versus Neptune 22
The icy interface: a perilous mixture 23

3
OVERALL ERUPTION: VOLCANO DISTRIBUTION
Atmospheric vents: above the tides 26
Submarine vents: below the tides 26
Explaining volcano distribution 29

4
POST-ERUPTION: DATING VOLCANOES
Radio-carbon dating: for the younger volcanoes 32
Thermo-luminescent dating: glows from the past 33
Potassium-argon dating: for the older volcanoes 34
Radioactive kit 35
Other isotope dating: for troublesome volcanoes 36
Fission track dating: scars from the past 38

5
EPI-ERUPTION: VOLCANIC ENVIRONMENTS
Volcano economics: credits and debits 42
Volcanoes and climates: coolhouse, greenhouse and ozone 49
Tracking major eruptions 52
Extinction volcanoes 56

6
FUTURE ERUPTION: VOLCANO WATCH
Monitoring volcanoes: many methods 60
Predicting eruptions and hazards: life saving 63
Vesuvius and Pompeii 66

7
SUB-ERUPTION: PLATE TECTONICS
Profile of a plate: its rocks, motions and melts 70
Mid-ocean ridge volcanoes: spreading floors 72
Subduction volcanoes: plate consumption 78
Rabaul eruptions 80
Within-plate volcanoes: rifts and hotspots 88
Flood basalts: ultimate volcanoes 95
Beyond plate tectonics: alternative views 96

8
VOLCANIC FORMS AROUND AUSTRALIA
Volcano landscapes: Australia and New Zealand 101
Young vents 101
The artist's view 104
The worn vents 113
Tombstones and burial mounds 115
Ancient burial grounds 119
The lava flows 123

9
VOLCANIC DISRUPTIONS AROUND AUSTRALIA
Drainage disruptions 132
Volcanic tombs 137

10
VOLCANIC MINERALS AND ROCKS AROUND AUSTRALIA
The minerals 146
Primary volcanic minerals 148
The rocks 150
The zeolites 152
Passenger minerals and rocks 157
Diamond mines 158

11
DYNAMIC VOLCANOES AROUND AUSTRALIA
The great migration 168
Volcanic boomerangs 169
Anatomy of volcanic boomerangs 170
Tasman Line of Fire 172
Gondwana's last volcanic fling 174
The pseudo-subduction line 177
The last subduction 180
New Zealand's volcanic climax 180

12
FUTURE VOLCANOES AROUND AUSTRALIA
The southern forecasts 184
The northern forecasts 187
Other forecasts 189
Combating future Australian eruptions 190

13
TRANS-TASMAN VOLCANO SPOTTER'S GUIDE
Victoria–South Australia 196
Tasmania 198
New South Wales 200
Lord Howe Island seascape tour 201
Queensland 203
South Island, New Zealand 207
North Island, New Zealand 210

APPENDICES
1
Maps of Australia and New Zealand Volcanic Areas 214
2
Some significant eruptions 224
3
Typical minerals of volcanic rocks 226

BIBLIOGRAPHY 228

GLOSSARY 232

INDEX 238

ACKNOWLEDGEMENTS

Many helped in the eruption of this book, especially Gayle Webb, Division of Earth and Environmental Sciences, Australian Museum, who assisted with many facets. Sue Folwell helped notably in processing the script. Other aid came from Ross Pogson, Nan Maeder, Betty Speechley, Alex Ritchie and Robert Jones. Rick Bolzan and Carl Bento, Photographic Services, provided reproductions. Dr Hal Cogger, Deputy Director, and Jan Barnett, Community Relations, liaised with the Australian Museum Trust and University of New South Wales University Press. Rex Parry, Science Publisher, guided publication.

Further anthropological, botanical, historical and volcanological material came from Dr Lesley Head, Dept Geography, Univ. Wollongong, NSW; Dr Mary White, Balgowlah, NSW; Kelvin Grose, Bowral, NSW; Dr Ivan Mumme, Port Hacking, NSW; Maude McBriar, Medindie, SA; Bernie Joyce, School of Earth Sciences, Univ. Melbourne; Dr Bill Birch, Museum of Victoria, Melbourne; Professors Ian Smith and Kerry Rogers, Dept Geology, Univ. Auckland, NZ.

Reproductions of original artworks came through Shay and Gill Docking, Paddington, NSW; Ranald Anderson, Tower Hill State Game Reserve and Dept Conservation and Natural Resources, Victoria. Colour artwork figures were made by Greg Sommers and Glen Ferguson, Dept Exhibitions, Australian Museum; black and white figures were drawn by Kathy Hollis, Trentham, Victoria; Tom Trnski, Sydney, NSW. Peter Kennewell, Cluff Resources Pacific, Sydney, NSW, supplied a diamond pipe diagram.

Undersea scenes were supplied by Dr Rodey Batiza, Dept Geology & Geophysics, Univ. Hawaii at Manoa, Honolulu; Dr Dan Fornari, Dept Geology & Geophysics, Woods Hole Oceanographic Institution, MA; Dr Rachel Haymon, Dept Geological Sciences, Univ. California–Santa Barbara; Dr Mike Perfit, Dept Geology, Univ. Florida; Dr Don Hussong, SSI Seafloors Surveys Inc., Seattle, WA; National Geographic Magazine, Washington, DC, USA.

Additional photography, artwork and images were provided by Dr Robert Coenraads, Frenchs Forest, NSW; Jane Takahashi, Photo Archives, Hawaiian Volcano Observatory, Hawaii; Dr Peter Rickwood, Dept Applied Geology, Univ. New South Wales; Anne and Tom Atkinson, Ravenshoe, Qld, and James Cook Univ., Townsville, Qld; Dr Jane Barron, St Ives, NSW; Sally Robinson, Ross Pogson, Greg Miller, Howard Hughes, John Fields, Dr Hal Cogger, Dr Alex Ritchie, Erica Hepburn and Judy Thompson, Australian Museum; Jim Frazier, Dural, NSW; David Barnes and Harvey Henley, Dept Mineral Resources, New South Wales; John Milne, Australian Information Services and Philippines Volcanological Survey, Manila; Dr Wally Johnson and Indonesian Volcanological Survey, Bundung; Pat Kelly and Geotrack International, Univ. Melbourne, Victoria, Dr Hugo Corbella, CONICET and Argentine National Meteorological Service, Buenos Aires; Dr Patrick Quilty and Australian Antarctic Division, Kingston, Tasmania; Rick Stevens, Fairfax Press, Sydney, NSW; Argyle Diamond Sales, Perth, WA; Professor John Lovering, Flinders University of South Australia, Adelaide; Rudy Weber, Gem Studies Laboratory, Sydney, NSW; Dr Chris Jenkins, Ocean Sciences Institute, Sydney; Dr Mike Aubrey, Technical & Field Surveys, Sydney, and The Silent Picture Show, Mullumbimby, NSW; The Royal Society of Victoria, Melbourne; Drs Jean-Paul Descoeudres and Derek Harrison, Dept Archaeology, Univ. Sydney, NSW.

Finally, to my family, Anne, Heidi, Lin and Dirk, for putting up with neglect during the preparation of the book.

PREFACE

△ *Pu'u 'O'o volcano in eruption, February 1994. Aerial view looking south. Steam column rising on coast marks lava entering the sea.*

FL SUTHERLAND

THE *Volcanic Earth* describes the life and times of volcanoes and their vital role in Earth's environment. Many volcanoes from around the world illustrate the book, but it concentrates particularly on the Australasian– Southeast Asian– Pacific region. Volcano shapes, eruption styles and erosional forms are followed from birth, through life to dormancy and extinction. The global consequences of major eruptions for climates are emphasised. The creation of volcanoes, from dynamic upwellings in the Earth's interior and the movements of its outer plates, is explained from modern research.

After setting the general volcanic scene, the book examines Australia's and New Zealand's past volcanoes, their present land forms and the possibilities of future eruptions. It also contemplates how people might react to and deal with an Australian eruption. The final section describes places where people can view volcanic features in Australia and New Zealand, and provides maps and details of selected easily accessible areas.

The book is intended for general readers interested in nature and geological sciences, senior secondary school students, beginning undergraduate students, and teachers. It provides a comprehensive survey of Australasia's volcanic landforms and their underlying geology, as well as a glossary and bibliography. The bibliography is organised according to various aspects of volcanoes, with each reference graded into three levels.

In the text, the names of volcanoes and volcanic features are highlighted in capitals the first or principal times they appear.

This book explores the continent's volcanic heritage. It asks whether new volcanoes await us, like the ones that once roared and spat in the very landscape we now domesticate and relax within. Australia is called the timeless land. We will see that its eastern edge is young volcanic land, ready to reawaken.

1

PRELIMINARY ERUPTION

◁ *Lava fountain, 450 m high, Pu'u 'O'o vent, Hawaii, 19 September 1984. Similar eruptions have occurred in the last 20,000 years in Australia.*

VOLCANOES create vivid events. Fountains of molten lava, bursts of gas and showers of rocky debris explode from their vents. Glowing avalanches or streams of lava pour down their slopes. Eruptive clouds may rise to stratospheric heights. Landscapes can be buried, rivers disrupted, lakes overwhelmed and local peoples endangered. Activity may last a few hours or continue over a decade. Some volcanoes erupt once only, others again and again. Vents can lie dormant for long periods between eruptions and eventually become extinct.

Long-active volcanoes, erupting for over a million years, build up huge mounds of solidified lava. The immense volcanoes on Hawaii rise 8,700 m above the sea floor. Lines of volcanoes may link into long ridges. Under the seas these extend for thousands of kilometres across ocean floors. Dense concentrations of volcanoes may coalesce into extensive lava plains and plateaus, as once spread across 1 million km² in Siberia. Such past lava floods apparently occurred in periods that coincide with the abnormal extinction of certain animal and plant species. These staggering statistics show that volcanoes are vital entities of Earth's environment and so justify intensive study.

THE HUMAN SIDE

Mankind has lived with volcanoes since earliest times. Remains of ancestral hominids in Africa are found in sediments between volcanic ash layers 1.3 to 2 million years old in Kenya and 2.9 to 3.4 million years old in Tanzania and Ethiopia. Trails of footprints left in wet volcanic ash uncovered at Laetoli, Tanzania, demonstrate that hominids were walking upright 3.6 million years ago. Prehistoric human footprints preserved in a hardened volcanic mud flow near Managua, Nicaragua, were originally buried under beds of volcanic ash 4 m deep. A village 1,400 years old buried under ash was discovered at Ceren in El Salvador in 1976. Excavations suggest the indigenous inhabitants were warned by an earthquake

△ Visitors at volcano displays, Thomas A. Jagger Museum, Hawaii Volcano National Park, Kilauea, February 1994.

▷ Tree splintered by 1980 blast, Mt St Helens. Inside Active Volcanoes Exhibit, Natural History Museum, Smithsonian Institute, Washington DC, US, July 1989.

◁ Hominids leaving footprints in 3.6 million year old volcanic ash, Laetoli, Tanzania. Diorama, Tracks through Time Gallery, Australian Museum, Sydney.

and fled, leaving their personal possessions behind. Artefacts made from glassy volcanic rocks from the Talasea region of New Britain are recovered from archaeological sites dating back 10,000–20,000 years ago. These sites extend across the Western Pacific and indicate that

obsidian was transported over 8,000 km.

Unusual effects recorded in ancient Chinese chronicles, such as unseasonal cold, clouds of dust, highly coloured twilights and strange images of the sun may mark past eruptions. Some scientists correlate those times with large European and Icelandic eruptions, such as ETNA in 44 BC, HEKLA between 1050 and 1150 BC and SANTORINI ISLAND between 1480 and 1600 BC. The term 'volcano' came from Vulcan, the Roman god of fire and the island of Vulcano off Sicily was his forge. Early descriptions of volcanoes were largely narratives describing their activity. Such writers included Plinius the Younger, who described the investigations and consequent death of his uncle Plinius the Elder during the AD 79 eruption of VESUVIUS. Since then, the study of volcanoes has grown into the exciting but exacting science of volcanology. Volcanologists, the scientists who study the behaviour and interactions of these moody landforms, use a wide range of disciplines and sophisticated technology to investigate their inner workings.

As an elemental force, volcanoes shape the lives of people who live near their realms. Many fearsome sites were worshipped as deities, sometimes to be appeased by human or other sacrifices. Pelé remains the goddess of fire to the Hawaiian people. Sightseers and travellers are attracted to volcanic vistas and their thermal playgrounds, which often support tourist and geothermal industries. The continuing restorations of Pompeii, Herculaneum and Stabiae, the towns overwhelmed by fallout and flows of debris and steaming mud from the AD 79 eruptions of Vesuvius, hold a special fascination for visitors. Besides the remaining buildings, many of the archaeological treasures retrieved from them are preserved in the Antiquarium at the Pompeii excavations and in the National Museum of Naples. Some of the numerous eruptions which have caused significant human casualties, evacuations or property damage in the last few thousand years are listed in Appendix 1.

THE DOCUMENTARY SIDE

The activities of volcanoes have stimulated a huge range of literature describing the more devastating eruptions, analysing archaeological and historical records and detailing many scientific studies. Only a fraction of this reading can be cited here (see Bibliography). With their rocky charms and climatic effects, volcanoes have also inspired the works of many artists and poets. William Ashcroft depicted the atmospheric effects seen in England from the 1883 KRAKATOA eruption in far-off Indonesia. He made 533 colour sketches in the three years following the event. Volcanic designs appear on postage stamps of many countries. These include active

EARTH VOLCANIC 4

CHAPTER 1

PROFILE OF AN ERUPTION

KILAUEA volcano, Hawaii, began erupting 3 January 1983 from its East Rift Zone and now, 12 years later, it is still erupting. This rift zone is often injected by the sideways migration of molten lava from the main magma chamber, which then bursts out in a flank eruption. The new Pu'u 'O'o cone and flows were joined in 1986 by a companion vent, Kupaianaha, which erupted until February 1992. Over 1 million m³ of lava has poured from the vents, much of it flowing in tubes under the lava surface. The flows have cut the coastal highway, destroyed about 200 homes, historic compounds and a visitors' centre, filled in a bay and added to Hawaii's coastline. Light airplanes and camera-equipped helicopters are kept busy giving tourists splendid views of the vent, flows and steam column where the lava exits into the sea. Eruptions like this one took place in western Victoria during the last 25,000 years.

JD GRIGGS/US GEOLOGICAL SURVEY

EARTH

VOLCANIC

CHAPTER 5

Pu'u 'O'o vent eruption, Hawaii, June 1984. Lava fountain (background) and stream of a'a lava forming (foreground).

▽ Flow from Kupaianaha vent near Kamoana campground, Hawaii, December 1989. Formation of pahoehoe lava with top of flow breaking loose and lava skin crinkling up at nose.

△ Steam column rising from lava flowing into sea, discharging from lava tube fed from Pu'u 'O'o vent, south coast, Hawaii, February 1994. Helicopter view.

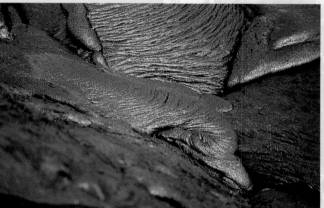

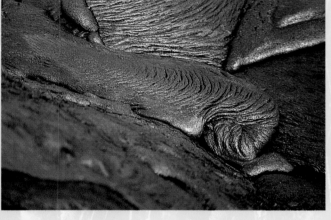

△ Further movement of skin and bunching up of nose. Same flow as shown above.

▷ Clifftop view of lava from Kupaianaha vent flowing into sea, near Kapa'ahu, Hawaii, June 1991.

island chains such as Japan, the Philippines, Indonesia, Tonga and New Zealand and the active Pacific continental margins of northern, central and southern America and the previous USSR. Isolated volcanic islands, such as Iceland, Ascension and Tristan da Cunha in the Atlantic have issued some fine stamps and even the French Antarctic Territories produced a special set depicting volcanic structures in Antarctic islands.

As photography and film-making techniques advanced and aeroplanes, helicopters, satellites and space flights made greater overhead coverage possible, the visual records of volcanoes improved. Notable luminaries were the intrepid volcanologist Haroun Tazieff, and Maurice and Katia Krafft, two inspired recorders who met an untimely death while filming the 1991 MT UNZEN eruption in Japan (see photograph, p.8). The dramatic eruption of MT ST HELENS in Washington State, USA, on 18 May 1980 was captured by many cameras, and in some cases these were the last scenes taken by photographers overwhelmed in the fast-moving blast. Movie films for mass audiences from novels based on historic eruptions appear from time to time. The film *Journey to the Centre of the Earth* based on the pioneer science fiction writer Jules Verne's book, featured an Icelandic volcano and ABC–Cinerama produced the epic film *Krakatoa, East of Java* in 1969.

The public appeal of volcanoes is evident in many natural history museum exhibits. Displays extend to elaborate reconstructions, such as a walk-in volcano in the Natural History Museum of New Mexico in Albuquerque, USA. A special exhibition 'Inside Active Volcanoes' at the Smithsonian Institute Natural History Museum in Washington, DC, in 1989 attracted 515,000 visitors in 3 months before touring the USA. One of its features was a communications link to real tremors being recorded below the Hawaiian volcanoes. A modern museum for volcanophiles is the Museum of Volcanoes built on the active IZU-OSHIMA VOLCANO, off the Japanese coast 120 km SSW of Tokyo. This museum was built after the island's inhabitants returned from evacuation during the 1986 eruption. The displays exhibit the growth of the local volcanoes, as well as volcanoes in other lands. They are illustrated with spectacular pictures, specimens, models and interactive techniques. A well-appointed theatre screens a film showing the human impact of the 1986 eruptions. In the future, developments in new computer technology that creates scenes of virtual reality will project people into volcanoes without any physical danger or hazardous effort.

THE GEOGRAPHIC SIDE

Some parts of Earth are liberally endowed with active volcanoes, other regions appear devoid of volcanic breath. This applies both to land and sea. Many ocean islands, such as Hawaii and Iceland and island arcs in Indonesia, the Caribbean and the Aleutians support the most active volcanoes. Sea floors are also quite volcanic, but this submerged domain has little impact on human awareness. Excluding undersea volcanoes, some 1,300 volcanoes have erupted in the last 10,000 years. Understanding the origin and life cycles of these volcanoes requires a close scrutiny of their structures and underlying chambers. Their birth is linked to heat flow and the movement of rocks deep in the Earth's interior. To solve their secrets completely requires intensive study not only of their active vents, but also of their past activity. Volcanic rocks in exposures and drill holes remote from any volcanism reveal the enormous role volcanoes play in Earth's evolution. An ultimate perspective of Earth's volcanoes requires their comparison with extraterrestrial volcanoes on other planets and moons.

All continents, except Australia, possess active volcanoes (see map, inside front cover). In some continental slivers, such as New Zealand and Papua Niugini, volcanism is prominent. However, many regions now destitute of activity reveal, under closer scrutiny, a strongly volcanic past. Australia is such a place. Although no obvious volcanism has ruffled the scene since European exploration and settlement after the 17th century, there are Australian Aboriginal legends about volcanic events and the continent's landscapes contain craters and lava flows scarcely touched by erosion. The more scientists examine Australia's rocky surface, the more Australia's volcanic make-up emerges.

ON OTHER WORLDS

JPL/NASA, USA

△ *The 8 km high Maat Mons volcano on Venus may be active — the dark areas on its slopes could be fresh lava. This is a computerised image from radar and elevation data sent back by the 1990–94 Magellan Mission and though it mapped tens of thousands of volcanoes and numerous long lava channels, no conclusive evidence of present activity has been found.*

OTHER planetary bodies, like Earth, have volcanoes. Our Moon was volcanic long ago. When its interior melted after impacts by huge meteorites, dark basalt lavas flooded into the craters. Moon basalts are 3–4 billion years old, but appear fresh with no atmosphere to weather them. Tiny green beads of glassy basalt found by Apollo 15 invoke volcanic fire fountains like those of Hawaii.

Mercury is cratered like the Moon, but its craters show no lava. Instead, areas between craters suggest basalt lava plains formed early in its existence. However, volcanism soon switched off in smaller rocky bodies such as these.

Venus, in contrast, has extensive volcanic terrains because a great deal of molten material could rise to a very hot surface compared to Earth. SIF MONS and GULA MONS, two large shield volcanoes, extruded longer lava flows than those on Earth. Pancake-like mounds, in the Alpha Regio highlands, may be lava domes or chambers. There is an impact crater called Cleopatra in which lava breached the crater rim.

Mars has massive shield volcanoes. OLYMPUS MONS is the largest in the Solar System. All the Hawaiian volcanoes would fit into its 600 km girth and 25 km height. Its summit crater is 65 km across and 2 km deep. The Martian shields rose over hotspots and unlike Earth, where surface plates move across hotspots to form volcanic chains, Mars has no moving plates to cut off volcanic growth. The Mariner and Viking Missions identified many volcanic features. Large, flat, old volcanoes like ALBA PETERA reveal explosive flows capped by lavas. Small, steep volcanoes, like THARSIS THOLUS, poke through floods of lavas and numerous flows wind across the lowland plains. Flood-lavas cover 10,000 km^2 in the Cerberus Plains and seem to be only a few hundred million years old. The Viking Lander photographed basalt boulders showing bubbly gas cavities.

Some unusual meteorites found on Earth may be Martian rocks, flung out by large impacts. Others may come from basaltic asteroids orbiting between Mars and Jupiter. They represent the oldest lavas in the Solar System at 4.6 billion years.

Io is the most active volcanic body in the Solar System. Caught between giant Jupiter and its other moons, Io is heated by enormous tidal forces. Its average heat flow is 20–30 times that of Earth. The Voyager Mission spotted volcanoes such as LOKI erupting umbrella-like sprays, which fell back in haloes around the vents. Longer lived, cooler jets rise up to 120 km high and form brightly coloured deposits around equatorial vents. Shorter lived, hotter jets reach 300 km high and form dark red deposits around vents in a longitudinal belt. Sulphur dioxide gas and sulphur deposits fuel Io's eruptions and produce yellow and orange surface colours, but other flows resemble silicate lavas.

Icy moons such as those of Jupiter and Saturn are not necessarily volcanically quiescent, as some may erupt water or brines.

CHAPTER 1

CHAPTER

KIYOTO NISHIKAWA

2

MAIN ERUPTION: THE VOLCANOES

VOLCANOES build up over openings in the ground and these vents expel the raw materials for their construction. Great variations in superstructures arise from each volcano's individual activity. Some volcanoes stand as true mountains, dominating their surroundings. MT FUJI rising to 3,777 m in Japan, MT KILIMANJARO topping 5,930 m in eastern Africa, MT ETNA standing 3,308 m above the sea in Italy and MT EGMONT peaking at 2,518 m in New Zealand are well-known examples. Some volcanoes combine sea and land forms. Although over half is submerged, Hawaii still rises to 4,206 m above sea-level. Such giants contrast with the extinct volcanoes scattered across the plains of western Victoria. MT ELEPHANT, the largest cone shown overleaf, is only 250 m high and 1 km across its base. There may be little surface expression remaining for fissures that once opened and poured out their covering lava flows. Surface depressions mark some vents, formed by explosive blasts. HYPIPAMEE CRATER in northern Queensland is an exceptionally deep example in Australia, with narrow walls 135 m deep blasted out of granite.

◁ *Collapse of lava dome, Mt Unzen Volcano, Japan, July 8, 1991. This eruption killed volcanologists Maurice and Katia Krafft.*

EARTH

VOLCANIC

EXPLOSIONS AND AVALANCHES: BLASTS FROM HELL

Avalanches of hot fragmented rocks sweep down the sides of volcanoes in violent explosive eruptions. These include deposits from nuées ardentes. This term for glowing clouds became fashionable after French scientists such as Lacroix studied such eruptions at MT PELÉE, Martinique, after one of them wiped out St Pierre in July 1902. The typical Peléean activity comes from explosions which often disintegrate growing lava domes. The collapse of a lava dome sends a dense, fast-moving mass of hot fragments and gas down valley slopes at speeds of up to 100 km/hour. A higher, accompanying layer of less dense ash may sweep over the hill slopes and such an 'ash cloud hurricane' may have struck St Pierre. An umbrella cloud of fine material also rises up above the moving train and falls as a fine ash deposit.

In some volcanoes, the collapse of a volcano side during eruption causes large debris avalanches. The 1980 MT ST HELENS eruption was an example, but these eruptions were labelled Bezymianny activity

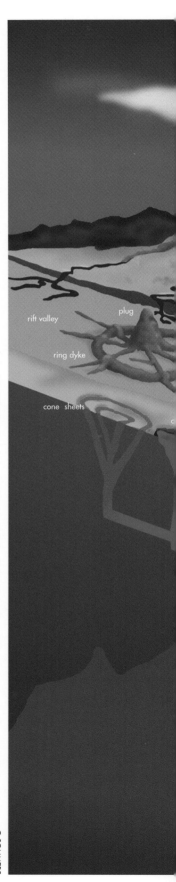

▷ Structures of volcanoes — surface and underground features. This generalised volcanic scene shows active, dormant and extinct volcanoes. The sideways spread of the eruption cloud at the top indicates the prevailing wind direction. The volcanoes lie along a fracture zone, marked by varying fault movements. Typical underground chambers, invasive sheets and feeders to the volcanoes are illustrated in the side sections cutting down through the crust. The way erosion dissects and exposes older volcanoes is depicted in increasing effects away from the active volcanoes. The volcanoes lie in the floor of a large volcanic collapse (caldera), with part of its remaining rim forming the background.

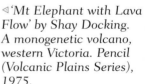

◁ 'Mt Elephant with Lava Flow' by Shay Docking. A monogenetic volcano, western Victoria. Pencil (Volcanic Plains Series), 1975.

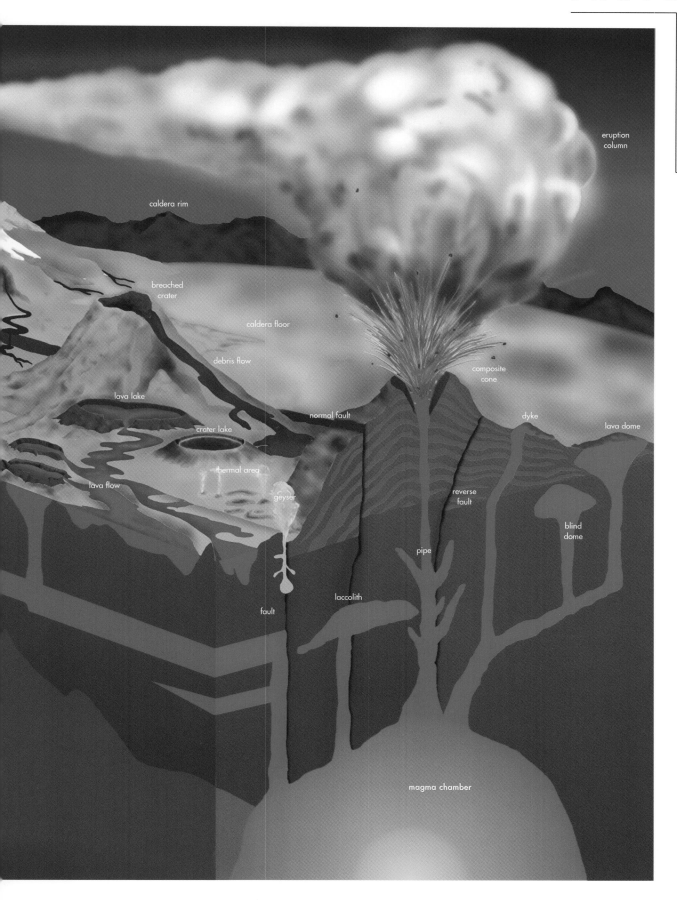

after the 1956 blow-out of the BEZYMIANNY VOL-CANO in Kamchatka. After an initial blast a full-scale eruption of exploding lava can build up, with the rapid ascent of a high eruption cloud. This activity is called Plinian, after its description by Plinius from the AD 79 eruption cloud of MT VESUVIUS. He likened the cloud to a Mediterranean pine tree, with a trunk and crown of branches. Plinian eruptions form two types of pyroclastic deposits: air falls of pumice and ash, and ground-travelling surges. Both types of deposits helped bury Pompeii, but the surges killed most inhabitants.

Ignimbrites form when denser parts of eruption clouds collapse under gravity and transfer their energy of fall into fast-moving ground flows. Their mobility comes from a fluidised mixture of larger fragments, fine particles, hot gas and air. They form layered deposits. A thin underlay, a pyroclastic surge rich in crystal fragments, may appear. The main deposit passes from a base of finer-grained pumice and rock fragments into thicker material of coarser-grained fragments. A layer of fine ash may settle above this. In many ignimbrites pumice fragments become stretched and flattened by the weight of overlying material or flowage. Thick, inner parts which remain hot become fused and form harder, welded tuffs and breccias.

CRATERS: LEFTOVER HOLES

CRATERS excavated by explosive eruptions feature in many volcanoes. Parts of the encircling rim may become destroyed by an eruption, giving a BREACHED CRATER. Some volcanic peaks exhibit both SUMMIT and FLANK CRATERS. Craters held within low embankments are called TUFF RINGS or MAARS when excavated more deeply by multiple explosive eruptions. The term 'maar' comes from the Eiffel volcanic field in Europe, but some of the best examples come from the young Australian lava fields. Craters erupted along a fissure form CRATER ROWS and are often elongate rather than circular structures. A dramatic line of elongate craters formed in the 1886 eruption of MT TARAWERA in New Zealand. They extend for 13 km along a

▷ Different kinds of volcanoes and volcanic eruptions on land and at sea. This busy volcanic scene is not based on any actual landscape, but illustrates a range of volcanoes, their scales and styles of eruptive activity. Volcanoes erupting in water are shown centre left. A large volcanic collapse structure (caldera) with its underground extension and new growing volcano is shown in section, centre front.

EARTH

VOLCANIC

▷ Mt Tarawera fissure crater, New Zealand, February 1986. Looking west towards Ruawahia dome. Note darker layers of basaltic lapilli deposits of 1886 eruption near top overlying white rhyolite ash layers of 700 year old eruptions.

▽ Catastrophic eruption of Mt Vesuvius, 24–25 August AD 79. Pompeii, the city buried by the eruption, lay downwind behind the eruption cloud. Since AD 79 Mt Vesuvius has regrown in the caldera. Reconstruction based on Rittman 1962.

Composite cone before eruption

Plinian eruption and characteristic eruption cloud

Caldera collapse after eruption, leaving remnant forming Mt Somma on the north (left) side.

FL SUTHERLAND

CHAPTER 2

14

▽ *Erosional caldera of Haleakala volcano, Maui, Hawaiian Islands, February 1994. View north from summit near the visitors' centre, across young lavas and cinder cones on the crater floor.*

FL SUTHERLAND

NE–SW fissure, but in detail their alignments are slightly staggered rather than following a continuous fissure. The CRATERS OF THE MOON in the National Monument in Idaho, USA, present another easily visited row of craters. Aligned along the Great Rift Zone, they were used as an astronaut training ground for moon missions.

GIANT CRATERS

Supersized craters appear in many large volcanoes. These incomplete volcanoes show huge gouges made by extremely powerful explosions. Some were watched as they formed and the 1980 Mt St Helens eruption was exceptionally well documented. This dormant 2,950 m high peak showed within minutes a gaping hole on its north face as exploding steam propelled an outward avalanche of pulverised rocks. A huge horseshoe-shaped crater 600 m deep, 2 km wide and 3 km long took 400 m height off the summit. Other volcanoes have erupted so much material that they collapsed into a huge depression, called a

◁ Mt Schank volcano, South Australia, dry craters, March 1974. Aerial view looking north.

▷ Blue Lake, a water-filled crater, Mt Gambier volcano, South Australia, March 1974, looking west.

CALDERA. This Portuguese term for kettle or cauldron was given to the CALDEIRA VOLCANO in the Azore islands. New volcanoes sometimes grow within these calderas. Lava domes successively grow and disintegrate within the Mt St Helens crater. The present Mt Vesuvius grew in the shattered shell of the old volcano, after the catastrophic AD 79 eruption. The old side forms Mt Somma and the growth of a second volcano is termed the Somma stage.

True calderas in older volcanoes require careful distinction because prolonged erosion can produce similar topographic depressions. Australia's largest extinct volcano, the TWEED VOLCANO, has an inner depression surrounding the central peak of MT WARNING. However, the 20 million years since its activity have produced an erosional caldera, not an eruption-formed caldera.

LIQUID FILLS

Craters of active volcanoes may fill with upwelling liquid lava. These LAVA LAKES overflow from time to time, crust over in quiet periods or vanish in later eruptions. A lava lake filled a new crater on MATAVANU VOLCANO in Savai'i Island, Western Samoa, and poured out abundant lava from 1905 to 1911. It was described by the Australian geologist, H.I. Jensen. The active lava lake in the pit crater of NYIRAGONGO VOLCANO in Africa became well-known after Haroun Tazieff led expeditions there from the late 1950s into the 1970s. It suddenly drained out catastrophically in

1977, devastating the lands in its path, killing 70 people and burying herds of elephants. Another famous example is the HALEMAUMAU LAVA LAKE in Hawaii. This formed in the pit crater of KILAUEA VOLCANO from 1823 to 1924, a site overlooked by the Volcano House hotel. An inaccessible lava pool resides in the summit crater of snow-clad MT EREBUS on Ross Island in Antarctica.

FLOODING WATERS

Many volcanic craters eventually fill with water. Sometimes this is a seasonal or irregular event, but crater lakes are extra attractions in volcanic terrains. The two youngest craters in South Australia provide a hydrological contrast. MT SCHANK has empty craters, whereas MT GAMBIER has four water-filled craters: BLUE LAKE, VALLEY LAKE, LEG OF MUTTON LAKE AND BROWNE'S LAKE. Crater lakes in active vents may have heated waters. The summit lake on RUAPEHU VOLCANO in New Zealand is a poignant example. The lake waters breaking out on Christmas Eve 1953 created a runaway avalanche of mud and rocky debris that swept down the Whangaehu River and destroyed the railway bridge. An oncoming express train partially plunged into the river, killing 157 people. TAAL VOLCANO in the Philippines sits in a caldera lake, a potentially dangerous situation. Water entering an erupting vent produces a violent steam explosion, and when this happened at Taal in 1969 about 150 lives were lost.

Craters and calderas lying within volcanic

islands or on coasts may become breached by erosion and flooded. Rabaul Harbour in New Britain was formed when the sea entered the southeastern side of the RABAUL CALDERA, which is 14 km long and 8 km wide. Drowned calderas must be distinguished from older eroded volcanoes. LYTTELTON, AKAROA and DUNEDIN VOLCANOES in the South Island of New Zealand, like Rabaul, include fine harbours. However, with no activity for over 5 to 10 million years, the New Zealand volcanoes are dissected structures.

VOLCANOES: VARIATIONS ON A THEME

From small to large, volcanoes are endlessly varied. Small one-off vents are called MONOGENETIC VOLCANOES. They can include both explosive deposits and lava flows. Some volcanoes are largely built up from explosive deposits, which are termed PYROCLASTICS, signifying 'fire' fragments. Pyroclastic volcanoes form CONES, tuff rings and maars. Maars usually show lower rims and more gentle outer slopes than tuff rings. Other volcanoes are built up from solidified lava and form a LAVA VOLCANO. Some of these feed long LAVA FLOWS, if they are made of fluid lava, and some merely well up into a LAVA DOME, if they are made of viscous lava.

Rows of craters, cones and lava volcanoes aligned along faults and rifts form FISSURE VOLCANOES, of which Iceland has many. The LAKI FISSURE erupted along a 16 km section in 1783, spreading 13 cubic km of lava over an area of 565 km² in 5 months. More than 115 old vents can be seen along the 27 km long Laki fissure. ELDGJÁ FISSURE nearby poured out 9 km³ of lava from a 30 km line in AD 950. These structures may be active over hundreds of years. Most large centrally sited volcanoes contain alternations of pyroclastic deposits and lavas, along with their intruding feeders.

These COMPOSITE VOLCANOES are also termed STRATOVOLCANOES. They have active lives of up to 100,000 years. The larger volcanoes that suffer a major collapse become caldera volcanoes and then become SOMMA VOLCANOES if a new cone grows in the caldera. The active life for caldera and somma volcanoes may extend over a million years. Volcanoes that build up gently dipping radial flanks are called SHIELD VOLCANOES because their shape resembles the front of a circular fighting shield. Two or more large volcanoes growing together form COMPLEX VOLCANOES. Large shield and complex volcanoes may have activity spans of 1–10 million years.

LARGE VIEWS

The largest volcanoes are usually a complex family, grown together from many different vents. Mt Etna, for example, has four summit craters and three zones of active fissures. Hawaii Island has two enormous summit volcanoes, MAUNA LOA and MAUNA KEA; active and historically active flank volcanoes, Kilauea and HUALALAI; an extinct volcano, NINOLE, buried under lavas from Mauna Loa; and an eroded extinct volcano, KOHALA. A submarine baby volcano, LOIHI, is growing on its southern side, 1 km below sea level. These volcanoes have fissure zones which radiate out from

▷ Mt Bromo volcano, one of the new Somma-stage cones rising from the caldera floor of the former Tengger volcano, Java, Indonesia, October 1983. An inactive lapilli cone, vegetated and furrowed by runoff, lies in the centre; the active vent of Mt Bromo is on the horizon at left.

▽ Kilauea pit crater and caldera walls on flank of Mauna Loa shield volcano, snow-capped on far horizon, Hawaii, looking west, February 1974. The distance is deceptive as the summit of Mauna Loa is nearly 3 days' walk away and nearly 3,000 m higher than the basalt-filled pit crater.

the main vent and spread the volcano's growth. Mauna Loa, the largest, has four fissure zones. Holding 40,000 km^3 of rock, it is Earth's most voluminous volcano. By contrast, Mt Elephant in Victoria is a single vent containing about 0.1 km^3 of rock.

A good variety of volcanoes can be viewed on airline flights between Australia and the neighbouring lands of New Zealand, Papua Niugini and Indonesia. Small volcanic craters, cones and shields are seen over Auckland and the large active composite cones of NGAURUHOE and RUAPEHU are sometimes visible on flights across the North Island of New Zealand. Tourist flights from Rotorua allow close views of the composite cratered cone of WHITE ISLAND, the 1886 craters along the TARAWERA FISSURE and surrounding geothermal fields. New Britain flights sometimes pass over the cratered cone of PAGO VOLCANO, centred in the caldera of WITARI VOLCANO, and over the large cone of ULUWAN VOLCANO, locally known as The Father. They provide close views of the drowned caldera and parasitic cratered cones of Rabaul volcano. Flights via Indonesia give views of many active volcanoes, such as the many cratered cones in the caldera of BATUR VOLCANO in Bali.

EVENTFUL EVOLUTION: VOLCANIC PROFILES

Active volcanoes have ephemeral shapes, constantly modified as eruptions destroy or add features. Those built up over a well-established central vent may achieve a classical conical shape. Such volcanoes become revered for their symmetrical beauty, particularly in populous areas. Mt Fuji in Japan and MAYON VOLCANO in the Philippines are famous profiles, although Mayon's slopes are sometimes disfigured by devastating eruptions, such as those in 1993. Lesser known, but strikingly beautiful conical volcanoes include the 3,100 m snow-clad KRONOTZKY VOLCANO in Kamchatka, the 2,857 m SHISHALDIN VOLCANO in the Aleutian Islands, the 5,825 m EL MISTI VOLCANO in Peru and the 6,000 m LICANCÁBUR VOLCANO in Chile. Caldera-forming eruptions drastically alter volcanic profiles, although a new volcano growing within the caldera may rebuild a conelike form.

VOLCANIC CRISIS

A dramatic alteration was seen after the 1815 eruption of TAMBORA VOLCANO on Sumbawa Island in the Indonesian archipelago. This 4,000 m high volcano lost a kilometre in elevation and gained a 6 km wide caldera. The event probably killed 10,000 people directly and over 80,000 people indirectly by consequent famine and plague in Sumbawa Island and Lombok Island. Even more profound alterations are evident in the TOBA CALDERA COMPLEX in Sumatra, Indonesia. Historical records are lacking but this great collapse site is 100 km long by 30 km wide and up to 1.7 km deep. It marks four major collapses since the original volcano was built 1.2 million years ago. The collapses occurred

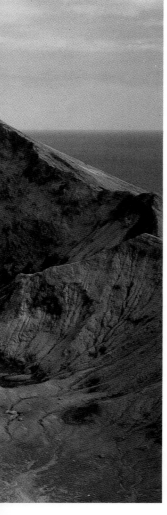

◁ White Island crater erupting, after collapse of the 1978 crater, Bay of Plenty, New Zealand, January 1979. Note breach in crater wall, southeast side.

34,000 to 43,000 years apart, culminating in the stupendous eruption 74,000 years ago when 2,800 km³ of material was ejected. The caldera is still active and lava domes rise from LAKE TOBA which lies within its steep walls.

THE MISSING SIDES

Great gaps show in some volcanoes, where huge avalanches removed enormous volumes of rock. Some avalanches are triggered by injections of molten rock. This creates earthquakes which release exploding gas and steam. A 30° segment of Mt St Helens was lost this way in the 1980 eruption. SOCOMPA VOLCANO in northern Chile lost 70° of its circumference, when an avalanche ten times that of Mt St Helens' tore out the mountainside and underlying basement rocks. Avalanches can also start just by earthquake, groundwater or gravity effects. The 1984 collapse of ONTAKE VOLCANO, the 1888 failure of BANDAI-SAN VOLCANO and 1792 slippage from MT UNZEN VOLCANO are Japanese examples. The 1792 avalanche swamped the adjacent bay, killing 14,500 people in the wave of displaced water.

The great bulk of Mt Etna, resting on 'feet' of clay, has destabilised its foundations. This helped form the VALLE DE BOVE, a 7 km deep valley left by land slippage towards the Ionian Sea. Large volcanic islands such as Hawaii or RÉUNION show major collapses on their flanks, where extensive

▷ Krakatoa caldera, Sunda Strait, Indonesia, on the 100th anniversary of its catastrophic submarine caldera collapse, 27 September 1883. Looking southeast to caldera remnant of Rakata Island from Anak Krakatoa (see text, p.23).

trains of rocks descend in huge submarine slides. These slides are caused by molten injections into the structure which displace and increase stress on the volcano's unstable flanks. Some Hawaiian slides travelled over 200 km, carting 5,000 km³ of rock. RÉUNION VOLCANO shows successive major slides as the volcano grew southeastwards.

AN OLD STAGER

Many old collapses and avalanche deposits mark the north sector of Ruapehu volcano in New Zealand; construction and destruction proceed together. When volcanoes become inactive for long periods, erosion makes increasing inroads into their structure. Only parts of the original surface slope may remain, then eventually only inner structures and finally just the underlying conduits.

THE WATERY INTERFACE: VULCAN VERSUS NEPTUNE

Island and coastal volcanoes face additional destruction, as the relentless sea attacks and undermines their structure. Australian examples include extinct volcanic islands in Torres Strait, and Lady Julia Percy Island off Victoria in Bass Strait. Some volcanoes struggle above water during active periods, only to disappear when activity declines. The COOK submarine volcano in the Solomon Islands nearly surfaced. It was found in 1964 when a British navy ship struck it with towing gear. Submarine explosions were recorded the next year. Historic births include GRAHAM ISLAND, also called Ferdinandea, which surfaced in the Mediterranean in 1831 and FALCON ISLAND, which rose in the Tongan islands between 1867 and 1892. MYOZIN-SYO VOLCANO off Japan, 420 km south of Tokyo, emerged briefly in 1952. It became notorious after a suspected submarine explosion destroyed an investigating scientific research vessel, leaving only pumice-charged wreckage. TULUMAN VOLCANO in the Admiralty Islands, north of Papua Niugini, was born without warning: it rose from the sea between 1953 and 1957 through submarine explosions and lava flows from several cones. CAPELINHOS VOLCANO emerged in an explosive fanfare off Caldeira volcano in the Azores in 1957. Sea flooding into the new craters created extremely violent eruptions. Eventually the waters were shut out after erupted material joined the volcano to the adjacent island. Another violent marine birth was witnessed 33 km off southern Iceland, as SURTSEY ISLAND emerged between 1963 and 1967.

▷ Mt Erebus erupting, 28 January 1841. Sketched by R McCormick from HMS Erebus and Terror during the 1839–43 Antarctic voyage with Captain James Ross.

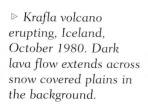
▷ *Krafla volcano erupting, Iceland, October 1980. Dark lava flow extends across snow covered plains in the background.*

PC RICKWOOD/UNSW

SUNDA SEA STORY

Far more violent was the destruction of KRAKATOA ISLAND in Sunda Strait in August 1853. The eruption dismembered three conjoined volcanoes, RAKATA, DANAN and PERBOEWATAN. A giant wave 30 m high was expelled as the sea flooded 23 km² of collapsed land. Only small outlying islands and a third of Rakata were left. A lethal blast of pumice skated over the sea, while the travelling wave killed over 36,000 people. The sea could not quench the volcano. In 1931 explosive submarine eruptions marked the birth of ANAK KRAKATOA, the child of Krakatoa, rising from the watery womb of the drowned caldera (see photograph, p.21).

THE ICY INTERFACE: A PERILOUS MIXTURE

Snow-capped and ice-capped volcanoes are typical in high latitudes and even appear in more tropical regions where peaks reach high elevations. Eruptions under ice cause melting and these volcanoes grow in lakes of meltwater. Such volcanoes are common in Antarctica, Iceland, Alaska, British Columbia in Canada and the high Andes of South America. The last region includes COTOPAXI at 6,039 m and NEVADO OJAS DEL SALADO, the highest active volcano in the world at 6,885 m elevation. Subglacial volcanoes were more common during the last glacial periods, when ice sheets were more extensive.

Subglacial volcanoes either form flat-topped, steep-sided tabletop mountains called TUYAS or sharp ridges called MÖBERGS where lava became chilled before flowing very far. Eruptions of subglacial volcanoes can cause dangerous large floods that sweep all before them in a rush of sediment-laden water. Such tremendous discharges are called JÖKULHLAUPS in Iceland. The KATLA eruption in 1918 discharged twenty times the flow rate and the GRIMSVÖTN eruptions in 1934 and 1938 discharged five times the flow rate of water discharged by the Amazon River.

SOUTHERN ICE

Glaciers covered high parts of southeastern Australia and Tasmania in the last ice ages. However, Australia's active volcanoes were elsewhere, so subglacial volcanoes are missing from the Australian scene. BIG BEN is an active ice-capped volcano on Heard Island, an Australian possession in the Southern Ocean. It displays subglacial effusions on its 2,745 m high summit. Mt Melbourne, at 2,732 m in eastern Antarctica, has densely spaced subglacial volcanoes around its flanks, while Mt Takahe in Marie Byrd Land, western Antarctica, at 3,460 m sits on a subglacial volcanic base. James Ross and Vega Islands off Graham Land, western Antarctica, display classic exposures of eroded overlapping tuya volcanoes. Other Antarctic subglacial volcanoes are found in the Hudson Mountains in Ellsworth Land, at Beethoven Peninsula on Alexander Island, and at Gaussberg on the Wilhelm II coast.

3

OVERALL ERUPTION: VOLCANO DISTRIBUTION

◁ *Pillow lava, summit of seamount 6c, East Pacific floor, about 102° 42' W, 12° 42' N, September 1957. Note elongate and bulbous form and surface striations of pillows.*

VOLCANOES are not random pockmarks, but usually lie in clusters, lines or broad bands, as shown on the map inside the front cover. They are visible in a variety of settings, such as island arcs, uplifted continental margins, continental rift valleys and oceanic islands. A few volcanic lines cross the continent–ocean boundary. The CAMEROON VOLCANIC LINE extends from BIU VOLCANO 800 km inland in west Africa to MT CAMEROON near the coast. It extends for another 700 km into the Atlantic Ocean along the island volcanoes of BIOKO, SAO TOME and PAGALU. The Ross Sea volcanoes along the eastern Antarctica margin extend over 1,200 km to the Balleny Islands in the Southern Ocean.

The oceans hide far more volcanoes than erupt on land. Some underwater vents release pumice on the surface, but most erupt in dark seclusion. Whereas thermal areas on land spout geysers and boiling mud pools, equivalent undersea regions discharge clouds of sulphide minerals from chimneys called 'black smokers'.

ATMOSPHERIC VENTS: ABOVE THE TIDES

A great cavalcade of volcanoes sweeps around the Pacific Ocean margin. Starting from rift valley and oceanic island volcanoes along the Ross Sea, this volcanic swirl meets the continental edge of New Zealand. It then follows the curve of island arcs passing through Kermadec, Tonga, Fiji, Vanuatu, the Solomons, Bougainville, New Britain and New Ireland. From there it engages the uplifted margin of Papua Niugini and then branches northwards through the Halmahera, Sulawesi, Sangihe, Philippines, Mariana, Ryuku, Japan and Kurile island arcs. Entering continental Kamchatka peninsula, it then extends eastwards along the Aleutian island arc into the uplifted continental margin of Alaska. After this, the zone swings down the uplifted continental margins of northern, central and southern America. It returns to Antarctica via the Shetland island arc and continental peninsula of Graham Land. After tracing the west Antarctic continental margin into the rift valley setting of Marie Byrd Land it finally comes back to the Ross Sea.

THE GREAT VOLCANO COUNT

The great PACIFIC VOLCANIC RIM boasts nearly 800 volcanoes with recorded activity in the last 10,000 years. Some 365 volcanoes are presently active or have erupted within historic time. As a volcanic showcase, this zone holds around 60% of the world's 1,300 visible volcanoes. In contrast, oceanic island volcanoes contained within this Pacific perimeter represent only 3% of observed volcanoes. The Indian Ocean, with 11% of visible volcanoes, is dominated by Indonesian island arc volcanoes, which make up 10%. In contrast, the Atlantic Ocean, with 8% volcanoes, is dominated by ocean island volcanoes. These make up 6%, largely due to Iceland's volcanoes. The Caribbean and Scotia island arcs contribute the remaining 2%. Continental Africa contains 8% of the volcanoes, mostly in rift valley settings. The Middle East (4%) and Asia (3%) include both uplifted continental and rift valley volcanoes. Europe, with under 2% of volcanoes, is famous for its classical continental margin volcanoes around the Mediterranean and Aegean Seas. Continental Australia comes last, with only a few volcanoes active in the last 10,000 years.

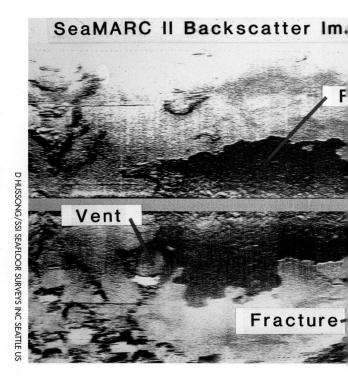

SUBMARINE VENTS: BELOW THE TIDES

The Galapagos islands are shield volcanoes with summits that break the Pacific waters south of Central America. Many similar volcanoes never meet the surface. More and more are found as bevies of scientists explore these ocean floors. They use human or robotic-driven submersible craft as well as surface vessels. Underwater cameras and video systems, sea floor mapping using echo-sounding techniques such as Deep Beam, or towed side-looking echo-sounding equipment such as Gloria, and hot water detectors, all help locate sea floor vents. Dredging and deepsea drilling from stabilised vessels gather samples from the oceanic volcanoes. Volcanic seamounts are also detected by gravity measurements made by space satellites such as SeaSat.

The prevalence of deepsea volcanoes was realised after a 1980 survey revealed over 200 centres between the East Pacific Rise and the Peruvian coast. The largest, 9.5 km across and 1.5 km high, showed a gaping summit caldera. Ocean floors exceed continental and submerged continental blocks in area, so that submarine volcanoes form a substantial, but silent majority of Earth's volcanoes.

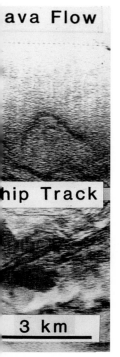

◁ Volcanic vent and lava flow on the sea floor. SeaMarc II back-scatter image from a ship-borne imaging system.

▽ Pumice sea voyage from 1962 South Sandwich Islands undersea eruption, across Southern Ocean to strandings on Australasian shores, 1963–64.

THE CENTRE RIDGE

Sea floor volcanoes usually form in groups, lines or bands. The great majority are the MID-OCEANIC RIDGE VOLCANOES. These form bands hundreds and even thousands of kilometres long, in undersea rift valley settings. They occupy dropped floors of older volcanic rocks within the sidewall crests of the ocean ridges. Strongly active sections form long ridges up to 600 m high and 40 km long, but less active parts merely produce some small volcanic cones. In other parts the median rift valley is poorly developed and volcanically inactive. The mid-ocean ridge volcanoes help create the ocean floors, but do not form uninterrupted lines. Large fractures, called TRANSFORM FAULTS, offset the ridges sideways by a few, tens or even hundreds of kilometres.

An active ocean ridge lies between Australia and Antarctica along the middle of the Southern Ocean (see map, p.29). It starts in the Indian Ocean at a junction of ridge systems, 400 km east of Malagasy Island (Madagascar). Passing between Australasia and Antarctica, it meets another junction of ridges in the eastern Pacific, 600 km off the Chilean coast. Large transform faults break up the line south of Tasmania and New Zealand, but this single volcanic ridge system spans three oceans across half the globe.

Other volcanoes populate the mid-ocean ridges, but outside the rift valleys. They often form in lines that branch away from the ridges, usually over tens of kilometres. The most closely studied ones, lying on the Pacific Rise, are the LAMONT SEAMOUNTS, a 50 km line of five shield volcanoes named SASHA, MIB, MOK, DTD and NEW. Small cones and lava flows surround their flanks and the main volcanoes rise up to 1 km. They have flat summits, calderas and old crater lava lakes. MIB and MOK have large and complex calderas.

◁ This computerised image of the Pacific Ring of Fire is generated from Global Positioning System (GPS) satellite navigation signals. Large earthquake centres (red dots, earthquake magnitudes of over 5 since 1980) clearly show the strong activity along the island arcs and continental margins around the Pacific. The active andesitic volcanism in these earthquake zones forms the Ring of Fire. The yellow lines mark plate boundaries in this global view centred on the Pacific Plate, with the Australian plate at bottom left. Other colours on land and sea indicate approximate elevations relative to sea level: green–yellow–orange–red, increasing elevations 0 to +5,000 m; pale blue–blue–violet, decreasing elevations from 0 to -9,000 m.

THE SIDE FLOORS

Other ocean floor volcanoes unrelated to the ridge systems have irregular distribution. Some are smaller versions of the ocean island shield volcanoes. Central Pacific examples include active seamounts north of Pitcairn Island and the MACDONALD SEAMOUNTS east of the Austral-Cook Islands. Other volcanoes close to the Pacific margin are related to the uplifted continent and island arc volcanoes. THE RUMBLES (I, II, III AND IV) are seafloor volcanoes lying between New Zealand and the Kermadec Islands and are very familiar to the volcano-conscious New Zealanders. The Woodlark Basin, northeast of Papua Niugini, has similar young submarine volcanoes. Australian, Papua Niuginian, Canadian and Russian scientists investigated these volcanoes from a submersible craft called *Mir*. FRANKLIN SEAMOUNT adjoining the mid-ocean ridge and continental margin has special interest here, as hot waters from its flanks are precipitating silver and gold.

UNDERSEA ACTION

Vast numbers of undersea volcanoes erupt unseen, but sometimes their activity triggers seismic shocks. An eruption in progress was seen by chance from the submersible *Alvin* on the East Pacific Rise in April 1991. The eruption made the bottom waters murky, the number of hydrothermal vents discharging hot fluids and mineral particle smoke increased, and lavas destroyed some animal communities established around previous vents. Spectacular shipboard observations of a massive undersea eruption were made on the Juan de Fuca ridge off western America in 1993, after scientists picked up seismic waves coming from the area.

Some submarine volcanoes reveal activity when they liberate large quantities of pumice. This light-coloured, silica-rich lava is highly charged with volcanic gas. The gases in the rock produce a porous or cellular structure that greatly reduces its density. Being lighter than seawater, it bobs up in large masses over the vent site, sometimes forming huge quantities of floating pumice called rafts.

A large pumice raft in the Southern Ocean was intersected by a British naval vessel off the South Sandwich Islands in March 1962. This raft dispersed westwards with the strong West Wind Drift currents and pumice washed up on sub-Antarctic Islands and southern coasts of Australia and New Zealand in late 1963 and early 1964 (map, page 27). The pumice circumnavigated the Southern Ocean, passing South America after three-and-a-half years. Its marathon voyage may have even included southern Australia and New Zealand, as finer pieces of this distinctive pumice washed up from late 1965 until early 1967. Pumice strandings also appear along east Australian coasts. These drift in on warm southerly flowing currents entering the Coral and Tasman Seas. Much of this pumice seems to come from eruptions in the Tongan Islands, with rarer deliveries from Indonesia.

EXPLAINING VOLCANO DISTRIBUTION

The distribution of Earth's volcanoes marks dynamic regions of plate tectonics, the theory discussed in chapter 7. The theory successfully explains volcano distribution. The andesitic clan of volcanoes marks collision zones where plates descend into the mantle, forming island arcs and active continental margins. The basaltic clan is abundant in stretched and ruptured regions within the plates. Large basaltic volcanoes lie over hotspots in the mantle, at active ends of volcanic chains that mark plate movements. Special basalts erupt under oceans in long lines that mark mid-ocean ridges.

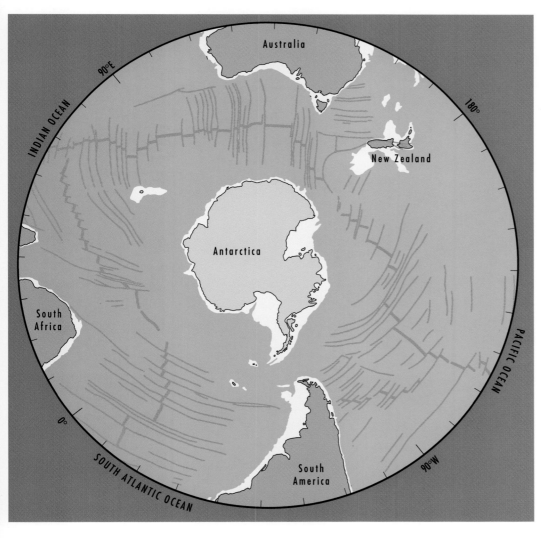

◁ *The great mid-ocean ridge volcanic line separating South Africa, India, Australia and New Zealand from Antarctica and South America. The displacements are fracture lines (transform faults). Southern hemisphere polar projection to 20° S.*

CHAPTER

RE POGSON

4

POST-ERUPTION: DATING VOLCANOES

◁ *Summit of Hanging Rock (Mt Diogenes), Victoria, September 1986. Outcrops of trachyte rock overlook the surrounding countryside. The rock has been dated as 6 million years old.*

VOLCANOES show their true colours when they erupt or declare likely activity with steam and gas emissions from small escape holes called fumaroles. Reports of distant eruptions in remote regions are unreliable, as local effects such as wind vortices in polar regions may simulate eruptive clouds. Many eruptions can be dated from historic records and legends, but only as far back as a few thousand years. Whether a volcano is extinct or merely slumbers is often hard to judge. MT PINATUBO in the Philippines had not erupted for 600 years when it devastated the countryside north of Manila in 1991 (see photograph, next page). Dating a volcano's past eruptions helps gauge its likely future performance. Usually volcanoes with eruptions estimated as younger than the last ice age (10,000 years ago) are best considered dormant.

Several 'tricks' of nature can be used to date past volcanic eruptions. One trick is isotope dating. Isotopes are varieties of the same element. They differ only in the number of uncharged particles, called neutrons, held in their atoms. Otherwise, the atoms have a balanced set of positively charged protons held to negatively charged electrons. Some isotopes contain too many neutrons and their structure is unstable. These break down by discharging rays and particles through radioactivity. The remaining atoms may then form a stable isotope of that element or may break down further, shedding more particles and finishing up as a different, less complex element. The breakdown for each radioactive isotope always takes place at a set rate. By carefully analysing the proportions of isotopes remaining in the host material, the length of time for its radioactive breakdown can be calculated.

RADIO-CARBON DATING: FOR THE YOUNGER VOLCANOES

The radio-carbon clock detects the point in time when living matter was caught in an eruption. It does this by analysing plant materials, usually turned to charcoal, or shells, corals and other skeletons in which carbon resides in calcium carbonate secreted by the host organism.

For carbon, three isotopes exist in nature. Carbon 12 has 12 protons balanced by 12 electrons. Carbon 13 has an extra neutron. Carbon 14 has two extra neutrons, which makes it unstable, and its breakdown rate is such that half the carbon 14 disappears every 5,700 years. The 'trick' with carbon 14 dating is that this isotope is made in the atmosphere as incoming cosmic rays collide with nitrogen atoms. Living bodies breathing air ingest the isotope, but when they die it is no longer renewed and the carbon 14 decays away. Because the amounts of carbon 14 are small, it soon becomes too difficult to measure accurately. For this reason, carbon 14 dating is useful only for materials younger than 50,000 years, or 70,000 years if special measurements are made.

YOUNG AUSTRALIAN DATES

Carbon dating has been used to date Australia's youngest volcanoes. In South Australia plant remains embedded in volcanic deposits at MT GAMBIER are at least 4,300 to 4,600 years old. Charcoal buried under explosive deposits at MT SCHANK is 18,000 years old. This merely gives an older limit to the volcano's activity, since the wood may have died well before the eruption preserved the remains. In western Victoria the carbon dating of sediments in the craters of TOWER HILL VOLCANO suggest that it last erupted about 24,000 years ago. Near MT ECCLES VOLCANO, the carbon dating of mud and peat deposits from a swamp suggests that the HARMAN VALLEY FLOW blocked the drainage about 27,000 years ago. In northern Queensland plant stems cooked under the TOOMBA FLOW as it descended the Burdekin Valley near Charters Towers give carbon dates of 13,000 years. So carbon dating shows that at least three areas in eastern Australia had active volcanoes in the last 30,000 years, a timespan well within that of Aboriginal occupation.

YOUNG NEW ZEALAND DATES

Radio-carbon dating is used for many of the prehistoric North Island eruptions, particularly in the voluminous and complex TAUPO VOLCANIC ZONE. Plant remains found in peats, soil horizons and vegetation buried by volcanics are typically used. At LAKE TAUPO VOLCANO numerous pumice, ash and ignimbrite deposits were dated to an enormous eruptive event 18,000–20,000 years ago and correlated with outlying ash beds up to 200 km away at Auckland. Radio-carbon results also help date old rock slides that avalanche off the large volcanoes. One fixed a collapse on Ruapehu's north flank at over 9,500 years ago.

Dates from MAYOR ISLAND VOLCANO (Tuhua) in the Bay of Plenty suggest major eruptions over the last 50,000 years, with the last of several caldera collapses 6,350 years ago and its last activity under 1,000 years ago. Around Auckland, reliable radio-carbon dates give eruption ages of about 29,100 years at CRATER HILL, 26,630 years at PANMURE HILL, 9,160 years at MOUNT WELLINGTON and 620 years on RANGITOTO ISLAND. This last date suggests the field is dormant rather than extinct. A radio-carbon date of 225 years for a tree supposedly overwhelmed by the lava flow on the island is considered a doubtful claim.

THERMO-LUMINESCENT DATING: GLOWS FROM THE PAST

This method uses the glow of light that some mineral grains show when heated. The glow is due to electrons which bombard the mineral from natural radioactivity and become trapped inside the crystal structure. Sudden heating drives out the loose electrons in a burst. The stronger the glow, the more electrons trapped inside, and this indicates how long the mineral was exposed to radiation.

Quartz is the best mineral for measuring thermo-luminescent glow. When a lava flow overruns sand, silt or rock containing quartz, the sudden heating drives out the trapped electrons. This sets the thermo-luminescent clock back to zero, like a stopwatch. When the quartz sample is collected and heated in the laboratory the strength of the glow indicates how much time has elapsed since the eruption. The thermo-luminescent clock gives a useful check for ages measured by the radio-carbon clock. It is doubly useful because it can measure time back beyond the radio-carbon limits.

GLOWING REPORTS

Thermo-luminescent dating pinpoints the age of Mt Schank volcano more closely than the maximum 18,000 years radio-carbon charcoal date could estimate. Samples tested from an old sand dune baked by a lava flow from the volcano showed that Mt Schank was erupting around 5,000 years ago.

Thermo-luminescence of plagioclase feldspars was used as an alternative to quartz as a dating technique in the Auckland volcanic field, New Zealand. It was used to confirm and extend the radio-carbon results and showed that some eruptions extended back at least 60,000 years, possibly to over 140,000 years.

▽ *Partly collapsed buildings near Manila, Philippines, after ash-fall from the June 1991 Pinatubo eruption.*

△ *Undara crater, McBride basalt field, north Queensland, looking west. The now vegetated vent was the source of the 190,000 year old lava flow dated by the potassium-argon method. The crater is 340 m across; flow extends from crater towards right background.*

POTASSIUM-ARGON DATING: FOR THE OLDER VOLCANOES

The potassium-argon method has dated thousands of volcanic rocks around the world. It uses the element potassium, which is fairly abundant in many minerals and rocks. One isotope, potassium 40, is radioactive. It is found with the common isotope, potassium 39, but in proportions of less than one hundredth of 1%. However, it breaks down very slowly — half of it takes 1,250 million years to decay — so it can still be accurately measured in most rocks. Potassium 40 breaks down and releases the gas argon, as the argon 40 isotope. From the breakdown rate, the age of the host material can be calculated. When molten lava is erupted, the argon gas escapes from its host into the air. As the lava cools and solidifies, new argon 40 accumulates in the freshly formed material and sets off the potassium-argon clock. The amounts of argon 40 are very small and difficult to measure at the beginning of the breakdown. For this reason potassium-argon dating overestimates ages for rocks under tens of thousands of years old. This was noticed, when the potassium-argon dates from the young Auckland lavas did not match the radio-carbon dates.

MT SURPRISE STORY

The potassium-argon clock has unravelled the ages of many Australian volcanic rocks. This is illustrated from a lava field containing several volcanic craters near Mt Surprise in northern Queensland.

RADIOACTIVE KIT

SEVERAL radioactive isotopes, besides carbon 14 and potassium 40 (see previous pages), can be used to date rocks. These isotopes break down into other isotopes of the same or different elements, and their proportions are measured against an unchanging, or stable, isotope of that element. Examples of these isotopes are given below.

Rubidium–strontium isotopes are the most widely used in dating rocks. Samarium-neodymium dating is very useful for altered rocks as these elements resist removal. Samarium 147, however, decays relatively quickly and the method is mainly used for old volcanic rocks. Thorium-lead and uranium-lead dating is used for minerals and rocks containing these elements. Zircon, a widespread, resistant zirconium silicate mineral, can incorporate the large atoms of uranium and thorium into its structure. Although these radioactive elements are present only in trace amounts, the isotopes involved can be measured with great accuracy.

FL SUTHERLAND

△ *Analytical laboratory equipped with mass spectrometer to measure isotopes of argon gas released from fused rock in potassium-argon dating method. Amdel Laboratories, Adelaide, South Australia.*

ISOTOPE	DECAYS TO	RATIO WHICH GIVES THE AGE
RUBIDIUM 87	STRONTIUM 87 (Sr87)	Sr87: Sr86
SAMARIUM 147	NEODYMIUM 143 (Nd143)	Nd143: Nd144
THORIUM 232	LEAD 208 (Pb208)	Pb208: Pb204
URANIUM 235	LEAD 207 (Pb207)	Pb207: Pb204
URANIUM 238	LEAD 206 (Pb206)	Pb206: Pb204

The youngest-looking crater, Kinrara volcano, gave an age of 50,000 to 70,000 years, whereas long lava flows which poured out of UNDARA CRATER gave an age of 190,000 years. An old-looking Mt McBride cone gave 1.7 million years. Overall, the potassium-argon dating showed that this volcanic field had erupted in bursts over the last 2.7 million years. The 50,000–70,000 date for Kinrara volcano is most likely its maximum age, because there are problems with the potassium-argon dating of very young lavas. The fresh form of its crater and flows suggest that it is more likely to be 20,000 years old and that the field is dormant rather than extinct.

AUCKLAND APPROACH STORY

Several young volcanic fields extend from Auckland in New Zealand for 250 km south. Detailed potassium-argon dating in these fields shows an intriguing march of decreasing age, going north from Okete (1.8–2.7 million years), through Ngatutura (1.8–1.5 million years) and south Auckland (1.6–0.5 million years) into Auckland itself (140,000–500 years). From the general pattern of eruptions over 0.3–1 million years in each field, it is clear that Auckland has barely started its volcanic story.

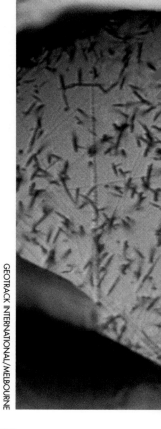

▷ Fission tracks in zircon grain from Mt Dromedary, New South Wales, dating the intrusion at 99 million years old. The zircon is about 0.3 mm long and the fission tracks are up to 0.01 mm long. The polished zircon grain is etched to show the tracks and viewed in transmitted light. The slight scratches are due to the polishing process.

OTHER ISOTOPE DATING: FOR TROUBLESOME VOLCANOES

Potassium-argon dating is unsuitable for rocks with little potassium, or for weathered or glassy rocks from which potassium or argon has escaped. These cases often give ages lower than the true ages. The argon-argon isotope method uses artificial irradiation to overcome such problems. The rock samples are bombarded with neutron particles, which converts the main potassium 39 isotope into argon 39. The proportion of argon 39 to argon 40 in the rock then measures its age. This dating technique was

◁ SHRIMP, an Australian-designed Special High Resolution Ion Micro Probe, Research School of Earth Sciences, Australian National University, Canberra. SHRIMP measures uranium and lead isotopes to date minerals such as zircon. It has dated 4 billion year old Australian zircons and much younger volcanic zircons. Operator P Kinny is beside the sample target, which is struck by a beam of charged particles (ions).

used on rocks dredged from volcanic seamounts off eastern Australia in the Tasman Sea. The argon 39–argon 40 ages were consistently 0.5 to 4.5 million years older than the measured potassium-argon ages. The latter were probably low due to alteration effects in the rocks.

AUSTRALIAN APPLICATIONS

The rubidium-strontium method helped date one of Australia's most celebrated volcanic features, HANGING ROCK (see photograph, p.30). This small peak near Macedon in Victoria formed the backdrop for the award-winning movie *Picnic at Hanging Rock,* made in 1975. The film's story, based on the 1967 book by Joan Lindsay, recounts a mysterious disappearance of schoolgirls climbing the rock. The rubidium-strontium dating shows the rock is 6 million years old and one of the earliest volcanoes in the central highlands of Victoria.

Old zircons give the best lead-uranium isotope results, such as an incredible 4 billion years old for a zircon dated from Western Australia. However, at the young end of the time-scale, it was found that the uranium 238-lead 206 isotope ratios were best for age measurements. By using this method, zircons from BULLENMERRI CRATER in western Victoria were estimated to be about 400,000 years old. The spectacular volcanic peaks around Rockhampton in Queensland are obviously much older volcanoes. Minute zircons extracted from these rocks showed ages of about 75 million years.

NEW ZEALAND APPLICATIONS

The South Island is known for its faulted and uplifted country forming the Southern Alps along the Alpine Fault. The Southern Alps, between Haast River and Lake Wanaka, show many dykes, sills and breccia pipes of a variety of alkaline rocks, which cut through the country schists. These are former injections of molten rock from a complex magma chamber of an eroded volcanic field. Potassium-argon dating of the dykes gives a spread of ages of hundreds of millions of years in those near the Alpine fault, suggesting that the uplifts and alterations caused argon to leak from the rocks. Rubidium-strontium dates on one intrusion and lead-uranium isotope dates on a zircon crystal from another intrusion both agreed with ages between 24–25 million years, considerably narrowing the previous age range.

SHORT-LIVED DATES

New ways of dating rocks by isotopes are continually being devised, especially for rocks less suited to conventional dating methods. These include young lavas from ocean-ridge volcanoes, which are difficult to date for ages under 800,000 years. Use is made of rare short-lived isotopes formed during chain reactions as uranium isotopes break down into the final lead isotopes. This method measures isotopes such as proto-actinium 231 which forms from uranium 235; thorium 230, which forms from uranium 238; and polonium 210, which also forms from uranium 238. When applied to lavas from the Juan de Fuca, Gorda and East Pacific Rise ridges, these method showed consistent rates for mid-ocean eruptions over the last 130,000 years. Some lavas were found to have erupted only months before the samples were collected. The young volcanoes active along the mid-ocean ridge between Australia and Antarctica are now ripe for detailed dating.

FISSION TRACK DATING: SCARS FROM THE PAST

Small scars called fission tracks (see photograph p.36–7) are left in minerals or glassy rocks that contain radioactive elements. Minerals such as zircon, apatite and sphene are found in many rocks and usually have traces of uranium and thorium. Particles shoot out as radioactive atoms break down and they damage the host material as they pass through it. These small tracks are visible under the microscope, especially after the material is etched with acid. The number of tracks measures both the content of radioactive elements and the length of time they have discharged particles. The radioactive breakdown takes place at a constant rate, so by analysing the quantity of radioactive material and counting the number of visible tracks the starting age can be calculated.

One complication is that the tracks become healed and disappear after moderate or strong heating. The tracks really only record how much time has elapsed since the last heating, but this does not always indicate the time of formation, which may be earlier — even hundreds of millions of years earlier. For unaltered volcanic rocks, the last heating is usually their eruption time.

AUSTRALIAN TRACKS

Fission track dating helps greatly in eastern Australia, as zircon and apatite appear in many of the volcanic rocks and old volcanic cores, such as Mt Dromedary. Zircon, in particular, remains a resistant mineral even when the rocks are altered and difficult to date by other means. It survives reheating to 700–800° C before fission tracks heal, but most eruptions will reset the fission tracks. Zircons from Weldborough in Tasmania provide a good example. The crystals wash down from the volcanic rocks capping Blue Tier. Fission track dates give an age of 47 million years, the same eruptive age for the capping lavas measured by potassium-argon dating. However, the uranium-lead isotopes of the zircons show they first formed 200–300 million years ago. They had remained in a deep chamber for 150–250 million years before being carried up in molten lava. All the previous fission tracks were erased.

Apatites helped solve another volcanic story at Ebor-Dorrigo in New South Wales. Here, the New England Tablelands form a spectacular steep and rugged escarpment to the coastal plains. Lava flows on the tablelands form the eroded remains of a large shield volcano. This must have once covered 1,000 km^2, according to the pattern of streams radiating away from the likely centre. As the escarpment retreated over time, only lavas on the western side were left and the east side exposes only underlying basement rocks and an isolated volcanic pipe. The EBOR-DORRIGO LAVAS have been estimated as 19 million years old from potassium-argon age dating. To check whether the pipe was once the volcano's mouthpiece, small apatites were recovered from the rock. Their fission tracks gave the same age as the lavas further inland. This indeed was the hub of the Ebor shield volcano and its radial run-off.

NEW ZEALAND TRACKS

Fission track dating was used around Lake Taupo in the North Island to investigate the volcano's complex explosive deposits. This showed that ignimbrite deposits west and northwest of Lake Taupo were distinctly older than those of the widespread 18,000–20,000 year old Taupo eruption. The fission track dates on zircons and volcanic glass from these older Whakamaru ignimbrites gave ages between 230,000–330,000 years and clearly came from a different vent to the Taupo volcano.

▷ *Basalt flows, Ebor upper falls, New South Wales, August 1985. These lavas are part of the eroded Ebor–Dorrigo volcano, dated by potassium-argon and fission track methods.*

CHAPTER

FL SUTHERLAND

5

EPI-ERUPTION: VOLCANIC ENVIRONMENTS

◁ *Geyser erupting, Whakarewarewa geothermal reserve, New Zealand, February 1986.*

VOLCANOES are a fundamental part of the environment. Without them Earth's surface would be less dynamic, more sterile and probably devoid of human diversity. The early volcanism of Earth provided atmospheric gases and water for rivers and seas. From there life evolved and expanded over time to form the present biosphere. Some scientists consider that volcanic vents on primitive sea floors incubated bacterial life. Descendants of such archaeobacteria still survive in heated waters at temperatures up to 115° C and possibly even over 150° C. Some archaeobacteria dwell in unoxygenated, sulphur-fed acidic environments. A further theory suggests life began from organic molecules coating crystals of pyrite, a typical vent mineral. A volcanic birth for life is only one of many models. However, experiments show that volcanic activity can produce water-soluble phosphate compounds, an essential requirement for evolving life forms.

Volcanoes present many environmental aspects, including human exploitation, climatic effects and hazards for local populations.

VOLCANO ECONOMICS: CREDITS AND DEBITS

Volcanoes offer products that can be farmed, quarried, mined or harnessed for energy. Their tourist value is less tangible and harder to cost. It ranges from the price of fresh lava, stamped with the day of eruption from MT ETNA, to a national tourist income, for who can imagine Japan without its volcanoes?

SOILS

Volcanoes produce soils, although not immediately after eruptions. Then, the land is laid barren with ash and cinders, rivers became choked with debris and mudflows and lavas present intractable rocky carapaces. However, the often fragmented and fine-grained volcanic rocks allow a relatively quick breakdown of these new coverings, particularly under tropical weathering. Studies demonstrate the speed at which vegetation and life return to devastated areas. Over time, volcanic terrains develop deep, fertile soils. The weathered rocks contribute many elements, such as potassium, to promote plant growth in soils.

Highly productive soils in Australia have developed on widespread basalt lava fields. In timbered areas, cleared lands often outline pockets of rich volcanic soil. Fertile red, brown and black basalt soils prevail in northwest Tasmania, the Monaro and New England regions in New South Wales and the Darling Downs, the Bowen Basin and Atherton Tablelands in Queensland. In some areas, as around Kingaroy, Childers and Atherton in Queensland, severe weathering leaves little fresh rock in the soils. Soils washed off parent rocks spread their richness in extended haloes. One drawback of these soils is swelling clay, which absorbs much water and makes the terrain difficult to traverse after rain. A peculiarity of the Monaro volcanic soils is the lack of trees in a region well below the snowline. This is unrelated to land clearing, but may result from a

▷ Thick red soils of a laterite horizon passing down to a pallid clay zone developed on basalt exposed in erosion gully, near Hivesville, Queensland, September 1978.

FL SUTHERLAND

cracking effect on larger plant roots caused by repeated shrinking and swelling in clay minerals.

Fossil soils occur in volcanic layers, but differ from the loose enriched topsoils used for farming. As basalts become leached by groundwater fluctuations, altered zones develop above the parent rock. In this pallid zone most elements are leached away, leaving a residue of clay minerals. A hard cap overlies the pallid zone, where evaporating groundwaters precipitate more insoluble iron and aluminium hydroxide minerals. These caps form laterites or bauxites, depending on whether iron hydroxides or aluminium hydroxides dominate. Laterites are redder and darker than bauxites. When thick and extensively developed, these profiles can be mined for clay, iron or aluminium ore. Such horizons can develop over other rocks, but old soils, laterites and bauxite in volcanic layers usually indicate time lapses between eruptions. In the Monaro volcanic field weathered layers reach 12 m and bauxite tops reach 3 m thick, suggesting periods of 0.5 to 5 million years between local eruptions. These zones are often concentrated around volcanic vents where circulating thermal waters had increased breakdown and precipitation processes.

QUARRIES

The Western Plains of Victoria contain large tracts of young lava little touched by weathering. Aptly named stony rises, they provide local properties with abundant blocks to construct stone walls and fences. More solid outcrops are quarried for building blocks, railway ballast or crushed aggregate for roads and concrete works. In Melbourne, basalt blocks used in buildings and pavements help give the city a solid strength. In Sydney and its environs a trachyte rock quarried from Bowral was sometimes used instead of the more common sandstone. The majestic Queen Victoria Building, opposite the Town Hall, is particularly instructive for viewing this stone. It employs rough-hewn, part-worked and polished trachyte for three quite distinct architectural effects. Some volcanic rocks are quarried for special properties. The rare leucitite lavas in central New South Wales are very alkaline. They are quarried at Flagstaff Hill, because the crushed rock bonds well with limestone.

Naturally fragmented volcanic rocks, such as scoria deposits, need little crushing for use as

EARTH
VOLCANIC
CHAPTER 5

▷ Giant tuff statues, Anakena, Easter Island, Pacific Ocean, July 1993. The heads and trunks are from Rano Raraku crater, capped with topknots of red tuff from a different site.

RR COENRAADS

△ Crushing rare leucitite volcanic rock, Flagstaff Hill quarry, central New South Wales, October 1993. The crushed aggregate is used to bind limestone.

FL SUTHERLAND

▷ Rano Raraku crater lake, Easter Island, Pacific Ocean, July 1993. Carved statues stand in front of quarried tuff, foreground.

RR COENRAADS

aggregate. Scoria cones in the younger Australian volcanic fields become quarry targets, raising issues of their commercial use as against conservation as landscape features. Scoria also provides heating stones for fireplaces and outdoor cooking. Pumice is extensively mined in some places, as in the immense quarries on Lipari Island, off Italy. However, pumice is not an important extractive material in Australia.

Certain volcanic rocks are sought for making carvings or stone implements. The giant statues with elongated heads and trunks made by early inhabitants on Easter Island are well-known. Made of yellow-grey tuff from the extinct crater of Rano Raraku volcano, some partly finished statues still remain in the quarry. A red oxidised tuff found elsewhere on the island was quarried to make the top-knots placed on the statues' heads. Obsidian is a favoured rock because of its glassy nature. It was extensively used and traded for blades and axes in New Britain, for arrowheads by indigenous tribes in North America and for sharp tools in Europe and central America. Some varieties of tuff, such as from Bali, Indonesia, are soft yet compact, and ideal for artistic carvings.

MINES AND ORES

A common deposit around volcanic vents is sulphur. This bright yellow mineral is mined as an important base mineral for the chemical industry,

particularly in the manufacture of sulphuric acid and the synthesis of phosphate fertilisers. Large sulphur deposits are mined around volcanoes in Chile, Japan and Italy. Sulphur mines in the Central Andean volcanoes near 6,000 m elevation are the highest working mines in the world. They hold an estimated 100 million tonnes of sulphur. Extensive sulphur deposits are mined near Mt Etna in Sicily, but are not volcanic as they formed from the bacterial breakdown of sedimentary beds of the sulphate minerals, gypsum and anhydrite. Sulphur mining at WHITE ISLAND VOLCANO off New Zealand suffered a setback when 11 miners were killed in a volcanic avalanche in 1911.

Heated waters circulating around volcanoes can dissolve large amounts of metals from local rocks and then precipitate them in metallic ores. A valuable deposit was discovered in LUISE VOLCANO on Lihir Island off Niugini in 1982. By 1990 exploration proved up to 600 tonnes of gold. The ores formed in the volcano's caldera after it exploded between 15,000 and 35,000 years ago. As such volcanoes become eroded, the deeper ore deposits from their activity become exposed. In Papua Niugini very large deposits of gold-bearing copper ores are mined from altered volcanic cores. The Ok Tedi deposit in the Star Mountains, formed over a million years ago, held

▽ *Section of Broken Hill orebody, New South Wales, showing present land surface and weathering zone.*

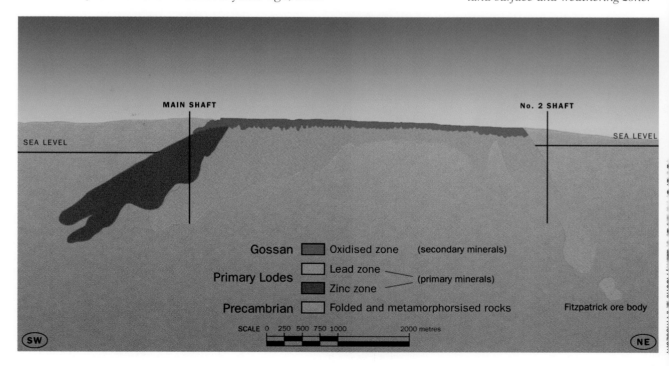

◁ Sulphur Banks, Kilauea, Volcanoes National Park, Hawaii, Febraury 1994.

▷ Argyle diamond mine, Western Australia, August 1986. Open-cut workings: with high grade diamond pipe just below (foreground), and treatment plant (middle distance), looking east.

F.L. SUTHERLAND

3 million tonnes of copper. The Panguna deposit in Bougainville Island, formed 3 million years ago, yielded 4 million tonnes of copper. The Ok Tedi mine, however, encountered environmental problems and Panguna mine closed due to local political fighting.

A huge variety of ore deposits pervade old volcanic structures around the world. Detailed studies of modern volcanoes help to understand the origin of these ores and to search for other deposit sources. These include the sea floor, where metal deposits presently form in the Red Sea, in Okinawa Trough off Japan, in Lau Basin behind Tonga, and in Woodlark Basin between Papua Niugini and the Solomon Islands. A very famous Australian orebody, the Broken Hill silver-lead-zinc deposit in New South Wales, held 250 million tonnes of ore. It probably formed around hydrothermal vents on a sea floor some 1,800 million years ago. The original deposits then became greatly reorganised during later earth movements, which lifted them up and obscured their original nature.

GEMSTONES AND ZEOLITES

Gemstones are attractive aspects of volcanoes and yield dividends to many mining operations. (chapter 10) Diamond, sapphire, ruby, peridot, garnet, feldspar, quartz and a host of rarer gem minerals are delivered in eruptions. Some gems are born within the deep chambers of the volcanoes. Such an example is peridot, the official gemstone of Hawaii, which greets visitors in jewellery and souvenir shops there. Other gemstones are accidental gifts plundered by volcanic activity from the underlying basement rocks. Diamond pipes are much sought by exploration companies, with the De Beers company achieving world dominance and household recognition. Most volcanic gem deposits are mined from older eroded volcanoes. The world's largest diamond mine, in Western Australia, recovered an annual yield of 39,000 carats, valued around $500 million, (mid 1994) from the 1,200 million year old ARGYLE PIPE.

Gemstones are often washed from their eroded sources and become concentrated in surrounding alluvial deposits. Sapphire deposits in central

Queensland and northern New South Wales yield about $20–25 million a year in production earnings (mid 1993–94), from about 65–75 million carats weight of stones. In some volcanic fields secondary gem minerals come from veins and cavities in altered volcanic rocks. The basalt lava fields of Brazil and Uruguay produce a seemingly never-ending stream of amethyst for gem and specimen markets around the world. More locally, Australia produces fine agates from its volcanic rocks, often in the form of 'thunder eggs'.

Last, but not least, are zeolite minerals mined from volcanic deposits (see pp. 152–3). They serve many industrial and environmental purposes. These aluminino-silicate minerals allow loose water, other molecules and metal atoms to move easily in and out of their framework. This prompts a wide variety of uses — as slow-release fertilisers, dietary supplements in animal nutrition, deodorisers, moisture control, desalination and water purifying agents. Zeolites form readily from volcanic glass, so they become widespread alterations in volcanic ash beds and sediments. They also form in geothermal areas, where heated waters alter surrounding rocks.

The Japanese islands are a volcanic zeolite province and one special use is in fertilising golf courses. A more unusual use of zeolites comes from MT EGLON volcano near the Kenya-Uganda border. Kitum cave contains pockets of zeolites in an altered basalt flow. Elephants exploit these zeolites as a nutrient source and their traffic has worn a precipitous path up the mountain cliff. Australia has potential for commercial zeolite deposits in old volcanic horizons in sedimentary basins. A mining venture in a 300 million year old deposit at Werris Creek, New South Wales, already exports zeolites to New Zealand and southeast Asia. An economic sideline of zeolites is their demand as attractive mineral specimens for collectors and shop-window displays. The superb zeolite crystals from the Deccan lavas in India are justly celebrated and grace many museum displays.

GEOTHERMAL WEALTH

Many volcanic areas show visible signs of underlying heated waters. The most spectacular are GEYSERS, named after an Icelandic term, geysir. These volatile discharges erupt when underground water collecting in a chamber heats up and begins to boil. The pressure release causes a rush of boiling water and steam which burst out in jets up to 100 m high. OLD FAITHFUL GEYSER in Yellowstone National Park discharges fairly regularly; other geysers are more erratic. Thermal springs, boiling mud pools, steaming waterfalls and hissing ground are other expressions of underground thermal activity. Drawing on these hot waters for heating and cooking is a great boon to local inhabitants, especially in cold climates like Iceland. Thermal waters are considerable tourist attractions, particularly where bathing in mineralised water is promoted as a health activity.

△ Specimen (about 10 cm across) of the zeolite mineral mesolite, from Deccan basalt lava flows, near Poona, India; displayed in The Planet of Minerals Gallery, Australian Museum.

New Zealand boasts a strong domestic and tourist development in its geothermal areas. According to legend these thermal sites are places where supernatural spirits from Hawaiiki arrived to bring warmth to a freezing Maori explorer. However, their excessive use around Rotorua also heated up the local political climate. Domestic use came into conflict with the world renowned thermal features. In 1982 geologists from the New Zealand Geological Survey warned that only 15 of the original 130 active geysers remained. The famous WHAKAREWAREWA GEOTHERMAL FIELD and its major attraction Pohutu Geyser would die, they predicted, unless geothermal wells in Rotorua were curtailed. The scientific findings came under attack by vested interests. Rotorua Council introduced some controls to improve thermal activity. However, in 1993 the controversy widened into a dispute over geothermal ownership between indigenous Maori people and the government. Australia, in contrast, has only a few thermal pools left in its volcanic regions with only minor tourist development.

△ *Geothermal tourist attraction, Hell's Gate, Rotorua, New Zealand, February 1986.*

The industrial use of geothermal energy is another source of economic return. First tried in 1904 at Lardarello in Tuscany, Italy, it proved a successful long-running development. Other fields were developed or tested in places like Wairakei in New Zealand, Pathé in Mexico, The Geysers in California, USA, Patoa in Hawaii, Sao Miguel Island in the Azores, Kamchatka in Russia, and Meager Mountain in British Columbia, Canada. The plan is to tap groundwaters trapped in permeable layers and heated to 200–300° C at depths of 1–2 km. The released steam then generates electricity. At Lardarello and The Geysers higher temperatures are tapped which generate steam directly, saving its separation from water. However, even the hydrothermal surface waters can be used, as at Kamchatka, where hot waters transform a liquid chemical into gas to run a turbogenerator.

New Zealand obtains 6% of its electricity from geothermal energy, largely from powerstations at Wairakei and Broadlands–Ohaaki. An early bore at Wairakei ran out of control and dissipated its energy. Detailed studies of the geothermal fluids, their mineral and gas contents and their reactions with reservoir rocks, help these plants operate most efficiently. At the Broadlands–Ohaaki plant hot, high-pressure waters were found to be precipitating a scale containing heavy metals such as copper, silver, gold, zinc and lead — gold depositing at over 100 grams a day.

VOLCANOES AND CLIMATES: COOLHOUSE, GREENHOUSE AND OZONE

The world's climates reflect many influences, including planetary mechanics related to the earth's rotation, inclination and orbit relative to the sun. Geographic factors encompass the distribution of continents, oceans and icecaps, and the circulation of warm and cold ocean currents. Geological factors uplift mountain ranges and separate continents by sea ways, but these slow processes have gradual effects. More recently, human activities have contributed pollution by burning forests and discharging industrial gas.

Volcanoes as natural polluters play a climatic role, but unlike present human activity they have contributed since Archaean time. Large volcanic eruptions provoke climatic consequences, particularly when eruptive clouds ascend into the stratosphere, the playground of high altitude winds. Some scientists link huge past volcanic outbursts to giant impacts on earth by extraterrestrial bodies, as both cause and consequence. This would create and reinforce a dramatic climate change, which could then trigger mass extinctions evident in the fossil record. Some scientists think that volcanic outbursts could cause such extinctions without the space invaders, while others attribute space strikes alone to this effect. These hypotheses are much debated and the volcanic side is explored later.

▷ *Eruption cloud rising from Mt Pinatubo, Luzon Island, Philippines, June 1991.*

ERUPTION CLOUDS

Volcanic explosions pollute and darken the surrounding atmosphere. Pulverised rocks, mineral particles, gases and liquid droplets called aerosols can produce severe local climates. These precipitate heavy rains and often spectacular lightning flashes. An awesome crackling discharge of volcanic lightning was filmed during the eruption of SAKUARAGIMA VOLCANO in Japan in November 1987.

Very large eruptions expel materials to stratospheric heights and help to cool the lower atmosphere. The MT PINATUBO eruption in June 1991 was a climate-changer. The eruptive cloud reached nearly 30 km in height and contained 10 billion tonnes of dust and gas. The dust screens out a certain amount of sunlight and initiates slight climatic cooling. Even more potent is the gaseous brew, which in Pinatubo's case included 22 million tonnes of sulphur dioxide, which combined with atmospheric water to form sulphuric acid. These acid droplets formed a shiny haze. This, with dust particles, reflected the sun's rays thereby reducing surface sunlight by an average 2% around the globe. It induced a stratospheric heating, but an inverse surface cooling effect. Many argue over how quickly cooling sets in and not all eruptions discharge much sulphur dioxide. Mt Pinatubo's cloud was exceptionally rich in sulphur, as pieces of pumice around the volcano contain crystals of anhydrite, a hydrated calcium sulphate mineral.

Many climatic variations can arise from eruption clouds. Some effects depend on the size of the eruption relative to other climatic controls and on whether the cloud stayed in the troposphere or reached the stratosphere. Other factors include the volcano's geographic location in northern or southern climatic belts, its eruption time relative to the El Niño or La Niña phases of the Southern Climatic Oscillation, and whether the eruption was an isolated event or one of a series. The relatively clear, warm atmospheres of the 1920s–1930s were untroubled by large eruptions, but since 1982 the large eruptions of EL CHICHÓN in Mexico, GALUNGGUNG in Indonesia, REDOUBT in Alaska, Pinatubo in the Philippines and MT HUDSON on the Chile-Argentina border all took place within a decade.

The El Niño–La Niña oscillation, which can reinforce or dampen the climatic effects of large volcanic eruptions, may itself reflect volcanic influences. This concept calls upon earthquake swarms and large submarine lava eruptions near the East Pacific Rise, which seem to coincide with El Niño. Other models suggest that large land eruptions can induce an El Niño effect due to the imbalance in land and ocean masses between the two hemispheres. Depending on the eruption site, the global climate system reacts in different ways and may induce strong or weak El Niño events. There is much to learn about these complex climatic interactions. Meanwhile, Australians can blame volcanoes for the drought El Niño has induced!

The progress of large volcanic clouds is followed by using laser-ranging light detectors, high-altitude jet flights, balloons and space satellites.

Satellites detect volcanic ash particles from their effects on radiation by using an Advanced Very High Resolution Radiometer (AVHRR). Ash-rich clouds from the 1982 Galunggung eruption were observed to cross Western Australia this way. Sulphuric acid aerosol is detected by a Total Ozone Mapping Spectrometer (TOMS). Ash and sulphur dioxide wash out of the cloud system after a few weeks, but lower aerosol contents can be traced by a satellite that carries a Stratospheric Aerosol and Gas Experiment instrument (SAGE). This method uses sunsets and sunrises to measure how much aerosols reduce the sun's radiation. Different temperatures in eruptive clouds can be detected by infra-red imaging. The clouds lose heat as they spread away from the vent and climb to higher altitudes. The box on the following pages gives examples of the tracking of eruption clouds.

'GREENHOUSE' VOLCANOES

'Greenhouse' gases are now notorious as agents of atmospheric warming. Since volcanoes pump out greenhouse gases regularly, they contribute to atmospheric temperatures. The 'greenhouse' gases were so named because their molecules trap longwave radiation coming off the earth's surface. This helps develop a 'hothouse' effect. Carbon dioxide, methane, nitrous oxides, chlorofluorocarbons, ozone and water vapour are typical greenhouse gases. Atmospheric temperatures are affected by a dual process: sulphur dioxide cools surface air temperatures through the aerosol effect, but greenhouse gases raise ambient temperatures. So the precise gas contents of large eruptions help sway the balance.

MT ETNA, the premier greenhouse 'volcano', delivers well over 25 million tonnes of carbon dioxide to the atmosphere each year. This does not depend on eruptions, as half the amount escapes from the volcano's flanks. In comparison, some 30–65 million tonnes of carbon dioxide escapes from submerged mid-ocean ridge volcanoes. All told, volcanic degassing probably donates 130–175 million tonnes of carbon dioxide a year to the Earth's atmosphere, compared to the 50,000 million tonnes supplied by human activities. Volcanoes previously kept the atmospheric carbon dioxide in balance, but humanity's effect on global temperatures has upset this balance.

'OZONE' VOLCANOES

Volcanic eruptions disturb the ozone layers 10–30 km above the Earth's surface. These layers are vulnerable to large eruptive clouds carrying ozone-destroying chemicals. Ozone, a special three-atom form of oxygen, O_3, is generated when the normal two-atom form of oxygen, O_2, absorbs ultraviolet radiation as it enters the atmosphere. Certain chemicals — erupted by volcanoes and produced by human activity — either react directly with ozone or act as catalysts for other ozone-destroying reactions. This depletes ozone concentrations, which are relatively low anyway (less than 10 parts per million). When the reactive chemicals become concentrated, ozone is destroyed more quickly than it is replenished by the ultraviolet light reaction. This leads to 'holes' in the ozone layer. Recent studies of high-level eruptive clouds from the El Chichón, Pinatubo and Mt Hudson eruptions now give us a better understanding of volcano–ozone relationships.

The ozone reactions are complex and influenced by seasonal, climatic and geographic factors. The most destructive chemicals for ozone are halogen compounds, containing chlorine, fluorine or bromine. Volcanoes erupt halogens mostly as hydrochloric and hydrofluoric acids (about 1,000 tonnes a year) and as certain chlorohydrocarbons such as methylchloride. The 1988 flank eruption of KLYUCHEVSKOY VOLCANO, Kamchatka, erupted high-temperature gases rich in water and halogens, but low in sulphur. However, volcanic halogens wash from the atmosphere quickly, so that human-made

TRACKING MAJOR ERUPTIONS

THESE techniques help to follow volcanic pollution in the global climatic belts and to evaluate the effects of large eruption clouds entering the atmosphere. The progress of four major eruptions since 1980 is described as follows:

1 The 18 May 1980 MT ST HELENS eruption cloud travelled east to cross southeastern Canada. Separate photographs taken by balloons on 7 May and 5 June 1980 showed that aerosols from Mt St Helens had reached the low stratosphere over Europe by June. They had increased the reflectivity of incoming sunlight threefold at a 15 km altitude. The eruption contributed relatively little sulphur dioxide, so that the aerosol content produced little surface cooling.

2 In contrast, the March 1982 EL CHICHÓN eruption spread westwards across the Pacific and circumnavigated the globe in 3 weeks. This 7 million tonnes eruption, rich in sulphur dioxide, was expected to produce a noticeable surface cooling, but did not do so because the eruption coincided with an El Niño climatic phase. El Niño phases are triggered by large upwellings of warm water in the equatorial Pacific which warm the surface atmosphere (cooler waters in the La Niña phase cool the surface atmosphere).

3 The June 1991 PINATUBO eruption, with three times the sulphur dioxide content of El Chichón, was easily followed as it spread across an equatorial zone. The SAGE instrument even located a patch of eruptive cloud that was torn off in a bout of bad weather and carried south over the Southern Ocean. An immediate cooling effect was noted, which increased through September. The eruption also coincided with El Niño, but still produced a near 1° C surface cooling. Away from lower-level El Niño effects, the stratosphere temperatures gave a more accurate picture of the aerosol effect. Here, in contrast to lower levels, temperatures rose and an observed 4° C rise matched the predicted value.

4 The MT HUDSON main eruption on 13 August 1991 was fairly poor in sulphur dioxide (0.3 million tonnes). However, the cloud was followed by TOMS, and after crossing the Argentinean coast under the westerly wind system it virtually circumnavigated the Southern Ocean by 21 August. It left a whiff of sulphuric aerosol off Tasmania as its nearest land point.

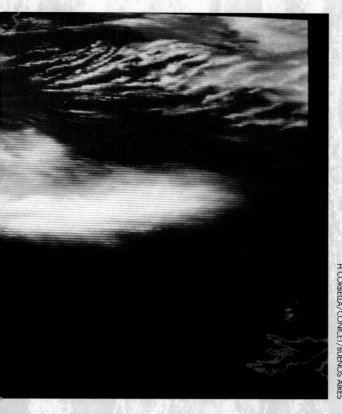

Satellite images of the Mt Hudson eruption plume of andesitic ash, from the meteorological polar orbit satellite NOAH II, and captured by the Argentine National Meteorological Service, Buenos Aires, 13 August 1991.

◁ The plume extends east from Mt Hudson, in Chile (left), crosses Patagonian Argentina (red line boundary, left centre) to beyond the coast (red line boundary), and heads over the Atlantic (right). The plume is mostly yellow and is driven east by a jet stream at more than 100 knots (185 km/hr).

▽ Image from the infra red band, with colours coded to different temperature intervals, of the volcanic plume top along its eastward path. The cloud cools as it rises, from –50°C (red) to –62°C (violet centre); some of it has passed through the troposphere to the stratosphere.

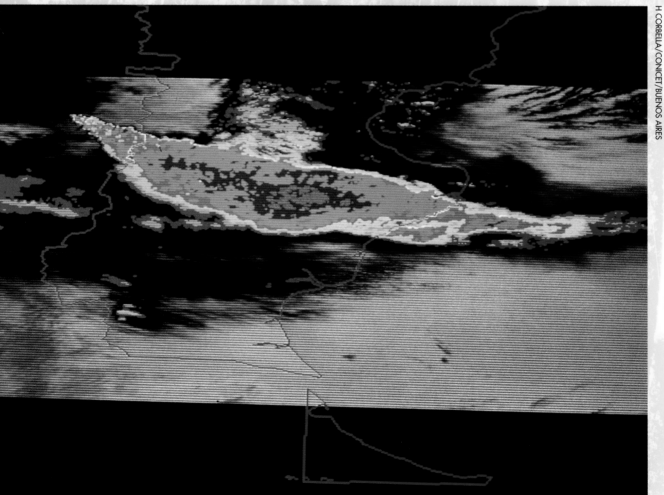

halogen compounds, like the chlorofluorocarbon compounds (CFCs), do greater damage.

Other volcanic chemicals such as sulphate acid aerosols assist ozone breakdown in several ways. They react with human-introduced chemicals to release their reactive halogens (50–170 million tonnes per year) and also react with atmospheric nitrogen to produce nitric acid aerosols. These promote further halogen-producing reactions. Also, by absorbing radiation these aerosols heat the atmosphere and alter the rate of ozone synthesis. This warming effect dominates ozone destruction in tropical regions, while chemical effects control ozone destruction in polar regions.

Northern hemisphere ozone decreased after the 1982 El Chichón eruption, where sulphuric and nitric acid contents increased at altitudes of about 25 km. The June 1991 Pinatubo eruption disturbed tropical ozone concentrations, which were affected by both sulphur dioxide gas and sulphate aerosols. Five months after the eruption the ozone layer had become disturbed up to 30 km in altitude and a 1.5° C rise in stratosphere temperatures had reduced ozone concentrations by 10–30%. The more southerly August 1991 Mt Hudson eruption caused an unusual ozone loss (near 50%) at 11–13 km above south polar regions by late September. From this level, the Mt Hudson aerosols were dispersed into higher ozone layers by a vortex of polar winds by 1992. Clearly, eruptions cause temporary dislocations to ozone layers. This, combined with continued human introductions of ozone-destroying chemicals, has led to the large ozone holes and record low ozone levels recorded over Antarctica in 1993.

'BAD WEATHER' VOLCANOES

The tracking of high-flying eruption clouds since 1980 has aroused awareness of volcanic effects on climates. These eruptions pale beside past disruptions. Just consider Indonesia's eruptions, which barely figure in this last cycle. KRAKATOA in 1883 erupted 10 billion m³ of ash, TAMBORA in 1815 discharged 100 billion m³ of ash, and the Toba explosion 74,000 years ago flung up an estimated 1–3 trillion m³ of lava and gas. The KRAKATOA eruption and its climatic effects are well documented and make absorbing reading. A recent evaluation of the atmospheric phenomena, however, suggests that the ozone layer vanished for a few years afterwards. A new estimate of the sulphur release gives 2 million tonnes, twice that previously determined. This helps to explain the observed climatic cooling and accompanying ozone loss.

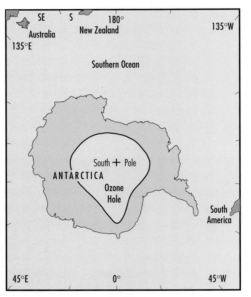

△ *Minimum ozone concentration (ozone hole) over Antarctica, 1994.*

The Tambora eruption preceded the severe 'year without a summer'. Other times of climatic misery may also correlate with large eruptions. The permanent fog in the northern summer of 1783 and subsequent severe winter of 1783–84 came after the extensive LAKI eruption in Iceland. These large eruptions leave their record in the form of thin sulphate deposits which become encased in ice sheets. This record extends back for thousands of years (see table, opposite page.) Sulphate concentrations in these bands from Greenland ice suggest that Laki, Tambora and Krakatoa each discharged over 50 million tonnes of sulphuric acid into the air. Acid aerosols from Laki, however, probably spread through the troposphere, rather than stratosphere, so that its climatic effects would differ. Toba probably ejected 3 billion tonnes of hydrogen disulphide and sulphur dioxide, 500 billion tonnes of water vapour and 2 billion tonnes of dust. This may have cooled climates by 3–5° C and raises the question whether exceptional volcanism can trigger an ice age. Some geologists see links between extensive Kamchatka and Aleutian Island volcanism before the northern ice age started, 2.6 million years ago. However, volcanism may only augment a prevailing cooling trend, rather than creating a sudden glaciation.

The known volcanic atmospheric contributions are dwarfed when past performances in the geological record are considered. Even more gigantic volcanic eruptions dominated the stage. The Deccan volcanism that took place in India about 65 million years ago erupted an estimated 1–3 million km³ of basaltic lava in a period of 1–3 million years. Even with such volumes of lava the calculated greenhouse effect from released carbon dioxide is only a 1° C drop in temperature. Even greater volcanism changed the face of the Pacific floor 100–130 million years ago. This was a time of warm greenhouse climates, but the volcanic carbon dioxide release is hard to determine. Another rapid, enormous volcanic outburst about 250 million years ago poured out over 1 million km³ of basalt lavas in Siberia. These were sulphur-rich eruptions, so that greenhouse warming from carbon dioxide would be cancelled out by sulphuric acid aerosol cooling. Geologists claim evidence of climatic chilling at this time. Australian geologists even identified a permafrost layer in 250 million year old sedimentary beds of the Sydney Basin.

YEARS OF HIGH SULPHATE FALLOUT IN THE GREENLAND ICE CORE AND SUSPECTED VOLCANO

Year AD	Volcano, Country	Sulphate level	Year BC	Volcano, Country	Sulphate level
1971	Hekla, Iceland	83	54	Sheveluch, Kamchatka, Russia	291
1969	Fernandina, Galápagos Islands	92	180	Volcano, Italy	93
1912	Katmai, Alaska, USA	67	413	Okmok, Alaska, USA	64
1902	Pelée, Martinique	41	585	Krafla, Iceland	132
1883	Krakatoa, Indonesia	46	737	Yantarni, Alaska, USA	74
1831	Babuyan, Philippines	68	1192	Bardabunga, Iceland	110
1830	Klyuchevskoy, Kamchatka, Russia	52	1454	Aniakchak, Alaska, USA	164
1815	Tambora, Indonesia	94	1623	Santorini, Greece	145
1781	Laki, Iceland	134	1991	Long Island, Papua Niugini	74
1641	Awu, Indonesia?	81	2310	Hekla, Iceland	80
1604	Monotombo, Nicaragua	61	2617	Black Peak, Alaska, USA	68
1587	Colima, Mexico	72	3258	Akutan, Alaska, USA	66
1478	Mt St Helens, Washington, USA	63	3518	Towada, Japan	174
1460	Kuwae, Vanuatu?	66	4267	Kikai, Ryuku Is	71
1344	Hekla, Iceland	54	4411	Avachinsky, Kamchatka, Russia	183
1285	Asama, Japan	44	4564	Masaya, Nicaragua	132
1259	El Chichón, Mexico?	349	4803	Mazama, Oregon, USA	141
1194	Oshima, Japan	61	4988	Hekla, Iceland	93
1175	Krafla, Iceland	148	5279	Kizimin, Kamchatka, Russia	404
1103	Hekla, Iceland	100	5521	Tao-Rusyr, Kurile Island	129
1026	Billy Mitchell, Solomon Islands	43	5688	Karymsky, Kamchatka, Russia	81
936	Eldgjá, Iceland	65	5954	Vesuvius, Italy	115
264	Ilopango, El Salvador?	48	6273	Pauzhetka, Kamchatka, Russia	150
152	Ibusuki, Japan?	45	6614	Bardabunga, Iceland	240
77	Vesuvius, Italy	95	6722	Towada, Japan	572

Relative sulphate levels are given in parts per billion. Other high levels cannot be matched with known eruptions, more so in the BC record. Only sulphate levels above 40 parts per billion are shown. More recent sulphur-rich eruptions not in the list include these: late 1974 Fuego (Guatemala), early 1976 St Augustine (Alaska), early 1980 St Helens (USA), late 1980 Uluwun (New Britain), early 1981 Aliad (Kurile Is) and Pagan (Mariana Is), early 1982 El Chichón (Mexico), mid 1983 Una Una (Indonesia), late 1983 Miyakeijima (Japan) and Pavlov (Alaska), late 1985 Ruiz (Colombia), early 1986 St Augustine and Pavlov (Alaska), mid 1991 Pinatubo (Philippines) and late 1991 Mt Hudson (Chile). At present Etna and Stromboli provide 10% of the global budget of volcanic sulphur dioxide. Note that ice core dates have an error of ±2 years, and those given here may not agree exactly with historical eruption dates — for instance, Laki erupted in 1781, Vesuvius in 79.

EXTINCTION VOLCANOES

DID the huge volcanic outbursts of the past produce climates too severe for most life forms to continue? This is one interpretation for sudden extinctions in the fossil record. When extinctions coincide with abnormal volcanism it excites suspicion of 'killer' eruptions that alter climates and place ecosystems under stress. This becomes more controversial when taken with evidence for giant impacts made by extraterrestrial bodies during some of these events. As with volcanism, such impacts also generate large dust clouds and acid rains. Such giant impacts could even trigger volcanism. As the earth's surface rebounds after impact, pressure is released on the hot interior mantle rocks, which then melt. Molten plumes rise up to erupt huge volumes of basalt lavas. Melting need not occur below the impact site, but could start at some distant site, after the great energy of the impact becomes focused as it travels through the Earth.

Large extinction events occurred at the end of the Cretaceous and Permian Periods. Both were times of large flood basalt eruptions, in the Deccan

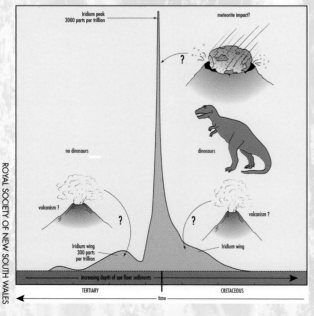

THE IRIDIUM ANOMALY

◁ *The extinction of the dinosaurs coincided with an abnormal deposition of iridium in old sea floor beds, thought to indicate a comet impact. The iridium peak is related here to a large impact in Mexico, and to Deccan flood volcanism in India which may have added to hostile environmental effects. NB: time flows right to left.*

◁ Baby Triceratops dinosaur hatching from its egg (photo of model). Dinosaur egg shells show increasing concentrations of elements such as selenium, and pathological damage to their proteins, leading to more fragile shells, even before the time of the meteorite impact. This suggests dinosaur survival was already threatened before the Cretaceous meteorite impact.

▽ Did volcanic outbursts assist a meteorite impact to cause many extinctions found in the fossil record 65 million years ago? This reconstruction shows Tyrannosaurus rex and Trachodon startled by surrounding events. Artwork by K. Gregg, after Z. Burian, Australian Museum.

province of India and in Russian Siberia. In the end-Cretaceous extinction, 65 million years ago, 75% of marine species and all known remaining dinosaurs on land disappeared, coinciding with the Iridium anomaly (diagram, left). It is claimed that a large crater of this age at Yucatan Peninsula, Mexico, was made by a meteorite, while the Deccan eruptions were already in full swing. So this may mark independent catastrophic events, rather than impact-induced volcanism.

The greatest end-Permian extinction 250 million years ago obliterated over 95% of existing species. This process coincides with the massive, explosive and sulphur-rich Siberian basalt eruptions, but so far evidence for any accompanying meteorite impact is indefinite. We can conclude that volcanoes sometimes help reshape life on Earth. Rather than referring to 'extinction' volcanoes, we could term them 'evolution' volcanoes.

CHAPTER

FL SUTHERLAND

FUTURE ERUPTION: VOLCANO WATCH

6

◁ Hotwater Creek and Frying Pan Lake, looking towards Cathedral Rocks, Waimangu hydrothermal area, New Zealand, February 1986. Formed in the 1886 Tarawera eruption, this area can erupt without warning.

VOLCANO surveillance in highly active and densely populated regions, is a life-saving service. Volcanoes can be checked in many ways: by taking ground, gas and water temperatures; by monitoring changes in volcano shape; or by listening in to earth tremors. However, these are only superficial checks of highly complex physical and chemical systems that evolve with time, and detailed studies of previous eruptions are also essential.

By closely mapping MT ST HELENS, geologists from the United States Geological Survey were able to predict 2 years before the 1980 eruption that eruption was likely to occur within 22 years. Developing local civil and national action groups, and raising awareness of volcanic risk among populations and ruling politicians are further necessary safeguards. The 1902 MT PELÉE eruption devastated St Pierre and killed 28,000 people. Newspaper editors and politicians, distracted by impending elections and uneducated about volcanoes, had declared St Pierre secure. When New Zealand geologists expressed concern in the 1960s about Lake Taupo's former dangerous eruptions, the local mayor — busy promoting tourism — declared this could never happen to Taupo. He was oblivious of nearby Mt Tarawera which erupted in 1886, killing 50 people or so, including a tourist or two.

MONITORING VOLCANOES: MANY METHODS

The main problem in volcano surveillance is predicting the exact time, scale and risk involved in an eruption. Methods for monitoring and predicting volcanic activity are becoming increasingly diverse and sophisticated. Investigating volcanoes is fraught with danger. Two volcanologists died monitoring KARKA VOLCANO off Papua Niugini in 1979. Three died in the 1991 MT UNZEN activity and six died in a small blast while testing monitors on GALERAS VOLCANO in Colombia in 1993. With 15 deaths of volcanologists since 1979, the International Association for Volcanology devoted a special session to volcanologists' safety at its 1993 Assembly in Canberra, Australia.

OBSERVATORIES AND NETWORKS

Many countries operate observatories to monitor the moods of their active volcanoes. For the more dangerous volcanoes, these are not necessarily permanent stations. Hawaii's observatory was set up on KILAUEA CRATER in 1912 'to take the pulse and blood pressure of the globe'. It renovated its present site in 1987, where an observation tower now gives a full-circle view of Hawaii's two active volcanoes. Volcanoes dictate observatory sites. After the shock of the Mt St Helens eruption, the Cascades volcano observatory was based in Vancouver, Washington, in 1980 to monitor the nine American west coast volcanoes.

A string of observatories monitor the highly active volcanic islands along the Australian–Pacific margin, extending from the North Island of New Zealand to Papua Niugini and Indonesia. Ground observatories link up with marine and aerial surveillance craft and the special advantages of earth-orbiting satellites. An eruption from a Kamchatka volcano was observed from the Space Shuttle

△ *Hawaiian Volcano Observatory, with observation tower, overlooking Halemaumau pit crater and Kilauea volcano, Volcanoes National Park, Hawaii, March 1987.*

in October 1994. A volcano space observatory, Vexuvio, is being designed for a scheduled launching from the Space Shuttle in 1996. Monitoring sea floor volcanoes is another frontier. Feasibility studies are already being made for setting up an underwater observatory on Lohi seamount off Hawaii and for siting remote-controlled laboratories on deep ocean ridges.

Japan is a world leader in monitoring threatening volcanoes and in developing predictive techniques to mitigate problems with evacuation. The nation has built up a strong collaborative network among scientists, civil government groups, law-enforcement bodies and emergency management agencies. Some developing countries with dangerous volcanoes need greater support for their observatories and monitoring programs. One such country is the Philippines, where 1335 died in the 1911 TAAL eruption, 500 in the 1951 HIBOK–HIBOK eruption, 300 in the 1991 MT PINATUBO eruption and 75 in the 1993 MAYON eruption. Only a third of the 22 active Philippine volcanoes are fully monitored and the country's 12 seismic observatories number well under the ideal — 100 — needed for such a volcano density.

HEAT AND FLUIDS

Changes in the temperature and chemistry of gas and waters emanating from volcanoes indicate unsettled interiors, even when the surface appears quiescent. Long-term monitoring is necessary for a full analysis of small fluctuations in the system. After 22 years of measurements at WAIMANGU CRATER LAKE in New Zealand, small fluctuations were attributed to changes in groundwater flow and local ground movements. However, the long-term trend was steady, without obvious volcanic influence. So the system born in the 1886 Tarawera eruption has become volcanically quiet. This contrasts to MERAPI VOLCANO, Indonesia, where a series of gas samples showed water contents dropping from 80% to 30% and sulphur dioxide contents increasing from 100 to 360 tonnes/day before the 1992 eruption. Measurements can be made on higher temperature gases escaping from fumaroles or on lower temperature waters in boiling pools, crater lakes and other aqueous discharges.

As molten material gathers and rises, temperatures usually increase and different gases come off during the advance. The relative contents of carbon dioxide, sulphur dioxide, hydrochloric acid and hydrogen in gases are useful signposts. High carbon dioxide signifies an early release from deep magma, sulphur dioxide escapes at higher levels, and high levels of hydrochloric acid and hydrogen indicate that magma is mixed with hydrothermal waters. Fluorine, boron and ammonium contents are also sensitive indicators for sources of fumarolic gas. Hydrogen, hydrogen disulphide ('rotten egg' gas) and methane are useful for indicating the origins of gases bubbling off boiling pools. Ammonia, boron and lithium contents help to categorise hot groundwaters. Other special measurements can be used, including isotope ratios of gases and the acidity or alkalinity strengths of waters. At Izu-Oshima and Mt Unzen volcanoes in Japan, isotopes of helium are used to monitor gas activity, as the amount of helium 3 released from magma is high compared to helium 4.

Special techniques are used for checking gases in volcanoes that are hazardous because of their difficult terrain or strong activity. Ground-based telescopes, fitted with spectral radiometers, detect sulphur dioxide/hydrochloric acid ratios in gases issuing from the dangerous lava dome of Mt Unzen volcano 1–3 km away. Similarly, remote-sensing spectrometers measured carbon, oxygen and sulphur ratios from a 900 m distance when the Galeras volcano emitted gases. Gases discharging in a continuous flux from lava lakes can also be checked. At MT EREBUS, for example, sulphur dioxide increases as the lava lake expands. Gas identification not only assists volcano surveillance, but also helps atmospheric investigations. Measurements of chlorine and fluorine escaping from EREBUS showed that this volcano contributes much of the halogen present in the pristine Antarctic atmosphere. An 8-legged robot, DANTE, was designed to sample gas from inaccessible fumaroles in the Erebus crater. Unfortunately DANTE failed in the operation, but represents a potential means of remote sampling.

Volcanic heat can be sensed by aircraft and satellites equipped with infra-red instruments. The Landsat Thematic Mapper (TM) was used to detect the growth and disruption of lava domes from satellite passes made between 1984 and 1993 over the LASCAR volcano, Chile's most restless but seldom visited volcano. This satellite technique can spot a range of volcanic activities, including fumaroles, crater lakes, lava lakes and lava flows. Aerial sensing even found a thermal high under ice, south of Ross Sea, thought to mark an active volcano under the Antarctic icesheet.

GROUND MOVEMENTS

Volcanoes change shape as they build up to or deflate from an eruptive mode. Such change can be caught by photography from the ground, air or space. Time-lapse photography, movie cameras and special slow-scan video television monitors have been used with success at Mt St Helens and in remote situations. Vertical and oblique aerial photography and side-scanning radar imaging in cloudy terrains all contribute to the documentation of volcanoes.

Geodetic levelling is a basic means of measuring changes in elevations around volcanoes. Sometimes crater lakes are used as giant spirit levels to record changes around shorelines. Tilt meters measure local deformation and can relay data by telemetry from remote sites to a base observatory. These results, however, need careful assessment in a controlled survey. Distance measuring and slope-distance networks can be set up within and around volcanoes and displacement meters can record changes in widths of cracks or movements on faults. Orbiting satellites can be used to monitor vertical and horizontal movements on volcanoes by means of the Global Positioning System (GPS) where radio signals are received between the ground and satellites.

SEISMIC ACTIVITY

Seismic recorders are widely used to pick up tremors in volcanoes caused by shifting rocks, magma movement or even gas release. Dramatic increases in seismicity often arise as volcanoes build up towards eruption. In HAWAII, seismic tremors mostly originate 2–3 km deep under the main craters and outline the tops of magma reservoirs. Seismic tremors also follow magma as it rises from 50 km to 60 km under Hawaii to enter the shallow reservoirs before an eruption. However, strong seismic activity does not always culminate in eruptions. At RABAUL VOLCANO, New Britain, the caldera began to deform and seismicity increased from below 1,000 tremors a month in 1983 to about 14,000 tremors a month by April 1984. On the threshold of eruption the volcano stopped short and the government emergency alert was lifted. This near-disaster proved a valuable forerunner for handling the September 1994 eruptions in the caldera.

In the North Island of New Zealand, seismic stations monitor tremors from both active and dormant centres. Stations set up near CRATER LAKE on RUAPEHU VOLCANO can detect two types of earthquakes. The typical low-frequency volcanic earthquakes come from cracks under the vent when they are invaded by molten magma. High-frequency earthquakes generated near the surface are probably caused by the sudden fracturing of ice, forming large crevasses. Around Auckland, three permanent seismic stations are spaced about 40 km apart on Waiheke Island, Hunua Ranges and Waitakere Ranges, forming a triangle to cover the volcanic field. It is difficult to detect any shallow earthquakes above 20 km, because of the background noise from the city and its environs.

Some volcanoes erupt with little seismic warning. New methods can now pinpoint molten chambers under volcanoes as future trouble spots.

Seismic tomography, a technique based on principles similar to those used in tomographical medical scanning (e.g. brain scans) uses waves generated by seismic shocks and earthquakes. Hot zones slow up seismic waves so that they arrive late at the recording stations and show as anomalous patterns. A 3–D velocity model of ETNA VOLCANO was made this way

◁ Recording drums for detecting seismic activity, Volcanological Observatory, Rabaul, New Britain, August 1970. More advanced seismic recorders are now installed.

▷ Thick volcanic pumice and ash beds formed in second phase of the Rabaul caldera eruption 1,400 years ago, Kobakaba road cut, near Rabaul, New Britain, August 1970.

and showed up potential magmatic reservoirs at depths below 10 km under areas of spatter cones. Seismic waves can be generated artificially from explosions set off around volcanoes. This was tried at the ice-covered KATLA VOLCANO in Iceland, and travelling waves slowed by hotter material located a large magma chamber 1–3 km under the volcano.

MAGNETISM AND GRAVITY

Magnetic surveys of volcanoes are useful since magnetism in the rocks gives indirect thermal information. The magnetism is acquired from the Earth's magnetic field, but lost when magnetic minerals in the rocks become heated above 580–680° C. The rocks reacquire the magnetism as they cool below these temperature points. Magmas and hot gases introduced into volcanoes will weaken magnetic fields over these sites. This magnetic effect has been monitored at WHITE ISLAND, New Zealand, since 1970. The magnetism rises and falls in sympathy with fumarole gas temperatures. It predicts not only the eruptive cycles but also which vent will erupt.

Gravity changes can chart a volcano's progress. For example, molten rock invading an

underlying reservoir distorts the superstructure and alters the gravity field. Gravity and ground tilt readings between 1988 and 1991 at ASKJA VOLCANO, Iceland, demonstrated a state of unrest under its caldera. Gravity readings around Mt Etna increased sharply 6 months ahead of the large lava outpouring in late 1991.

DUST AND LAVA

Volcanic ash clouds sometimes affect jet airliners at high levels — a new hazard besides aircraft hijacks and terrorist bomb explosions. Acid aerosols from the 1982 EL CHICHÓN eruption caused costly damage to aircraft flying on northern routes. Ash from the 1982 GALUNGGUNG eruptions stopped the engines of British Airways and Singapore Airways planes at altitudes of 9–11 km, forcing them into terrifying dives to start them again. Another plane met ash from the 1989 REDOUBT eruption over Alaska and lost power before managing an emergency landing. Many aircraft flew into ash from the 1991 Mt Pinatubo eruptions and on the ground the weight of ash on tail pieces tilted aircraft off their landing wheels. No eruptions have caused a crash, but the potential problem interests Australian scientists at the CSIRO laboratories. They designed an instrument which detects volcanic clouds by their infrared signature. Problems with development grounded the device, and prevented its immediate commercial use.

Even lavas, once eruptions start, can be used to estimate the likely volume and length of eruptions. This method is based on studies of lavas erupted during the prolonged activity at Unzen volcano in Japan and during large explosions from western United States volcanoes. Larger eruptions show subtle chemical differences in their oxygen, strontium and neodymium isotopes. Since these isotope values reflect a large influx of magma, knowing the isotopic signature of early erupted lavas makes it possible to foretell the volume still to come.

PREDICTING ERUPTIONS AND HAZARDS: LIFE SAVING

Forecasts of eruptions and their likely consequences must take into account all aspects of the volcano's behaviour, history and morphology. Even after eruptions start, decisions about evacuating local populations should be taken only after the severity and length of activity are fully assessed.

△ *Icelandic volcanologist, H Tryggvason, monitoring lava stream during eruption of Krafla volcano, Iceland, October 1980.*

ASSESSING AFTERMATHS

The eruption itself is not always the main hazard. As a tragic example of this, the small 1985 eruption of NEVADO DEL RUIZ VOLCANO in Colombia melted snow and ice on the summit, which triggered a series of mudflows. One of these, bursting out of a confining canyon, buried Amero township 40 km from the eruption site, killing 25,000 people. Detailed studies made since 1985 reveal that this volcano is very vulnerable to collapse because the dissolution of rocks by hydrothermal fluids along faults has greatly weakened its structure. Models of these potential collapse areas — susceptible to earthquakes as well as eruptions — now allow better evaluation of the volcanic risks and hazards.

Submarine eruptions are harder to predict than those of land-based volcanoes. Although not a direct threat to populations, underwater vents may pose special hazards if tsunami waves are generated. KICK 'EM JENNY, the most active volcano in the Lesser Antilles island arc, is such a vent. Computer models exploring its eruptive wave-making potential were used to assess its tsunami threat.

The likely consequences of a volcanic eruption depend on the exact nature of the volcano and its environment. The Auckland volcanic field in New Zealand lies within a city and harbour, so eruptions on land and under water have to be considered. A volcanic hazard assessment for Auckland's small volcanoes could involve initial earthquakes in increasing swarms over an impending volcanic site, but only minor earth deformation is likely. Blast shock waves radiating at supersonic speed would devastate areas up to 5 km from the vent and explosion debris could bury this zone up to 30 m deep. An eruption column would deposit ash and create problems for domestic and international aircraft using Auckland's airports. Accumulating ash could make roofs collapse, contaminate the city and surrounding country water supplies, and spoil crops. Rapid lateral surges of hot gases, steam and ash would devastate the land in their paths and could travel across water. Sea waves created by submarine explosive eruptions could scour the coast. The magnetic measurements of Auckland's volcanoes suggest that clusters of volcanoes may erupt together, thereby compounding their effects.

FINDING FORMULAS

Forecasting eruptions remains an inexact science, due to complex factors. Many dangerous eruptions are unexpected because the volcano's repose times (between eruptions) are either erratic or little understood. It is interesting that, of the World's 101 Most Notorious Volcanoes listed in 1981, not one figured in the major lethal eruptions recorded in the 1991 update of the list. Formulas for predicting eruptions have been devised. In the 1960s, for example, the geophysicist Blot drew on observations of earthquakes and eruptions in the Melanesian island arcs to link significant eruptions with earthquakes from intermediate depths in the Earth's mantle. From this he calculated an eruption migration time, and claimed an eruption probability of 80% within 15 days of the predicted date. This method, however, is confined to island arc settings.

Another predictive model was proposed in 1991 by two American geologists, Voight and Cornelius. This model considers the point at which materials fail under rapidly increasing stress, and uses seismic activity to do it. Seismic rates are measured in near-real time (i.e. without time delays to collect, process and calculate data) to detect imminent eruptions as soon as possible. The method involves implanting laser reflectors and tiltmeters over the volcano to register constant, instantaneous measurements, and Merapi, a dangerous volcano overlooking Jogjakarta in Indonesia, was wired up to test it. The theory best applies to volcanoes that erupt stiffer, more explosive materials. Whether the theory can fulfil predictions or applies to fluid volcanoes remains to be seen.

Predicting Australia's next eruption is a greater challenge (see chapter 12). Since its volcanic fields have no present activity for scientists to study, present knowledge cannot be defined, and repose times between eruptions range from several to tens of thousands of years.

▷ *Galunggung volcano erupting heavy ash cloud, Java, April–May 1982. Ash from this eruption stopped the engines of planes flying 9–11 km overhead.*

MOUNT VESUVIUS AND POMPEII

Vesuvius (Vesuvio to the Italians), is a volcano notorious for threatening settlements on the Bay of Naples. The present cone grew after a larger Vesuvius blew apart in AD 79. Mt Somma ridge remained on the north side, partly shaped by even earlier massive eruptions. One of the largest is identified in deposits that date to about 1365 BC. The mountain had been quiet for 600 years when it caught the settlements by surprise in its AD 79 outburst. Pompeii, the chief casualty, was 9 km downwind and unlucky because the wind direction was unusual on that day, recorded as 24 August.

The afternoon sky turned dark as white pumice rained down and the eruption cloud rose 20 km high. The pumice, a frothy phonolite rock, was erupting from the vent at a rate of 5,000–80,000 tonnes per second. By evening, grey pumice started to fall from discharges at a rate of up to 150,000 tonnes per second. Pompeii's streets and dwellings were filling with pumice and roofs began to collapse under its weight. People sought shelter or tried to flee as panic increased. The final blow came as collapsing eruption clouds sent surges of hot gas, ash and stones down the mountain slopes. One, ripping through Pompeii, knocked down many upper structures and killed the remaining residents. The nearby town and citizens of Herculaneum suffered a different end, being buried in a torrent of hot mud.

The burial of Pompeii left only a few higher structures protruding. Its position, layout and wealth of public and private creations were left in limbo until rediscovered during excavations in 1763. Later excavations revealed the style of dwellings and possessions and also hollows left by decayed corpses, still containing bones. These hollows, when filled with plaster, yielded casts of human forms and even details of their dress. Casts were also made domestic animals.

Figures found in 1991 inclu ed a man whose cloak covered t face of a pregnant woman, and tv small children lying farther awa From the archaeological view, husband is protecting his wife the last, the children havi become separated from them the dark. From the volcanologic view, the bodies are surrounded surge deposits containing aligne timber poles, which denotes high-energy transport event. T first interpretation is that the bo ies fell just where they were whe disaster overtook them. The se ond interpretation is that a sur swept the bodies into their restin place and the children bei lighter, were swept further alon This divergence of analyses sho the importance of co-ordinati detailed volcanological studi with such excavations.

◁ *Victims of the Vesuvius AD 79 eruption. The hollows in the pumice left by bodies (excavated in 1991) are now preserved as plaster casts containing skulls and other bones. Regio 1, Insula 22, Pompeii, September 1994.*

Vesuvius has recorded many eruptions since the AD 79 catastrophe. A large eruption is traced to AD 472 and ash from a destructive eruption in 1631 reached as far as Turkey within a day. Further large eruptions damaged surrounding parts of Naples, most recently in 1872, 1906 and 1944. Mud flows or lava flowing between Vesuvius and Mt Somma have spilled out of the old crater and destroyed the village of San Sebastiano four times in the last 150 years. As Naples and Vesuvius grow it becomes vital, for the city's safety, to understand the behaviour of the volcano's underlying chambers, which extend to 5 km below it and can hold up to 10–12 km³ of magma.

The need to monitor Vesuvius led to the founding of the first volcano observatory, Osservatorio Vesuviano, in 1841. This observatory and its equipment will become a museum in 1995. The studies will continue through a network of regional universities. Vesuvius is not the only Neapolitan threat, as Campi Flegrei, north of Naples, is a giant volcanic complex. In the past its calderas have erupted enormous ignimbrites and ash falls, which reached the Isle of Ischia to the east and Sorrento to the south. Naples would be devastated by such eruptions, but present activity seems to be lessening.

▷ *Crater of Vesuvius, Naples, Italy; one of the most dangerous volcanoes overlooking a large city. Taken during a 50-year quiescent period, September 1994.*

CHAPTER

PC RICKWOOD/UNSW

SUB-ERUPTION: PLATE TECTONICS

7

◁ Lava streaming from Krafla, Iceland, October 1980 eruption. Iceland sits over the oceanic join between the Eurasian and North American plates.

BELOW the surface eruptions of volcanoes lie the moving forces of plate tectonics. The rise and fall of volcanoes are not shallow, subterranean whims of the god Vulcan. They are expressions of deeper, more dynamic and grander thermal events, unimagined by the Romans. Volcanoes are part and parcel of global movements of the Earth's plates, known as plate tectonics. Plates are rigid segments of the Earth's upper layers, although they can grow and move sideways. They include continents and oceans within their boundaries. Unlike the continents, which behave like icebergs floating in denser material, plates are thicker units which behave more like conveyor belts.

The term 'plate' holds the key to volcanism in its letters. Plate, made of Lithosphere, rides over Asthenosphere creating Tectonism and generating Eruptions. Each plate is bounded by active geological boundaries. These can be mid-ocean ridges generating new sea floor, collision zones between interacting plates, or fault zones tearing plates apart. Plates can be made of ocean floor, as in the huge Pacific plate, or mostly continental rocks, as in the large Eurasian plate. Most plates include both oceanic and continental parts, as in the Indian–Australian plate. This last plate is of greatest interest here. Mid-ocean ridges bound its west and south sides in the Indian and Southern Oceans and tear faults and collision zones mark its east and north sides. The plate is moving northwards as it grows away from the Southern Ocean ridge. Consequently, the Australian plate is colliding with the northwest-spreading Pacific plate to the east of it, and with the slower-moving Asian plates to the north.

▷ *Basalt dyke cutting lower part of ocean floor rocks uplifted above sea level, northeast Macquarie Is, Southern Ocean, October 1981.*

▽ *A SW–NW section of the Australian Plate, through Tasmania, showing the crust and mantle structure from the Southern Ocean floor to Bass Strait. The low velocity zone below the base of the lithospheric plate is the source of molten basalt magma, rising here to form chambers and volcanoes for basaltic eruptions. Crustal layers melting at higher, earlier levels formed large granite and diorite magma chambers after subduction and folding. Most of these erupted rhyolite and andesite lavas have now eroded.*

PROFILE OF A PLATE: ITS ROCKS, MOTIONS AND MELTS

Plates contain two separate layers, a top crust and bottom mantle, each made of different rocks. Continental crust is rich in silicon and aluminium. This is typical of sedimentary beds such as sandstones and shales, igneous rocks such as granites and syenites, and metamorphic rocks such as gneisses and granulites. Oceanic crust is richer in magnesium and iron typical of basaltic rocks and coarser equivalents such as gabbros. Both continental and oceanic crusts overlie the mantle, typically made of denser peridotite rocks, richer in magnesium. In a sense, the crust floats on the mantle and the lighter continental rocks form a crust 20–60 km thick. By contrast, the heavier oceanic crust is usually between 5–10 km thick. The crust-mantle boundary, named the Moho after the geophysicist Mohorovičić, shows up as an increase in the speed of earthquake waves as they travel into the denser mantle rocks.

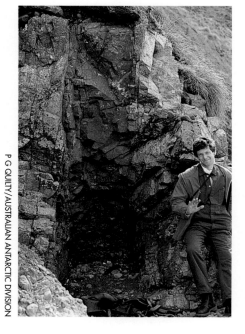

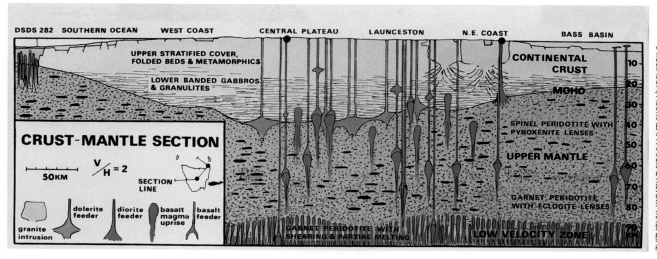

The bottom of the plate overlies the Low Velocity Zone (LVZ), where passing earthquake waves lose speed. This is thought to mark a fluid-impregnated region, where rocks start to melt. The LVZ acts as a mobile zone along which plates move sideways under geological forces. The cooler rigid plates move over a warmer underlying mantle, the asthenosphere. So begins plate tectonics.

tear apart — lateral slip faulting. Once subduction starts, gravity aids the process and the downgoing slab drags the oncoming plate towards its zone of descent. The downdragging force depends on the steepness of the descending plate, but other factors can reinforce the process. The downgoing oceanic slab, with a veneer of sea floor sediments, starts to lose fluids when these are driven off by increasing

PUSH OR PULL?
WHAT DRIVES PLATE MOVEMENT?

Plate motion seems to start at mid-ocean ridges, where molten basalt wells out of axial rift volcanoes. As each molten injection cools and crystallises, its feeder channel gives the sea floor this new width (diagram, foot of previous page). In this way the sea floor opens from a few and up to tens of centimetres a year. This accumulates into tens to hundreds of kilometres over millions of years. This pushing force moves plates away from the ridges and any continents on the plate drift along ahead of the enlarging sea floor. However, Earth is a confined sphere so that a moving plate will encroach on its neighbouring plates. Something gives under the slow collision of trillions of tonnes of rock. The interplay of colliding plates provides a complex geological equation involving relative plate speeds, angles of collision, and the thickness and resistance of rocks in each plate.

One plate may slide under the other — subduction — or may ride up over it — obduction — or may

△ *Crustal plate boundaries (yellow lines) shown in a Global Positioning System (GPS) computerised image generated from satellite navigation signals. The plate boundaries follow mid-ocean ridges and active continental margins. The Indian–Australian plate is at the lower left, the Pacific plate is centre left and the Antarctic plate is at the bottom. Colours indicate approximate elevations relative to sea level: purple/red +2,000 to +5,000 m, orange +1,000 to 2,000 m, yellow +500 to 1,000 m, green 0 to 500 m, pale blue 0 to –500 m, blue –500 to –3,000 m and dark blue –3,000 to –9,000 m.*

temperature and pressure. Deeper down, the basalt converts into still denser rocks, such as eclogite, increasing the downdragging force.

All these ridge and subduction forces can be calculated and balanced from estimates of the physical parameters. Many geologists consider the pulling force as the driving force in the ocean ridge to subduction zone movement. They assume that mid-ocean ridge volcanoes are largely zones of melt bleeding from the mantle, not major hearts of deep upwelling melt.

VOLCANO ORIGINS

Volcanoes spring up where rocks in the deeper lithosphere or asthenosphere begin to melt in earnest. Although plates move over a fluidised Low Velocity Zone, this does not normally generate enough melt to well up to the surface. Also, below the LVZ temperature gradients drop off as heat is carried away by convection in the asthenosphere. Under the high pressures in the asthenosphere the rocks remain solid, even though plastic. Extra triggers are needed to melt enough rock to form batches of rising magma.

Triggers that produce smoking barrels of volcanoes include:

1 upwellings of extra heat, which raise local temperatures and initiate melting;
2 upwellings of fluids (water, carbon dioxide), which pervade rocks so that they melt at lower temperatures than dry rocks do;
3 uplifts of the lithosphere, which reduce the pressures on rocks and permit melting to start at lower temperatures.

Some of these triggers may operate together. Lava is often finally erupted from shallow chambers. Recent studies of Italian volcanoes suggest their chambers take 10–100 years to refill and erupt only 1–10% of their contents each year.

VOLCANIC ROCK CHEMISTRY

Each layer in the lithosphere and asthenosphere produces a characteristic melt, called magma, from its mineral make-up. As the constituent minerals melt at different temperatures, the extent of melting also moulds the character of the magma. For example, a slight melting (up to 15%) of peridotite mantle rocks produces magmas that are deficient in silica (below 50% silicon dioxide) but moderately rich in the alkali elements, sodium and potassium (usually over 4% sodium and potassium oxides). Such magmas erupt as alkali basalts. A greater extent of melting allows more refractory minerals from mantle peridotite to enter the melts. These magmas have more silica (over 50%), but lower levels of alkali (usually below 4%) and volatiles substances, and erupt as tholeiitic basalts (see Glossary).

More subtle mineral entries into melts can give us a better idea of the source rocks of lavas. For example, the base of the lithosphere and top of the asthenosphere both consist mostly of garnet peridotite rock. When this mantle melts to make basalt, the garnet bestows its trace elements to the molten pool, so even a surface lava can show a chemical imprint of its deep original source. However, towards the top the mantle converts to a spinel peridotite, as spinel is more likely to become stable at these lower pressures than garnet is. Basalts erupting from melting at this level lack the pattern of heavy rare earth elements found in melts formed from the garnet peridotite zone. The change from garnet peridotite to spinel peridotite usually occurs at depths between 50 km and 70 km.

Lower crustal rocks have more feldspar and quartz, so that melting here produces magmas rich in silica and alkalis. These erupt in lavas such as trachytes and rhyolites. Analysing lavas to pinpoint their depths of origin is not always simple. Many magmas pause on the way up and disguise their original character by precipitating early formed crystals or by digesting crustal rocks from around their chambers. In this way, a basalt magma formed in the mantle can transform itself into a trachyte or rhyolite before erupting. When large amounts of water invade the mantle—released from subducting plates, for example—then basalt magmas become enriched with aluminium. They erupt as high-aluminium basalts, basaltic andesites or andesite, lavas typically found in volcanoes along island arcs and active continental margins. Such magmas may also transform in crustal chambers into dacites, trachytes and rhyolites before final eruption.

MID-OCEAN RIDGE VOLCANOES: SPREADING FLOORS

These tireless makers of ocean floors erupt huge amounts of lava. The swells they create on both sides of spreading ridges occupy about a third of the ocean floors. The ridges pour out a special basalt, called MORB (Mid Ocean Ridge Basalt). With fairly high silica contents and low alkalis and volatiles, MORB is a variety of tholeiitic basalt.

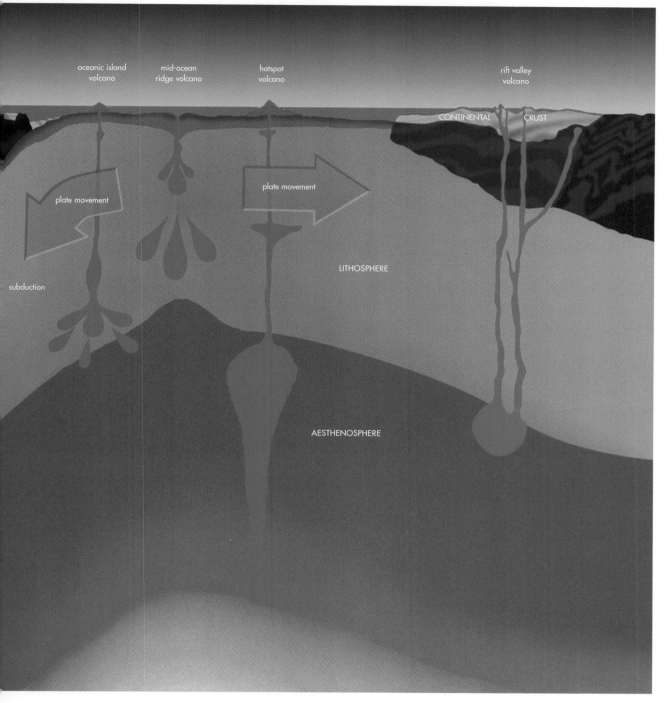

The lavas originate from copious melting of mantle peridotite under the ridges. Traces of uranium and thorium and their radioactive isotopes in MORB suggest that it originates from a wide melting region in the garnet peridotite layer about 80 km deep. The MORB magmas then rise at about 1 m/yr before erupting. In deep water, the lavas erupt without much explosive activity and quickly cool into small cones, lava domes and restricted lava flows and lakes.

△ *Section of mid-ocean ridge and spreading sea floor, passing down through the lithosphere into the asthenosphere. It shows the rise of magma and spreading motions according to plate tectonic theory. Depth of section 200 km.*

VOLCANIC EARTH

▷ Submersible exploring walls of pillow lavas in the slow-spreading north Atlantic rift during the FAMOUS north Atlantic ocean project.

METZER © 1975 AND NATIONAL GEOGRAPHIC SOCIETY

The crust created by mid-ridge volcanoes is revealed in old sea floors thrust up on land by earth movements. The upper layers, apart from sea floor sedimentary deposits, are largely pillow lavas. These quickly chilled basalts adopt lobe-like forms with glassy crusts on contact with water. A central layer is jammed with dykes, the channels that fed the overlying cascades of lava. A lower layer contains coarser rocks crystallised from basalt settling in magma chambers, forming bands of lighter gabbros and denser peridotites. This section overlies the typical spinel peridotite mantle.

An example of uplifted ocean crust forms Macquarie Island, an Australian possession south of Tasmania. This ocean ridge was formed over 12 million years ago before it tilted above sea level. The lower magma chambers and dyke sheets are exposed in the north end of the island (see photograph, p.70) and the overlying pillow lavas, massive lavas, glassy fragmental deposits and intermixed oceanic oozes are exposed in the southern end.

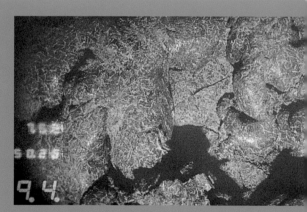

△ Pillow lava. Shiny, reflective crust indicates that it is recently erupted on the crest of the East Pacific Rise near 9° 31' N. Small white tubes on the pillow surface are surpulid worms and indicate a nearby hydrothermal vent system. Photo taken 1990 at 2,530 m depth with ARGO-I optical–acoustic imaging system operated by Woods Hole Oceanographic Institute—Deep Submergence Operations Group, Santa Barbara.

SLOW AND FAST SPREADERS

There are two extremes of volcanic spreading ridges:
1. slow-spreading ridges, such as the mid-Atlantic ridge, where plate floors grow at under 6 cm/yr;
2. fast-spreading ridges, such as the East Pacific ridge, where plate floors grow at over 6 cm/yr.

Volcanoes along both types of ridges form similar hills of pillow lavas. Away from ridges, the flanking volcanoes and structures of both kinds of spreading ridges show distinct differences.

Slow-spreading ridges have a more rugged ridge morphology. The ridge volcanoes lie in an inner rift valley, less than 10 km across, which also receives lavas flowing down from valley walls. The lavas vary in composition: valley side lavas retain fewer early formed minerals than the ridge MORB lavas. The variations in lava suggest that many small magma chambers fed the inner rift, in contrast with the large continuous chambers that underlie fast-spreading rifts. The outer rift has a jagged profile formed by fault blocks of new ocean crust and extends over a 25–30 km width.

Fast-spreading ridges are usually flanked by low-lying lava plains. Here, fast outflows of lava are confined as lava lakes within low-walled rift valleys. These broad, low-relief spreading structures probably overlie a large area of partially molten rock, perhaps 100 km across. More concentrated magma flows upwards under the ridge, to collect and mix the magmas in large continuous high-level chambers. These then supply steady eruptions of typical MORB basalts. Periodically, a slightly different MORB, enriched in elements such as barium and potassium, erupts. These magma fluctuations take place within distances of 15–40 km and within timespans of 10,000 years. Active hydrothermal vents line the ridge near the youngest lavas. Away from the ridge axis smaller batches of more enriched basalts are thrown up to form seamount chains. The plumbing that supplies fast-spreading systems is clearly complex and needs more study.

ABNORMAL SPREADERS

Besides varying in spreading rates, mid-ocean ridges also vary in their elevation. Iceland is an outstanding example because the mid-Atlantic ridge actually emerges on land. Like the north Atlantic ridge, Iceland is widening at a rate of 1–2 cm a year and is split by axial valleys in which fissures erupt voluminous amounts of tholeiitic basalts However, these basalts differ from MORB and range into more

△ Whorl of sheet lava, indicating rapid effusion rate, on the crest of the fast-spreading East Pacific Rise near 9° 17' N. This forms when fast-flowing lava meets a small obstruction in the pre-flow surface: the plastic crust deforms around it. Depth of photo is about 2,580 m; distance across it is about 5 m. Photo taken 1990 with ARGO-I optical–acoustic imaging system operated by the Woods Hole Oceanographic Institute–Deep Submergence Operations Group, Santa Barbara.

△ Sheet flow in the axial summit caldera of the East Pacific Rise crest near 9° 30.8' N. This flow (named the ODP Flow as it was drilled during Leg 142 of the Ocean Drilling Program) is extremely flat because it was ponded in a small collapse area along the western margin of the summit caldera. Depth 2,520 m; distance across is about 4 m; taken from the ALVIN submersible operated by the Woods Hole Oceanographic Institute 1991.

evolved basalts, including one called icelandite. Away from the axial zones, Icelandic volcanoes erupt alkali basalts. Under Iceland something extra is helping to raise its head above water and erupt diverse lavas from extensive magma chambers: a large thermal upwelling from the asthenosphere, called a hotspot. Iceland exists where two volcanic processes overlap. Even well away from hotspot upwellings, some MORB basalts are enriched by infiltrations of asthenospheric material.

In contrast, the Australian–Antarctic spreading ridge becomes exceptionally deep in the Southern Ocean. A rugged 500 km section between 120° and 127° E at 50° S, called the Australian–Antarctic Discordance, is 4,000–4,500 m deep and is one of the world's lowest ridge systems. The ridge volcanoes erupt MORB basalts, but two distinct chemical types. West of 126° E, the lavas are typical of Indian Ocean MORB; lavas east of this are typical of Pacific MORB. Pacific mantle flows in and invades Indian Ocean mantle under the Discordance, before melting to form MORB magma. The Discordance is only part of a general depression which extends into low-lying parts of southern Australia and Antarctica. It is thought that a downwelling of cold mantle is dragging down the mid-ocean ridge where it crosses this geological 'sink hole'.

AWAY FROM THE RIDGES

As new ocean crust forms, the rocks acquire a magnetism from the Earth's magnetic field as they cool and solidify. Their magnetic minerals take the north-south polarity of the magnetic poles. When older sea floor was examined, a very surprising feature emerged. Strips of the floor showed reverse magnetism, opposite to that of the present poles. These alternating strips of normal and reversed magnetism not only suggest that the magnetic poles habitually flip from north to south, but give geologists a very useful means of dating the growth of ocean floors. By matching the magnetic patterns against dated sea floor basalts, geologists started to map the history of the oceans. They found that many fractures which offset spreading ridges also displaced the magnetic stripes. These transform faults, once a part of ocean growth, record different rates of spreading in individual ridge segments.

As the ocean floor moves away from the hot seam of its birth, it gradually cools and subsides, and the magnetic stripes allow geologists to time this process. After 80 million years the floors have largely subsided and become covered with sedimentary deposits. A formula for estimating the subsidence of sea floor basalts since their formation at a ridge is:

$$\text{DEPTH} = 2{,}500 \text{ m} + 350 \sqrt{\text{OCEAN AGE}}.$$

Thus, ocean crust 50 million years old will subside to 5,000 m deep.

The magnetic pattern in the Southern Ocean reveals the flight of the Australian continent. Spreading gained speed in the last 10 million years and is moving Australia away from Antarctica at about 7 cm/yr. Before then, up to 43 million years ago, Australia moved north at about 5 cm/yr. Prior to that, the separation was slower than 1 cm/yr. So, after a long slow start, a steady stride and then a quickening, Australia's migration collided against the Indonesian and Melanesian borderlands.

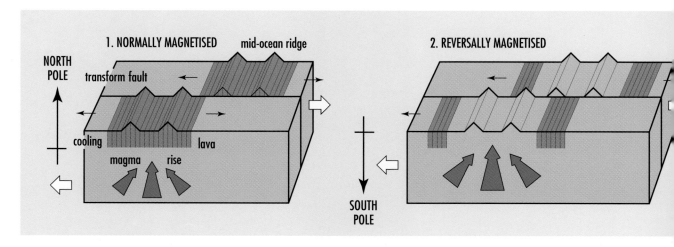

▷ Magnetic stripes on the sea floor around Australia. Normal magnetism shown as dense stripes, reverse magnetism as light stripes. Note different orientation and fewer stripes in the shorter-lived Tasman Sea floor and older sea floor southwest of Western Australia. The active spreading mid-ocean ridge is the dashed line, offset by transform faults.

▽ When magma rises and lava cools at the mid-ocean ridge, as it solidifies it takes on the prevailing magnetic orientation. Normal magnetism (North Pole up) is again shown as denser stripes. When the Earth's magnetic field is reversed, the solidifying lava is reversally magnetised (South Pole up, light stripes). As the sea floor spreads away from the mid-ocean ridges (small black arrows), it carries alternating normally and reversally magnetised stripes.

Transform faults at right angles to the ridge and stripes result in the latter being displaced laterally, shown by the blocks of sea floor sliding past each other (white arrows).

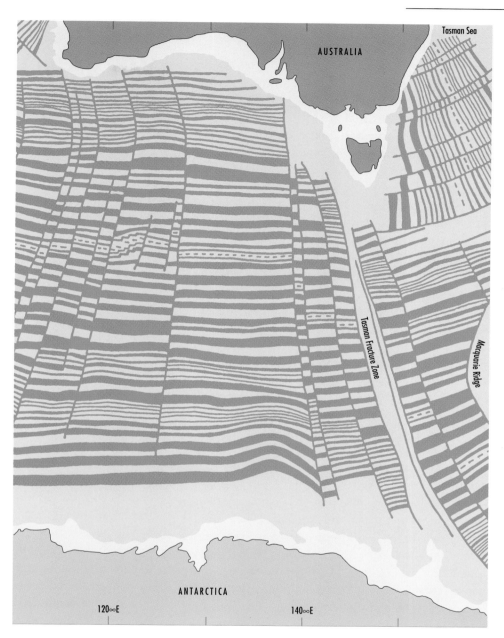

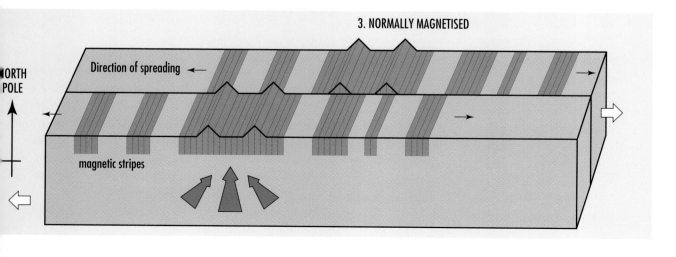

Eastwards, magnetic patterns in the Tasman Sea show an early spreading stage, which 85 million years ago started to move Lord Howe Rise and New Zealand away from eastern Australia. The spreading stopped 55 million years ago, leaving an extinct ridge and a fossil Tasman sea floor. Further north, magnetic patterns in the Coral Sea floor indicate that a slow, limited spreading started 65 million years ago and stopped with the Tasman Sea 55 million years ago. Magnetic patterns in the Indian Ocean floor, northwest of Australia, show that a much earlier separation — over 130 million years ago — split India off Australia.

The symmetric magnetic patterns lead back to the very inception of sea floors. By closing up the parallel stripes, we can join up the opposing sides of original rifts and reconstruct old continental landmasses. Such reassemblies show that many of the breakup points were triple rift junctions, where rifts' arms joined at about 120° to each other. Under stretching and uplift, continents tend to break into such triple points. As stretching continues, the hot mantle rises up under the rifts, often generating some volcanism and then finally the massive melting which creates MORB and the sea floor.

A triple point junction once joined South America, South Africa and Antarctica 135 million years ago before southern Atlantic spreading. Sometimes one rift arm fails to form sea floor unlike the other arms. West Africa, for example, has a failed rift. The two remaining Atlantic rifts separated as spreading floors between the outward bulge of Brazil and the inward curve of western Africa. Closer to Australia, the Gippsland triple point started Tasman Sea spreading: a northern New South Wales rift and a southern Tasmanian rift opened, and the western Gippsland rift remained as the Gippsland basin, keeping Australia and Tasmania together.

Parts of magnetic patterns become lost when later events such as subduction and obduction disrupt the sea floor. The wholesale subduction of sea floor, as it expands away from the mid-ocean volcanoes and collides with an adjacent plate, creates a second generation of volcanoes — the dangerous subduction volcanoes.

SUBDUCTION VOLCANOES: PLATE CONSUMPTION

As the sea floor sinks below an over-riding plate, powerful forces operate on the sliding, cracking slab. The sea floor is dragged down into a deep trench along this action front. These trenches may lie 5–8 km below the surface of the sea and are usually about 2 km lower than adjacent sea floor. As the slab and any overburden of oceanic sediments descend, brittle movements and then mineral changes set off earthquakes from shallower as well as deeper foci. These earthquakes outline the dynamic dive of the slab. This is the Benioff zone, named after the geophysicist Benioff. The angles and depths reached by the subducting slab depend on the relative speeds, directions and resistances of the colliding plates.

Some Benioff zones reach 200–300 km in depth, others descend to 700 km. New studies suggest that the seafloor off northwest Australia has penetrated down to 1,200 km under Indonesia. Steeper subduction is more typical of ocean-ocean subductions and shallow subduction is more common under ocean-continent subductions. Volcanism is part of the subduction process, but the more molten regions of subducted plates seem to miss the main earthquake zones. The hotter, more fluid conditions for magma production probably release stress more easily than do the colder, more locked parts of the slab.

At lower levels, the cooler slab material starts to heat up and release fluids into the overlying mantle. These fluids initiate the melting of mantle peridotite and basaltic magmas rise up behind the trench zone. The flush of fluids releases many elements originally incorporated into the slab by hydrothermal activity near the ridges. These enrich the character of magmas and subduction volcanoes erupt their own brands of lava. However, the overlying plate also adds extra chemical characters to the magmas as they rise and interact at higher levels.

An overlying oceanic plate offers less diverse interaction than a continental plate. Volcanoes supplied by oceanic subduction form lava suites that are simpler than those delivered by continental margin volcanoes.

OCEANIC SUBDUCTION — ISLAND ARC VOLCANOES

An early stage of ocean subduction may be taking place in the western Pacific. Here, a zone of buckling and earthquake activity extends in an arc across the sea floor south of the Marianas Islands to the Samoa Islands. In developing island arcs, volcanoes form over the subducting slab where it begins releasing fluids from the altered sea floor basalts. Water-bearing minerals, such as amphibole, break down at the temperatures and pressures that prevail in the asthenosphere at about 80 km depth. This water and other fluids become enriched by many trace elements. The chemical invasion causes extensive asthenosphere melting and the formation of tholeiitic basalt magmas.

The distance volcanoes develop from the trench depends on the steepness of the subducting slab. Lavas also change in chemistry both along and across arcs, as different parts of the slab melt. Besides tholeiitic basalt, arc volcanoes erupt many lavas rich in aluminium, such as andesites and dacites, which belong to a calcalkaline series. Other lavas richer in alkalis, particularly potassium, belong to an alkaline series. The earliest volcanoes in an arc and those closest to the trench usually erupt tholeiitic basalt. Later volcanoes, and those at greater distances from the trench, typically erupt calcalkaline and then alkaline lavas.

Superimposed on this chemical spectrum are the effects of magma evolution in chambers before eruption. This evolution produces lavas rich in large crystals of olivine, pyroxene, feldspar, amphibole or mica. Lower chambers in the mantle mostly feed low shield volcanoes which erupt basalts and basaltic andesites. Higher chambers in the crust tend to feed large cones and complex volcanoes, which erupt andesites and dacites. The volatile-rich arc magmas erupting from shallow chambers produce highly explosive eruptions, with extensive fallouts and avalanches of debris and common caldera collapses.

Island arcs border the northern Australian plate where subduction has become entrenched. The New Britain volcanoes lie behind the subducting Solomon Sea plate, north of the New Britain trench. The earthquake zone dips at 70° to over 560 km depth. Eruptions in the last 100 years are recorded from Rabaul, Uluwan, Loboban, Pago and Langila volcanoes. The volcanoes mostly erupt andesites and dacites and become more alkaline with increasing distance from the trench.

Another arc lies to the west, off northern Papua Niugini. Islands erupting in the last 100 years include RITTER, LONG, KARKAR, MANAM and BAM VOLCANOES. Their lavas are mostly basalts and andesites. No trench appears along this arc and earthquakes show patchy concentrations, with only some suggestion of a steep north-dipping zone east of Karkar Island. This arc differs from the more

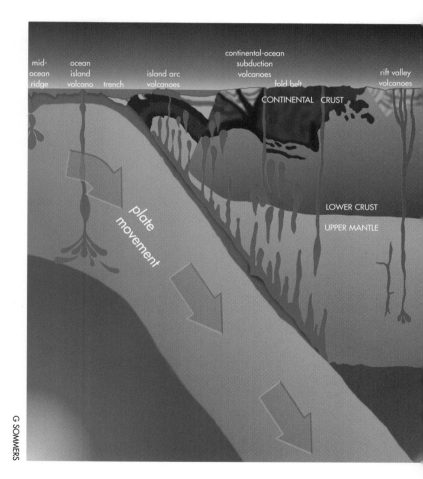

△ Section of a subduction zone showing a plate spreading away from the mid-ocean ridge and descending into the mantle. The sea floor melts and rises as magma producing an unstable continental margin and volcanoes such as those in island arcs, with a basin behind them. Depth of section 200 km.

RABAUL ERUPTIONS

RABAUL VOLCANO AWOKE in the early morning of 17 September 1994, the seventh known eruption in 230 years. Two satellite vents — VULCAN and TAVURVUR, on opposite shores of the harbour — erupted within 90 minutes of each other. The eruption column reached 20 km high and prevailing winds carried its plume southwest over New Britain. Pumice choked the harbour, ash and cinders rained on Rabaul and nearby villages and two large waves damaged islands and shores in Blanch Bay. Several new cones formed around the original Vulcan cone.

Rabaul was ready for this, after the near-eruption in 1984. A well prepared evacuation plan, new roads and an emergency airport site were in place. Five people died, but 30,000 people fled the town unharmed. The eruption had subsided by the end of September and people began to return to the damaged town. It was hardest hit in the southern sections.

Vulcan and Tavurvur had erupted together in 1937, when 500 people were killed. Tavurvur had erupted again in 1941–42, when Rabaul was occupied by the Japanese. These eruptions were dangerous, but not as devastating as those that created Rabaul caldera, before its settlement by the Tolai people. The initial caldera formed in a massive eruption about 6,000 years ago. Another major eruption about 1,450 years ago produced the caldera's present dimensions. Later parasitic vents, such as Vulcan and Tavurvur, acted as safety valves for the underlying magma chamber. Hopefully, the volcano is less likely to produce any more caldera-forming eruptions in its subsequent evolution. A volcano town like Rabaul needs constant vigilance over its surrounding vents.

◁ Aerial view of Rabaul Caldera and its related volcanoes, looking northeast, 1970. The vegetated cone of Vulcan Crater is seen in the foreground and the bare vent of Tavurvur Crater is partly visible across Simpson Harbour on the far right. The large volcano in centre top is Mt Kombiu (The Mother) and the vent at its foot is Rabalankaia Crater.

▽ Vulcan and Tavurvur volcanoes erupting together, watched by locals, Rabaul harbour, September 1994.

EARTH VOLCANIC

81

CHAPTER 7

typical New Britain arc. The lavas become more alkaline from west to east along rather than away from the arc front. This suggests that some volcanoes may pick up arc-like chemical imprints left in the mantle by older subductions, rather than being true subduction volcanoes.

The Indonesian arc volcanoes erupt over a belt 1,500 km long, where the Australian plate is subducting under the Java and Timor troughs. Ocean floor descends along most of this line, but Australian continental materials enter at the Timor end. The arc is about 70 million years old at the Sumatra-Java end and less than 10 million years at the Timor end. Sumatran volcanoes erupt through continental crust. Volcanoes on Java and Flores, the two other large islands, sit 90–190 km above the earthquake zone and mostly erupt tholeiitic basalts and andesites. More alkaline volcanoes sit 185–210 km above their sources and trace elements and isotopes in their lavas indicate that the underlying mantle is still little affected by subduction. The subduction of the old Australian crust in the east must be recent because volcanoes there still show no chemical imprints of the downgoing continental material.

COMPLEX OCEANIC SUBDUCTION — BACK ARC BASINS

Many island arcs, especially in the western Pacific, develop miniature sea floors within their arms. These Back Arc Basins seem to form where old, cold sea floor descends at steep angles. Upwelling hot mantle behind the descending slab splits the lithosphere under the arc and generates spreading ridge volcanoes. If mid-ocean ridges are first generation volcanoes and subduction zones second generation volcanoes, then back arcs are third generation volcanoes.

Back arc lavas resemble MORB basalts but also some subduction lavas. An extra lava type in some arcs is an unusual magnesium-rich andesite called boninite, after the Bonin Islands south of Japan. Boninites probably form when refractory peridotites melt in the lithosphere, as extra heat wells up during back arc formation.

Some island arcs are very complex. The Japanese arc was developed by the subduction of Pacific seafloor under an Asian continental margin about 70 million years ago. It became an island arc after back arc spreading formed the Japan Sea 30 to 15 million years ago. The southeast Japan arc lies behind the subducting Philippine plate and carries highly active volcanoes in Honshu, Kyushu and Ryuku Islands. An oceanic subduction between the Philippine and Pacific plates forms the Izu–Bonin trench and an active volcanic ridge. This joins Japan at a junction between the Philippine, Pacific and Japan Sea plates.

▷ *Cratered cones in Batur volcano caldera, Bali, Indonesia. Accessible to tourists, it is subject to periodic eruptions; taken in repose, August 1983.*

MT FUJI, Japan's largest volcano stands over this great tear in the lithosphere. Its large volume (1,400 km^3) is mostly basalt, with much explosive material. HAKONE VOLCANO, next door, is more typical of island arc volcanoes, erupting basalt to dacite lavas and showing a caldera collapse. North of this, Pacific plate subduction generates highly active volcanoes along northeast Honshu, Hokkaido and the Kurile Islands.

Back arc basins are actively forming along the Australian plate margin, behind the New Britain, New Hebrides and Tonga–Kermadec arcs. The Bismarck Sea Basin formed over 3 million years ago behind the New Britain arc and is opening at a very rapid 13 cm/yr. The Fiji basin, which began 10 million years ago, is opening at 7 cm/yr. The old arc has rotated a remarkable 60° from its original position since the basin began opening. The Lau Basin began 6 million years ago behind the Tonga and Kermadec volcanic chains and is opening at 8 cm/yr.

Studies of the Lau Basin floor have revealed an unexpected process. The original asthenosphere that gave a Pacific isotope signature to the arc and back arc volcanoes here is vanishing. Asthenosphere with an Indian Ocean isotope signature is

OCEANIC-CONTINENT SUBDUCTION — ANDEAN VOLCANOES

Ocean floors, subducting under continents, generate magmas similar to those erupted in island arc volcanoes. However, thick continental crust changes the amounts and complexities of the lavas that finally erupt. A major ocean-continental subduction forms the 10,000 km long Andean mountain chain. These mountains were raised as the eastern Pacific floor subducted below South America over the last 100 million years. Most of the present volcanoes form where the Nazca plate descends along the Peru-Chile trench (see photographs, next page). It dips around 30° under the Andes, but in some places becomes less than 10°

coming in. This reverses the invasion and replacement of Indian Ocean asthenosphere by Pacific asthenosphere in the Australian–Antarctic Mid-Ocean Ridge. The overall pattern hints at deep mantle flows operating under the Australian plate as it moves northwards.

△ *Fuming Mihara-Yama crater in Oshima volcano caldera, Izu-Oshima Island, Japan, September 1992. Lava fire fountains and flows from this crater in the 1986 eruption forced evacuations.*

◁ *Crater rim, with group descending into steaming crater, Villarrica volcano, Chilean Andes, 1989.*

▷ *Composite volcanic cones, Volcan de las Tres Virgines, Baja California, Mexico, August 1982.*

and too shallow to generate volcanoes. Moving mantle backing up behind the steep zone seems to be forced sideways to flow around the north and south ends of South America.

The amount of melting and volcanism caused by descending slabs depends on whether younger and warmer or older and colder ocean floor goes down. With warmer slabs, higher temperature gradients allow greater melting of overlying mantle and even the continental crust above. This last layer will yield voluminous eruptions of silica-rich lavas such as trachytes and rhyolites. The Andean margin is subducting relatively young Pacific floor and its volcanoes form in three main zones:

1. The northern volcanoes in Columbia and Ecuador (2–5° S). These erupt largely andesites and basaltic andesites.

2. The central volcanoes in southern Peru, northern Chile, Argentina and Bolivia (18–26° S). These erupt a great range of calcalkaline and alkaline lavas and their silicon, aluminium and potassium contents are higher than those found in the other zones.

3. The southern volcanoes in southern Chile and Argentina (33–45° S). These erupt basalts with minor basaltic andesites and dacites.

The central volcanoes reflect the presence

△ Looking north along Chilean Andes from Villarrica volcano to Osorno volcano, September 1990.

▽ A'a lavas from Volcan de las Tres Virgines volcano, Baja California, Mexico, August 1982.

and melting of an older continental block where the crust is 50–70 km thick. Old uplifted ranges like the Andes, where subduction began 100 million years ago, also expose large chambers and intrusions from previous volcanism, now bared by erosion. These bodies reach 300 km across in huge, elongated intrusions called batholiths or in smaller more symmetrical intrusions called stocks. They often exhibit complex structures of coarser-grained rocks such as granite, granodiorite, diorite, syenite, monzonite and gabbro. These are the deep equivalents of rhyolite, dacite, andesite, trachyte, high potassium alkaline and basalt lavas respectively. The Peru batholiths have four granite phases which invade and surround large blocks of earlier granites and volcanic beds, giving nested 'belljar'-shaped intrusions.

C H A P T E R 7

Similar continental subduction volcanoes and intrusions border the Pacific. The central American–Mexico belt, with over 30 active volcanoes, extends along the subducting Cocos plate and into the Baja California region. The Cascades Range and Southern Coast Range volcanoes extend from northwest America into southwest Canada. They lie inland of the Juan de Fuca plate subduction zone, north of the San Andreas Fault. The active Cascades volcanoes form well-spaced centres with large volumes of lavas and explosive deposits, but the Southern Coast Range volcanoes form a line of numerous small volcanoes. Alaska supports over a dozen active volcanoes, where the northern Pacific plate subducts under the Alaskan Peninsula. Nearly 30 active volcanoes and some huge calderas border the northwest Pacific plate where it subducts under Kamchata Peninsula.

An active continental belt passes under the high mountains of the Himalayas, Iran and Turkey into the Aegean and Mediterranean regions. It involves a complex continental-continental collision between the Indian, Asian, African and European plates, with subduction under back arc basins at its western end. This zone features an association of calcalkaline lavas with highly alkaline lavas, very rich in potassium. These high potassium lavas are well known at Vesuvius and around Rome.

OCEANIC-MICRO CONTINENT SUBDUCTION — AUSTRALIAN MARGIN

Active margin volcanoes fringe the Australian plate in the continental slivers forming Papua Niugini and New Zealand. In Papua Niugini there are 19 volcanoes in western Papua, 7 volcanoes in eastern Papua and 9 volcanic fields on the D'Entrecasteaux Islands off southern Papua. In New Zealand there are 4 active volcanoes, several geothermal fields and some dormant volcanic fields in the North Island.

Western Papuan stratovolcanoes have erupted basalts, andesites and alkaline potassium-rich rocks. Widespread explosive deposits around MT HAGEN and MT GILUWE suggest that caldera eruptions occurred 30,000–50,000 years ago. Eastern Papuan volcanoes include MT LAMINGTON, WAIOWA and MT VICTORY, all of which erupted in the last 100 years. Mt Lamington was inactive until a glowing avalanche eruption in 1951 devastated everything in its path and killed 3,000 people. The Australian volcanologist G.A.M. Taylor was awarded the George Cross medal for his brave surveillance of the Lamington eruptions. The D'Entrecasteaux Island volcanoes include active geothermal areas and those along Dawson Strait are noted for explosive rhyolite deposits.

◁ *Volcanic bomb in younger explosive deposits (yellow stained layer) overlying older explosive deposits (lower grey layer), Nevado de Toluca, Mexico, September 1993.*

RR COENRAADS

▷ *Eruption from crater lake, Ruapehu volcano, New Zealand, 8 May 1971. The steam explosion ejected breadcrust bombs of fresh andesite with old rocks and much mud, ash and water.*

▽ *Summit of Ngauruhoe volcano, New Zealand, showing loose andesite lava fragments, on the active cone's slopes, 1967.*

The Papua Niugini volcanoes erupt calcalkaline and alkaline potassium-rich lavas typical of subduction margins, although no clearcut subduction underlies the region. The rhyolites of D'Entrecasteaux Islands are chemically similar to those of continental rifts. The area is complex; it had early sea floor obduction along the Owen Stanley ranges, and subduction events took place before the present uplifts and strong lateral faulting. The volcanoes probably tap mantle that was chemically imprinted by earlier subduction events.

North New Zealand shows a classical suite of subduction volcanoes where the Pacific floor descends under the southern Kermadec trench. The TAUPO VOLCANIC ZONE, with 20 volcanic centres, extends northeast from TONGARIRO to WHITE ISLAND in the Bay of Plenty. The TARANAKI VOLCANIC ZONE, with 4 centres, lies 100 km west of Tongariro. The Taupo zone sits 250 km in from the subduction front and 80–120 km above the slab; the Taranaki zone is 350 km from the front and sits 180–200 km above the slab. The large TONGARIRO and TARANAKI STRATOVOLCANOES probably rise along a tear in the descending slab.

Andesites dominate the Tongariro massifs and active RUAPEHU and NGAURUHOE VOLCANOES. The adjoining low-lying LAKE TAUPO CALDERA does not immediately convey its extreme potential for catastrophic eruptions. However, the AD 186 eruption discharged 9 km³ of pumice in a stupendous upward blast, at a rate of 100,000 m³ per second, reaching a height of 50 km. It covered the North Island with barren white ash over 1,500 km². The WAIRAKEI GEOTHERMAL FIELD now operates on the volcano's north side.

Further north, the MAROA CALDERA contains the Broadlands–Ohaaki geothermal field; the OKATAINA CALDERA is linked to the 1886 TARAWERA RIFT eruption and to the WAIMUNGU and ROTOMA GEOTHERMAL FIELDS; and the ROTORUA CALDERA and GEOTHERMAL FIELD lies west of the Okataina caldera. All these calderas erupted rhyolitic lavas and contain lava domes as well as huge explosive deposits. A narrow belt of tholeiitic basalt, andesite and dacite volcanoes extends along the south side to WHITE ISLAND VOLCANO, but to the north the rhyolitic eruptions fill a basin opening up under the stress of subduction.

The TARANAKI VOLCANOES in contrast form 4 centres which become younger towards the southwest. They erupt potassium-rich relatives of the Taupo Zone andesites. MT TARANAKI is the youngest vent of EGMONT VOLCANO, which is less than 120,000 years old. The present cone erupted lavas 2,800–7,000 years ago, and 400–700 years ago, and last erupted in AD 1655. Despite its scenic charms, the steep slopes with abundant fragmental deposits make Egmont volcano a dangerous edifice. Its hazards are now under detailed surveillance.

A former volcano rises in solitary splendour off the South Island of New Zealand, SOLANDER ISLAND. It is an eroded andesite volcano, similar to Mt Egmont in nature and about 1 million years old. The island, 300 m high and 1.5 km across, is probably part of a large submerged volcano which is 15 km across at its base. It probably formed as oceanic crust from the Tasman Sea, subducted below this complex structural region.

▷ *Lapilli tuffs of young cone exposed at Herchenberg quarry, northern outskirts of Laacher See volcano, east Eiffel field, Germany, September 1990.*

WITHIN-PLATE VOLCANOES: RIFTS AND HOTSPOTS

Various breeds of volcanoes inhabit plate interiors. They range from small explosive craters of highly alkaline lava, like the LAACHER SEE VOLCANO in Germany, to huge and complex basalt volcanoes, like HAWAII. They also include rift volcanoes and hotspot volcanoes. Australia and its marginal sea floors form a plate interior setting, referred to as 'within-plate'. No vents are active, but extensive belts of former within-plate volcanoes embroider the east side.

RIFT VOLCANOES — STATIONARY LINES

Stretching, along with downwarping and doming, eventually ruptures the crust into fault blocks and intervening rift valleys. The lithosphere thins under the rupture zone and asthenosphere wells up. High heat flows may cause melting and volcanism. Some rift lines develop virile volcanism, others develop only a little, and many display none at all. In essence, volcanoes erupt periodically along the fault structures and volcanism stays within these confines. The CAMEROON VOLCANIC LINE, West Africa, probably overlies a deep fracture

△ *Rift valley wall, Bahia de los Angeles, Baja California, Mexico, August 1982.*

alongside an old rift and has erupted up and down this line, across continental and oceanic crust, for 66 million years.

The Rio Grande Rift in New Mexico, USA, is a strongly volcanic rift that evolved over 32 million years. After extensive volcanism, the mantle now lies 28 km below the rift. Magma is present in a thin pool in the crust and extends over a 1,700 km² area that is 20 km deep as well as in small pockets 5–10 km deep. The East African–Ethiopian rifts extend for 1,000 km, branch into separate rifts and have erupted 500,000 km³ of lava in 45 million years. Ethiopian crust has stretched by 30 km, eastern Kenyan crust by 10 km, and the western Tanzanian crust by 2–3 km. Each branch differs in its volcanic nature.

Rifts often show volcanic activity peaks which vary in their volume and in the chemistry of their erupted lavas. This reflects the rates of crustal uplift, extension and amount of mantle upwelling. Alkali basalts commonly evolve into trachytes, rhyolites and phonolites. Lavas also include silica-rich basalts and alkaline lavas that are very low in silica, such as nephelinites, melilitites and leucitites.

THE RISE OF RIFT VOLCANOES

Early rift activity marks the Eiffel volcanic district of Germany. These volcanoes erupted in the last 700,000 years with the uplift of the Rhenish shield, a large block of lithosphere cut by the Rhine River. Restricted melting and upwelling in

the mantle led to the eruption of small cones and flows, maars and tuff rings, and a few larger explosive complexes. A western group of 250 volcanoes erupted basanite, nephelinite, melilitite-nephelinite and leucitite lavas. A more evolved eastern group of 100 volcanoes erupted phonolite lavas in the main volcanoes, such as the 1 km wide Laacher Sea crater. This 11,000 year old eruption buried artefacts such as elaborate slate carvings from the Magdalenian settlement. Bones and tools recovered from older crater fills make it possible to reconstruct the life and environment of people who dwelt there 400,000 years ago.

The East African–Ethiopian rift system demonstrates progressive stages of rift volcanism. The western rift shows restricted volcanism, but includes the highly active NYIRAGONGO and NYAMURAGIRA VOLCANOES. Their lavas are commonly low in silica, highly alkaline and potassium-rich, and resemble the Eiffel lavas.

The eastern rift is highly volcanic, supporting large volcanoes like MT KENYA and KILIMANJARO and many smaller and unusual alkaline volcanoes. The main lavas include members of the basanite–phonolite, alkali basalt–trachyte and basalt–rhyolite series. OLDOINYO LENGAI is a most unusual alkaline volcano, the only known active carbonatite volcano. It erupts a sodium carbonate lava, similar to washing soda, which flows more freely at lower temperatures than normal silicate lavas do.

The Ethiopian rift volcanoes mostly erupt lavas that are more transitional to compositions of tholeiitic basalts. These lavas evolve into rhyolites and show influences of MORB basalts. This rift system has a greater mantle melting tendency near its junction with the Red Sea–Gulf of Aden triple point; sea floor spreading and eruption of MORB basalts has taken over there.

Australia has no active rift volcanoes but there are old examples along the breakup rifts that formed its continental margins. Rift volcanoes are found in Antarctica on the Ross Sea–Marie Byrd Land margins, but ice and snow make their detailed study difficult.

HOTSPOT VOLCANOES — MOVING CHAINS

These exciting volcanoes act as flares, marking the movements of plates. They involve a deep upwelling from the asthenosphere and a plate in

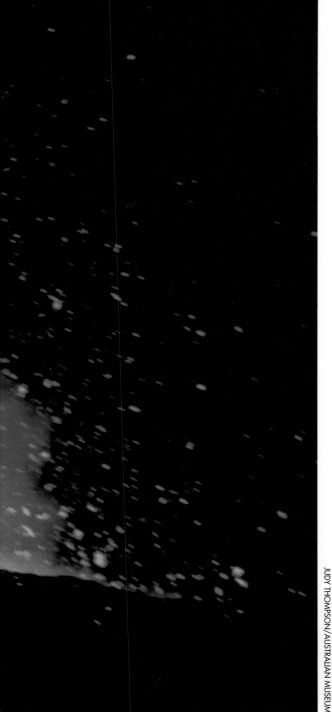

△ Nyamuragira volcano erupting at night, Virunga Park, Rumagaro/Tongo station, Zaire, August 1992.

▷ Formation of volcanic chains over a hotspot. As the continental plate moves to the left, the magma supply to the presently active volcano will be cut off and it will become extinct. A chain of volcanoes results from the plate motion, those further left being older, and more eroded.

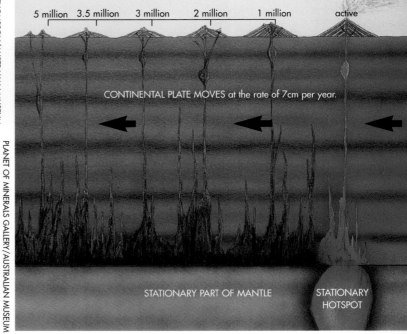

motion. The upwelling heat mobilises melting and large quantities of magma erupt to form a volcano over the melting zone. However, the volcano is eventually doomed. It cannot grow indefinitely because the plate carries the edifice away, choking off its magma supply. Another volcano takes its place, then another, and so on, in a process that produces a chain of volcanoes getting older and older and farther away from the hotspot.

The most famous hotspot volcano is HAWAII, but other islands of the chain once stood in active glory over the hotspot. MAUI was there 0.75–1 million years ago, MOLOKAI 1.75–2 million years ago, OAHU 2.5–4 million years ago and KUAI 5 million years ago. This volcanic sequence measures the Pacific plates motion over the Hawaiian hotspot — 8–9 cm/yr in a westnorthwest direction. The hotspot volcanoes vanish under the Pacific waters as they age and erode away, until only submerged stumps remain to trace the spread of the Pacific floor.

Some 3,000 km from Hawaii the volcanic chain, now 43 million years old, bends in a near-northerly direction. This marks a change in the Pacific spreading direction, with the chain continuing as the Emperor seamounts. MEIJI SEAMOUNT is the last volcano in the chain before the Pacific floor dives under the Aleutian trench, destroying the hotspot record earlier than 73 million years ago. LOIHI, the young volcano at the foot of Hawaii, will inherit the hotspot's mantle.

LIFE CYCLE OF A HOTSPOT VOLCANO

How do hotspot volcanoes grow? This question makes Hawaii the most studied hotspot volcano. Its beginnings are buried, but its offspring Loihi suggests an undersea start. A caldera full of active hydrothermal vents would have formed a summit from pillow lavas and flows up to 2 km long, with steep lower aprons of broken pillows and lava. The first lavas would have been alkali basalts, formed from small-scale melting starting in the underlying plate when it engaged the hotspot. Then tholeiitic basalts joined in as melting increased over the hotspot, escalating to massive rapid eruptions of tholeiitic basalts as shield-building brought Hawaii above sea level. As Hawaii moves off the hotspot, the reduction in melting produces more alkali basalts, with the stagnation and evolution of magmas in chambers leading to eruptions of some alkaline lavas such as trachyte. Chapter after chapter in Hawaii's future erosion as an extinct volcano can be predicted from what is now seen in the neighbouring islands as they increase in age.

As each giant volcano dies erosion sets in, stripping its bulk. After 1–3 million years, the unloading causes uplifts and small degrees of mantle melting. Last gasp, post-erosional eruptions break out, but differ from the previous basalts. Very poor in silica and rich in alkalis, the lavas form basanites, nephelinites and melilitites. After this final fling, the island slowly sinks as the ocean floor cools and subsides away from the hotspot. It is reduced to sea level, then crowned with its coral reefs to form an atoll. It sinks into a seamount, a far cry from its heady hotspot days. The birthplace of the hotspot itself is a deeper story.

Hotspots such as Hawaii may rise from the mantle-core boundary layer 2,900 km below. This thermal layer cools by releasing funnels of heat, which rise in plumes of solid hot rock. These plumes impinge on the lithosphere at temperatures of 1,300–1,550° C. A large zone of mantle melting forms and feeds the hotspot chain on the overlying plate.

Not all hotspots rise off the Earth's core. Some smaller hotspots form in lines, once linked to sea floor spreading ridges. These may come from a discontinuity 670 km deep, which separates the upper and lower mantle. Other hotspots may turn out to be 'wetspots', where mantle that is altered by fluid infiltrations melts more easily and produces the same effect as a hotspot. This has been suggested for the Azores in the mid-Atlantic.

Hotspots stir the Earth's mantle, but what processes are involved is still debated. Slabs of subducted ocean floor become stored at the upper-lower mantle boundary or even the lower mantle-core boundary. Eventually they ascend as hotspot upwellings hundreds of millions of years later. Hotspot lavas are spiced with traces of old chemistry, giving complicated stories for geologists to decipher.

▷ *Western flank, Mt Kenya, Kenya, January 1984.*

◁ Eroded hotspot volcano with fringing coral reef, Lord Howe Island, Tasman Sea, aerial view looking north to Old Settlement Beach, North Beach and Mt Eliza, April 1982.

Hawaii is one of many hotspot volcanoes and its pedigree can be traced across the Pacific plate. RÉUNION ISLAND hotspot in the Indian Ocean originated in northwest India, over 65 million years ago. HEARD ISLAND hotspot, southwest of Australia, can be traced back through KERGUELEN ISLAND and the Ninety-East undersea volcanic chain into northeast India, 115 million years ago. Many lesser hotspot chains extend across the Pacific, such as the Pitcairn–Tuamotu Island group and Macdonald seamount–Austral Island group. Some mimic the same bend in direction shown by the Hawaiian–Emperor chain. The Louisville chain crosses the southwest Pacific floor until it vanishes at the Tonga trench at the 65 million year benchmark.

Whole series of hotspot islands border the Atlantic ridge. THE AZORE, CANARY, CAPE VERDE, ASCENSION, ST HELENA, TRISTAN DA CUNHA, GOUGH and BOUVET islands trace the movement of Africa away from the hotspots during Atlantic opening. The Iceland hotspot traces its origins back across Greenland 60–65 million years ago and perhaps across the Canada basin back to 150 million years ago. The hotspot became anchored on the mid-Atlantic spreading ridge to build the ICELAND basalt volcanoes. JAN MAYEN ISLAND, further north, may trace its hotspot back to Siberia, 250 million years ago.

CONTINENTAL HOTSPOTS

Hotspot volcanoes also bob up in continents, the most prominent being YELLOWSTONE CALDERA in Wyoming, USA. A tremendous eruption 600,000 years ago ejected 1,000 km³ of rhyolitic ash, leaving a 70 km long caldera and spreading ash across 21 States and into Canada and Mexico. It was a far greater explosion than the 1980 Mt St Helens eruption. The caldera's present activity includes only the renowned geothermal plays of geysers and springs, but it overlies an 8–10 km thick chamber of hot rhyolitic rock. Below this, basaltic magma chambers lie above partially molten mantle descending to 250 km. The Yellowstone hotspot left a trail of widespread lavas across Idaho, Oregon and Washington States forming the Snake River and Columbia River basalts. These basalts age westwards to 17 million years and were formed as the North American plate drifted west over the hotspot.

Australia has a long-lived continental hotspot system. It lies under Victoria, Bass Strait, Tasmania and the Tasman Sea floor at about 40° S. Present activity is confined to seismic zones between Flinders Island and the Tasman Sea, and to deep gas discharges under Victoria and Tasmania. This hotspot system left chains of basaltic volcanoes as the Australian plate moved northwards across its line. Many of these volcanoes have large central eruptions of rhyolite and trachyte lava besides tholeiitic and alkali basalt lavas, resembling the Yellowstone situation. The Tasman floor volcanoes, however, are mostly tholeiitic and alkali basalts, resembling the Hawaiian situation. Hotspot magmas under continents are more likely to evolve and interact in crustal chambers than under thin ocean crust.

FLOOD BASALTS: ULTIMATE VOLCANOES

Some parts of Earth, continent and ocean, were once inundated with huge volumes of lava, much greater than those produced by normal hotspot volcanoes. Such volcanic outbreaks, not witnessed at present, appear every few tens of millions of years. Lavas from such wholesale melting are mostly tholeiitic basalts. The main continental regions affected over the last 250 million years include the Columbia River, Ethiopian, Deccan, North Atlantic, Rajmahal, Serra Geral, Karoo, Newark, Siberian basalt provinces and Antarctica–Tasmania. These basalt lava provinces reach 9 km in thickness and cover areas up to 2 million km^2. They also include oceanic equivalents, such as Ontong Java Plateau on the Solomon–Pacific margin and the Kerguelen Plateau in the Southern Ocean. The emplacement rate for Ontang Java lavas probably exceeded the whole present mid-ocean ridge output.

The Pacific super-swell was an enormous volcanic area in the central Pacific during the Cretaceous Period. Termed a 'superplume', it created many island and seamount chains, such as the Line Island and Musician seamounts. These formed over a hot upper mantle zone extending down to 450 km. Some scientists claim that these

▽ *Transantarctic Mountains, North Victoria Land, Antarctica, 1970; a region of eroded flood basalt lava remnants and exposed dolerite sill intrusions.*

are periodic outbursts with runaway volcanism occurring every 20 to 40 million years. Some even claim that extraterrestrial meteorite impacts triggered such exceptional melting. Others disregard visitations from the heavens and favour the theory that Earth's core shrugged off excess heat to produce supersized hotspots.

The hotspot models invoke plumes rising into the asthenosphere and flattening out below the lithosphere in a huge mushroom-like head 1,000–2,000 km across. This produces massive melting and a flood basalt province within a few million years. As the hot mushroom expands, it incorporates cooler lithosphere. In cross-section the plume axis carries an enlarged head with 'eyes' of cooler material on each side, giving the fanciful term 'tadpole tectonics'. The rising plume head forces up the lithosphere into a dome and the fractures allow ready access to waves of rising basalt magma.

Called 'starting plumes', these events can involve rifting and sea floor spreading, and they sometimes feed hot material into flanking mid-ocean ridge systems. After the plume-head generates flood basalts, the thermal system cools, but its conduit continues as a hotspot source. Thus, the Deccan plume basalts can be traced to the Réunion Island hotspot, the Kerguelen Plateau–Bunbury–Rajmahal plume basalts to the Kerguelen and Heard Island hotspots, and the Greenland plume to the Iceland hotspot. A proposed Cretaceous superplume is linked to many hotspot volcanoes in the Pacific. Australia may have a plume that once produced rifting and spreading in the Coral Sea before it continued as a hotspot system down eastern Australia and the Tasman Sea.

While the starting plume model accounts for many features of flood basalts, there are difficulties with applying it to certain cases. For example, the proposed starting plume for Greenland–North Atlantic basalts seems to connect to an earlier hotspot trace extending north along Baffin Island, Ellesmere Island and the Alpha ridge in the Canadian Basin. West Greenland basalts seem to erupt before the east Greenland–North Atlantic basalts, so that rifting may stimulate rather than cause the volcanism. Similarly, the Deccan plume may arise from an older incubatory plume. The Cretaceous superplume may just be a large area of hot mantle, not cooled by any subduction events for hundreds of millions of years. This illustrates the complexities in these large volcanic events.

BEYOND PLATE TECTONICS: ALTERNATIVE VIEWS

Plate tectonics theory explains many of the operations involved in volcanic eruptions, even some features that seem anomalous. For example, in the seamounts behind Marianas trench the slab temperatures would be too low to produce volcanoes. Investigations show these to be 'mud volcanoes' erupting cold, green serpentine muds. Fluid distilled from the subducting plate has changed the peridotite in the overlying plate into serpentine, which can then flow up under pressure.

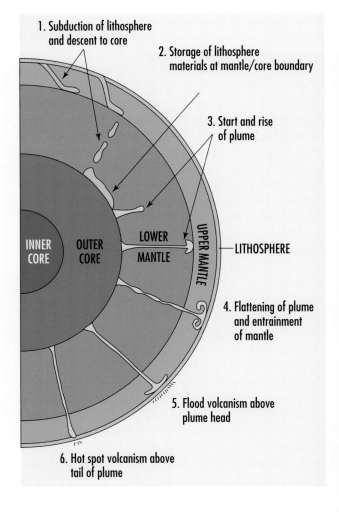

Plate tectonics assumes that Earth's plates all act as rigid bodies. However, a series of satellite measurements has indicated that some areas might deform in less rigid ways. Some volcanoes show little obvious connections to plate features and have enigmatic origins. Most plate tectonic theories maintain that these processes take place on a planet of consistent size — where all the new sea floor made by mid-ocean ridge volcanoes is swallowed up by subduction.

However, a few scientists suggest a different tectonic aspect — that Earth expands. Several expanding earth ideas have been proposed. For example, in the 1930s Hilgenberg considered continuous continents on an Earth 55% of its present diameter, and in the 1950s–1970s Carey, an Australian Professor of Geology in Tasmania, championed fast expansion over the last 160 million years to explain the present ocean floors. This view precludes ocean floor subduction. In the 1970s–1980s Owen, from the British Museum, favoured expansion, but with subduction, proposing that 180 million years ago Earth's diameter was 85% of its present size. In the 1980s Weijermers, in contrast, accounted for the Earth's plates by assuming a constant size Earth, but made some assumptions on ocean floor production and subduction.

Other scientists, such as Keith in the 1990s, have promoted alternatives to plate tectonics, basing their theories on mantle upwellings under continental margins. Here, they maintain, sea floor descends at mid-ocean ridges to cause volcanism, and decreases in the age of sea floors towards the ridges are due to contracting volcanism. They suggest that hotspot tracks come from the recycling of mantle from old rift zones into the asthenosphere, rather than plate movements over upwelling plumes.

All these different models illustrate the complexities involved in the moving Earth and the origin of its volcanoes. In this book, Australia's past volcanism is described according to the basic concepts of plate tectonics.

◁ *Diagram showing evolution of hotspot systems. The cycle illustrated from initial subduction to the rise and decay of hotspots may take hundreds of millions of years.*

VOLCANIC FORMS AROUND AUSTRALIA

*Breadknife,
 dyke,
nbungle
ains, New
Wales,
ber 1986.*

AUSTRALIAN Aborigines saw the volcanic land in action. Myths from northern Australia, where young scoria cones exist, can be interpreted in terms of volcanic eruption. The myth of Goorialla recounts how this ancestral being was cut open to free a pair of brothers after he had swallowed them. Awakened, Goorialla tore apart the mountain he had slept on and scattered it across the land. In their efforts to find cover, other ancestral beings took to the air, water and ground to form the creatures of the land. Goorialla returned into the ocean off Cooktown, but took on the form of shooting stars to keep an eye on the world.

Another myth involves three volcanic crater lakes. The breaking of a taboo by newly initiated men aroused the rainbow serpent. The earth under the campsite roared, twisted and cracked, the wind blew like a cyclone and a red cloud of unusual hue entered the sky. After forming LAKE EACHAM, similar events formed LAKE BARRINE and LAKE EURAMOO.

△ *Lake Eacham, a young water-filled volcanic crater, Atherton Tablelands, north Queensland, November 1971. A local Aboriginal legend attributes this feature to the Rainbow Serpent.*

In western Victoria, some Aboriginal place-names suggested that there was heat in the ground at a former period. This is reported by James Dawson in his 1881 compendium on the languages and customs of local tribes. There was little tradition for craters throwing out smoke or ash. However, there were memories of a fire that came out of 'Bo'ok', a hill near Mortlake, and of volcanic bombs in scoria at MT LEURA that resembled stones thrown out of the hill by the action of fire.

Legends of the local Aborigines around Mt Gambier in South Australia invoke the family of the giant Craitbut. They tried to camp and make ovens at MT MUIRHEAD and MT SCHANK but were frightened by a spirit. They escaped to Mt Gambier and made an oven, but water came up and put it out. They made others, until the four MT GAMBIER CRATERS were created.

A volcanic myth seized the imagination of European settlers after explorers discovered Burning Mountain near Wingen in New South Wales. Vent-like, cracked hot ground, constant smoke exhalation and encrusting yellow deposits of sulphur and other mineral sublimates fostered this myth, which prevailed until another theory — that water was acting on iron pyrites — surfaced in the 1840s. Conflicting views emerged in 1866. *The Illustrated Sydney News* proclaimed Burning Mountain as the only Australian volcano, while *Baillieres New South Wales Gazetteer and Road Guide* favoured a coal seam combusting at depth. Geological studies by Professor T.W.E. David from Sydney University confirmed a coal seam conflagration and the volcanic myth gradually died. It still lingered in some school texts in the 1920s and in advertisements for Winjennia soap, made of 'volcanic substances of curative powers'.

VOLCANO LANDSCAPES: AUSTRALIA & NEW ZEALAND

Many volcanoes, particularly those in eruption, have caught the artist's eye and become subjects of artworks. VESUVIUS was frequently painted over the centuries, the Japanese artist Katsushika Hokusai produced '36 views of Fuji' and Hawaii's volcanoes have inspired many artistic endeavours since 1794. While unable to capture an Australian volcano in its eruption finery, early Australian artists brought their talents to bear on volcanic landforms in the young lava fields (box, p.105). New Zealand not only offers artists splendid young volcanic scenery in the North Island, but also turns on eruptions every few to tens of years to guide artists' reconstructions of past activity.

Vivid artistic depictions of New Zealand's volcanoes have appeared in books by Geoffrey Cox, a professional illustrator and writer of natural history themes, including: *Slumbering Giants: the volcanoes and thermal regions of central North Island*; *Fountains of Fire: story of Auckland's volcanoes*; and *Mountains of Fire, the volcanic past of Banks Peninsula*.

△ *Large volcanic bomb with spindle tail, Hampden Shire Quarry, Victoria, September 1981.*

▽ *Burning Mountain, near Wingen, New South Wales, October 1987. Smoke and mineral deposits coat bare ground in contrast to the surrounding land.*

◁ *Volcanic bomb embedded in scoria, Hampden Shire Quarry, Victoria, September 1977.*

YOUNG VENTS

Well-preserved volcanic vents greet visitors to young basalt fields in Queensland, Victoria and South Australia. The western Victorian fields exhibit the greatest pride of vents. About 50% are dominated by scoria deposits, 33% are largely lavas, 10% are maars and tuff rings, and 3% are complex cones and maars. In contrast, the South Australian fields show 70% scoria vents, 25% maars and 5% lava volcanoes. Scoria cones and lava volcanoes dominate most Queensland fields and maars are only prominent around Atherton.

SCORIA, CINDER AND COARSER PYROCLASTIC CONES

Scoria is loose rubbly basalt, flung up in regular eruptive bursts. It falls around the vent, building up a cone with loose slopes. The typical size of lava fragments is 2–64 mm and they are often frothy from escaping gas bubbles. Near the vent, scoria usually contains large blocks of basalt called bombs (photographs, previous page), too heavy to travel far. These may be spindle-shaped or ribbon-shaped due to the rotation of plastic lava in flight. Breadcrust and cowpat bombs are self-explanatory. These deposits are typecast as Strombolian deposits, typical of activity at STROMBOLI VOLCANO between Sicily and Italy. Scoria grades into finer material called cinders (although they are made of rock, not burnt wood), or else it grades into coarser fragments, the size depending on the intensity and height of explosions and wind effects. Scoria cones usually build a circular vent with a summit crater and can form in clusters or rows. Many release lava flows during their activity.

Some cones include basalt spatter, lava fragments formed in a fast incandescent rain from explosive fire-fountaining. The pieces stick together since they are hot and plastic on landing, but are too stiff to form a trickle of lava. These deposits are typical of Hawaiian activity. Other cones contain numerous, larger pieces of basalt torn from older lavas of vent walls. These more explosive, irregular and mixed deposits are the trademarks of the unpredictable explosions of VULCANO ISLAND, near Stromboli Island.

An isolated, but accessible pyroclastic cone rises from the plateau west of Ingham in north Queensland. MT FOX, a cone 120 m high, has a shallow crater and a lava flow on its southern side that fills an old valley. This monogenetic volcano is probably about 100,000 years old. In contrast, the SEVEN SISTERS are eight small aligned scoria cones without craters, east of Atherton in north Queensland. Nearby, the prominent 170 m high MT QUINCAN CONE exhibits cinder beds and volcanic bombs in quarry operations. A swamp occupies a summit crater and the higher northwest side and northwesterly elongation of the cone indicates eruption during prevailing ESE winds.

Near Gayndah, in central south Queensland, MT LEBRUN forms a compound pyroclastic cone

▷ Crater rim, Mt Noorat, Victoria, looking over Hampden Shire Quarry and surrounding plain, September 1977.

▽ Hummock lava volcano, Bundaberg, Queensland, March 1991; from summit, showing basalt flows extending east to the coast.

that is 330 m high and contains two crater lakes, the COULSTON LAKES. These cones erupted long flows of lava which ran into the Burnett River over a 130 km distance. This basalt is about 600,000 years old. In contrast the small, shallow BRIGOODA CRATER 90 km to the southwest marks a brief explosive burst that occurred about 400,000 years ago with negligible cone construction.

Numerous cones relieve the monotony of the lava plains in western Victoria. MT ELEPHANT, the

breached-summit crater in a rampart of spatter bombs, both cowpat and breadcrust varieties. The large scoria cone at Mt Eccles rises above an elongate crater lake, LAKE SURPRISE, which fills coalesced craters in the lower lavas. Massive scoria beds in the cone contain some contrasting layers of more closely packed bombs and thin, more explosive deposits which punctuated the main Strombolian buildup. TOWER HILL is a cone complex built by both Strombolian and Vulcanian activity, but began by forming within a large maar volcano (see p.104).

The young South Australian volcano, Mt Schank, boasts a fine scoria cone built over a volcanic fissure. The fissure cleared its throat with some early explosive activity and lava escape. It then built the main cone in two stages, joining up with a small maar on its south side in the process.

LAVA VOLCANOES

These vents show little explosive activity. They form low cones or domes, from which lava can often flow down over considerable distances. Profuse lava may erupt from fissures, and vents may collect spatter from fire fountains. These vents are often buried by later flows and the landscape develops extensive lava plains and plateaus. In Iceland, such volcanoes typify Icelandic activity and examples appear in young Australian basalt fields.

UNDARA VOLCANO centred in the McBride basalt field, north Queensland, is celebrated for exceptionally long lava flows, which cover 1,550 km². One flow extends 160 km northwest into Einasleigh River and another travelled 90 km north to the Lynd River. The flows were fed by spectacular lava tubes. UNDARA CRATER is 0.3 km across and 40 m deep and has only minor pyroclastics. A long flow from TOOMBA VOLCANO, 100 km west of Charters Towers in the Nulla field, travelled west down to the Burdekin River. It erupted from a fissure with minor pyroclastic cones about 13,000 years ago. The HUMMOCK BASALT near Bundaberg flowed over an area of 215 km² and originally extended down beyond present sea level. Its cone contains considerable spatter and bombs, and remnants of a lava lake lie in a crater 350 m across. Tearooms near the basalt summit give a splendid vantage point for imagining the eruption 900,000 years ago.

largest, is a relatively simple cone made of massive, crudely stratified bomb-ridden scoria. The crater summit is large and its breach marks, perhaps, the collapse of scoria where a flow issued from its base (see p.11). MT NOORAT is noted for a pronounced crater that reaches right to the bottom of the cone itself. Two of the youngest cones, MT NAPIER and MT ECCLES, are more complex and have produced successions of flows, some travelling over considerable territory. MT NAPIER CONE has a shallow,

THE ARTIST'S VIEW

GEORGE F. Angus painted views of Mt Gambier crater lakes in 1844–45. Then came Eugen von Guérard, a Viennese who studied in Düsseldorf in Germany, before travelling to Australia in 1852.

He had painted Naples from Vesuvius and Mt Etna from Taormina in 1838; in Australia he added Victoria's volcanic landforms to his works — including 'Tower Hill' in 1855, which shows a splendid panorama of the scoria cone complex, lying within the water-filled maar. The Tower Hill State Game Reserve is presently restoring the landscape back to the natural state depicted by von Guérard. His other volcanic landscapes include 'Larra', a view of Mt Elephant in 1857; 'Basin Banks', about 20 miles south of Mt Elephant in 1857; 'View of Lake Bullen Merri' in 1858; 'Stony Rises, Corangamite' in 1857; 'Purrumbete' from across the lake; 'From the Verandah of "Purrumbete"' in 1858; and 'Pulpit Rock, Cape Schanck' in 1865.

A later artist who identified strongly with Australian and New Zealand volcanic features is Shay Docking. Rather than treating the volcano as part of the landscape, she elevated their vistas into isolated and emotional images. Her early volcano works date from 1954 to 1964 and encompass pencil, pastel, watercolour and oil depictions. They include Tower Hill, Landscape of Volcanic Plains, Western Districts Victoria and the Glass House Mountains Series. Other works include the 1970–85 South Seas Icon Series (Volcanoes of Auckland's isthmus, New Zealand), the 1972 Volcanic Peninsula Series (Banks Peninsula, New Zealand), the 1977–86 Volcanic Plains Series (Western District of Victoria) and the 1972–88 Volcanoes of NSW Series, as well as some South Australian, Lord Howe Island and Hawaiian volcanoes.

EARTH VOLCANIC

105

NATIONAL PARKS DIVISION/MINISTRY FOR CONSERVATION VICTORIA

△ 'Tower Hill', 1855. Oil painting by Eugen von Guérard, held in Warnambool Art Gallery, Victoria.

BENDIGO ART GALLERY VICTORIA

▷ 'Volcano, Town and Western Plains', Shay Docking, 1977. Pastel and acrylic, 109 x 121 cm.

CHAPTER 8

About 140 lava volcanoes are recognised in western Victoria and 80% lie in the central and eastern parts. Many lavas are tholeiitic basalts and icelandites. Lava volcanoes are common near Melbourne and Geelong cities and built the 2,000 km² lava fields of Werribee Plains. Around Ballarat, large lava cones and domes form MT BLOWHARD, MT HOLLOWBACK and MT PISGAH, and lava discs are found at Lawaluk and Mondilibi. Near Gisborne, the complex MT GISBORNE LAVA DOME erupted 9 lava flows from 3 vents. Further west, MT PORNDON VOLCANO erupted large flows in all directions, and 4 bursts of activity built a 200 km² apron of lavas. Although capped by scoria cones, Mt Porndon is basically a lava volcano.

TUFF RINGS, TUFF CONES AND MAARS

Instructive examples punctuate the Atherton Tablelands and western Victorian plains and form the many splendid crater lakes of southeastern South Australia. These structures develop in response to water in the volcanic system. This sometimes comes from volatile rich magma, but more often from encounters of erupting lava with groundwaters. Shallow surface water allows highly intense explosions to build a tuff cone. Sides are steep here, and inside the rim finely fragmented ash beds dip in towards the vent as well as reversing in dip outwards from it. Compacted beds of ash, called tuffs, are made of fragments smaller than 2 mm. Those with over 10% of basalt fragments or lapilli (little stones) are called lapilli tuffs.

Maars are created by greater water-lava interactions, which generate strong steam explosions. This is called phreatomagmatic activity, and it excavates deeper craters. Because of varied mixes of magma and water, maar deposits range in their textures and structures. Two types of deposits may form:

- air-fall material ejected in explosions;
- ground-hugging base surges, which exploded outwards from the vent.

The air fall deposits pass into scoria-rich beds, but are mostly fine-grained, ashy beds. Some deposits include balls of fine ash collected around fragments of scoria or basalt — these distinctive fall-out structures are called cored, accretionary lapilli. The base surges of fast-moving materials often build up structures resembling dunes where waves of material formed climbing ripples as they

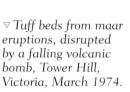

▷ *Tuff ring exposed in quarry operations near Camperdown, Victoria, September 1981. Pale tuffs from base surge maar eruptions are overlain by dark scoria deposits from Mt Leura cone eruptions.*

▽ *Tuff beds from maar eruptions, disrupted by a falling volcanic bomb, Tower Hill, Victoria, March 1974.*

swept over the ground. Sag structures formed under the ballistic impact of large bombs landing in soft and wet maar deposits are often seen.

Several maars lie within a 20 km² area east of Atherton in north Queensland. Far from spoiling the landscape, they provide picturesque settings. The largest maar at BROMFIELD SWAMP feeds North Johnstone River and is 1.6 km wide and 60 m deep. Lakes occupy maars at Lake Barrine (1 km wide and 120 m deep) and Lake Eacham (0.9 km wide and over 65 km deep). Pyroclastic beds in crater walls 30 m high dip 15° away from the vent. At Lake Barrine, the beds contain surge deposits and sizeable blocks of local schist torn from the bedrock. Lake Euramoo, the smallest maar lake, is perhaps the most charming. LYNCHS CRATER is a small swamp-filled maar containing old sediments. Vegetation deposited in the maar sediments ranges from 10,000 to 200,000 years in age and records some distinct floral changes in the region over this period.

A world-class series of structures show up in western Victorian maars and tuff rings. Their prevalence springs from aquifers in underlying limestones and sandstones of Tertiary age, which supply abundant water for explosive activity. The largest simple maar encloses LAKE PURRUMBETE and is 2 km wide. It exposes a fine succession of base surge deposits. Besides typical outward-dipping beds, it preserves some inward-dipping deposits that slumped into the active vent. Twin maars, 10 km WNW of Lake Purrumbete, contain LAKE

△ *Emerged basalt volcanoes, Murray Islands, Torres Strait, August 1974. Dowar Island (right) and Wyer Island (left) seen from the shore of Maer Island.*

BULLENMERRI and LAKE GNOTUK, with Gnotuk's waters noticeably more saline. Sediments in these lakes and in nearby LAKE KEILAMBETE MAAR show that these maars formed over 9,500–11,500 years ago. Bullenmerri probably has two overlapping craters and surge deposits dip 30–50° in towards the lake. Bullenmerri and Gnotuk are not identical twins, as Gnotuk exposes part of a tuff ring and some lavas, and it may be a collapse feature.

Between PURRUMBETE MAAR and BULLEN-MERRI-GNOTUK CRATERS, Mt Leura rises as a prominent scoria cone. Appearances are deceptive, as the base forms a well-defined tuff ring 2.5 km across, formed by early lava–water interaction (see p.106). The rim beds dip symmetrically into and away from the vent. Thicker beds with climbing dune structures probably mark steady and longer base surges. Thin, mantling ash beds probably formed 'co-surge' ashes when lighter, finer material became separated from denser surge material by rising currents of air or water.

Multiple maars reach their climax in the RED ROCK COMPLEX near Colac. Here 42 vents include 28 scoria cones and 14 maars. These explosive vents were formed after extensive lava eruptions. Since Lake Corangamite and Lake Colac were formed against lava flows, they are not maars, although several of the maars do contain lakes. Tuff ring and maar formation at MT ALVIE and maar and tuff cone eruptions at Red Rock preceded scoria cone formations. The rim at LAKE CORAGULAC is exceptionally high in scoria content, which shows that Strombolian and maar activity both took place. Large basaltic blocks of older and fresh-formed basalt create many sag structures in the Coragulac deposits. The Tower Hill complex near Warrnambool, in contrast, contains a scoria cone complex in a large maar 3 km long. Maar rims exposed in quarries show 13 intervals of Strombolian and Hawaiian style deposits which interrupted maar deposits. Water-charged eruptions finally lost out against lava-rich eruptions.

Mt Gambier provides South Australia's answer to Victoria's many maars. A fissure-controlled maar complex built up in two bursts of activity 4,300–4,600 years ago. In the first, tuff rings formed at LEG OF MUTTON LAKE, some lava issued forth and a scoria cone finished off proceedings near BROWNE'S LAKE. After a short break the main explosive activity produced a linear nest of craters now occupied by Browne's Lake, VALLEY LAKE, Leg of Mutton Lake and the famous BLUE LAKE (see p.17). Base surge beds left by radial blasts from the vents are common and accretionary lapilli feature in the Leg of Mutton Lake deposits. A last fling of fire fountains built up moulds of glassy spatter in BROWNE'S LAKE CRATER. A few final steam blasts blew out blowholes, such as DEVIL'S PUNCHBOWL, and the fireworks were over.

▽ *Cape Grim, looking north from Slaughter Bluff to Bass Strait, northwest Tasmania, November 1965. Dark beds of pillow breccia dipping into Cape Grim under pale fossiliferous marine beds (right) make an underwater delta built up by an eroded offshore volcanic centre.*

EMERGENT VOLCANOES

These underwater vents emerge in explosive eruptions more extreme than most groundwater interactions. Lake waters and shallow seas become battlegrounds as the vents surface, hurling material aloft in violent blasts and sweeping materials outwards across the surface through base surges. Deposits of highly fragmented pieces of chilled lava build up. These are Surtseyian deposits, named after the activity which created SURTSEY ISLAND in the sea off Iceland. Accumulations of such glassy fragments are called hyalotuff deposits ('hyalo' for glass). The glass commonly hydrates to a secondary material called palagonite. Fragments often become cemented by minerals which precipitate from solutions soaking through the rocks.

Where waters are deep enough, the volcano forms submerged mounds of pillow lavas before Surtseyian activity arises. Some vents create their own water bodies by erupting a large lava flow, which blocks a river and creates a water dam for subsequent emergent activity. Vents erupting under ice can also melt their cold covers and erupt into substantial pools of water, but not so much in Australia. As volcanoes build past the Surtseyian stage, they enter a deltaic stage. Underwater deltas made of broken lava build outwards as lavas erupted in air flow down the land into surrounding

waters. The chilled lava starts to form pillows, but is rent asunder by steam explosions. Broken-up pillows and glassy fragments avalanche down the submerged slopes and fan out as deltas. These deposits are called pillow lava breccias or flow foot breccias. Finally, the volcano builds up into a substantial open-air vent, producing lava flows and pyroclastic deposits of Strombolian, Vulcanian, Hawaiian or Icelandic nature.

There are several extinct emergent volcanoes off Australia's coast. The Murray Islands in Torres Strait, between Cape York and Papua Niugini, include 8 volcanoes which grew from the sea over 1 million years ago. MAER ISLAND, the largest, is 3 km long, rises 200 m from the sea and has 3 vents. Many of these islands expose palagonitic tuffs and base surge deposits. Fragments of coral indicate that eruptions blasted through underlying reefs. Other islands are made of basalt flows that erupted after their emergence. STEVENS ISLAND, one of the South Barnard Islands off north Queensland, is another emergent volcano. Its dipping pyroclastic beds include base surges and contain large blocks torn from the underlying metamorphic basement rocks. Basalt dykes have cut through the pyroclastic beds, and adjacent islands are made of basalt flows; both features indicate a former open-air surface on the volcano.

LADY JULIA PERCY ISLAND off southwest Victoria shows flow foot breccias under lava flows, marking a volcano which emerged in Bass Strait. BLACK PYRAMID, an isolated island off King Island, is rarely visited. Photographs show 30 m high cliffs of pillowy and columnar basalts, a mid-rib of horizontal tuff beds, and a 40 m high peak of flow foot breccias. Tasmania has outstanding examples of emergent volcanoes in coastal sites and in old inland lake systems, but these are older and more dissected structures. The coastal volcanoes rose when higher Bass Strait seas overlapped onto northwest Tasmania.

CAPE GRIM VOLCANO has a dramatic sea cliff cut into flow foot breccias, which dip into the cliff. This volcanic delta was built out from offshore vents near the Doughboys islands. From the volcano's emergence hyalotuffs appear in shore platforms north of Cape Grim. A deep channel filled with fossiliferous marine beds of Miocene age overlies the volcano on the south side. The full spectrum of an emergent volcano is exposed in FLAT TOPPED BLUFF VOLCANO to the south. Pillowy lavas and hyalotuffs appear in its northern cliff faces and flow foot breccias and capping lava flows form the southern cliff faces. TREFOIL ISLAND, STEEP ISLAND and ROBBINS ISLAND lying off northwest Tasmania also show flow foot breccias, hyalotuffs and pillowy lavas erupted by other emergent volcanoes.

This suggests that eruptions flowed into a sea that was over 70 m higher than its present level. More flow foot breccias overlie Miocene age marine limestones in the Marrawah-Redpa district of Tasmania. The volcanic structures are weathered and poorly exposed and can only be seen clearly in

quarries. The elevations of the limestones and volcanics indicate eruptions into seas 100–130 m higher than their present level. The volcanic layers have protected the limestones from the full effects of erosion since the seas retreated.

CENTRAL VOLCANOES

The central feature of these volcanoes is the cap of silica-rich lavas. This adds to the character of Australian volcanoes, as introductions of viscous trachyte and rhyolite lavas bring more violent explosive activity into play. Unfortunately, no young examples remain, so older structures must serve for illustration.

The early parts of such caps are made of basalts forming shields built up mainly by activity of the Hawaiian and Icelandic nature. The next stage builds the central complex of trachytes and rhyolites, erupted from large chambers below the core of the volcano. Lava domes, short lava flows, pyroclastic aprons and avalanche deposits are typical. The pyroclastic deposits resemble those of violent historic eruptions, such as occurred at MT PELÉE, MT ST HELENS and VESUVIUS. However, Australian central volcanoes are not subduction volcanoes, but within-plate volcanoes. They resemble smaller versions of Hawaiian hotspot volcanoes, but with more prominent trachytes and rhyolites. A further stage in some central volcanoes adds more basalts, so that their proportions of trachyte and rhyolite to basalt vary greatly.

The last central volcanoes formed in Australia are small examples in central Victoria, but after 6 million years most superstructures are gone and pyroclastic deposits are rare. The larger 11 million year old CANOBOLAS VOLCANO near Orange in New South Wales retains some pyroclastic aprons, is easily visited by road to a summit lookout, and makes a good case study.

The CANOBOLAS BASALT SHIELD covers over 800 km² and the central area of trachytes and rhyolites covers over 200 km². Over 50 vents occupy the central part. Some 20 lava domes on the northwest side intrude and rest on pyroclastic deposits. Crystal tuffs of trachyte composition contain abundant crystals of potassium-rich feldspar. West of Old Man Canobolas, air-fall and pyroclastic flow deposits are bedded together. Volcanic bombs formed sags in the trachytic tuff beds and the ignimbrites have stretched and distorted pumice and glass fragments. Volcanic breccias contain coarse angular pieces of trachyte or rhyolite in fine-grained matrix. Obsidian bands near Mt Towac grade into ignimbritic tuffs and mark strongly welded ignimbrites fused to glass. Plugs, domes and dykes of solid trachyte, phonolite and rhyolite intrude into the complex and one trachyte flow has descended from the volcano's south flank over a distance of 10 km.

The close of the activity of Canobolas volcano and other east Australian central volcanoes can be visualised by visiting extinct but younger volcanoes in Lake Turkana region in Kenya and the Massif Central region of France. Although similar to Australian central volcanoes, these are rift volcanoes rather than hotspot volcanoes.

CALDERA VOLCANOES

When large eruptions expel enough lava, underlying magma chambers collapse into a huge volcanic sink. No clearly visible calderas appear in young Australian volcanoes. However, Mt Porndon volcano in western Victoria may mark a buried caldera, according to gravity measurements made over the structure. These indicated a large disc of

oded summit of Mt
as' central shield
from Mt
as Rd, southwest
e, New South
October 1994.

lava embedded in
eccias, south side
Grim, northwest
a, November 1965.

basalt filling a crater structure 3 km long, 2–5 km wide and 330 m deep. This in turn indicates that a caldera collapsed after an eruption from the vent poured out widespread lavas to form the surrounding stony rises. Calderas commonly mark explosive Plinian eruptions and there may be examples in the big central volcanoes such as FOCAL PEAK VOLCANO near Barneys Peak in Queensland. However, caldera remnants became severely obscured by erosion in these older structures.

A large caldera filling remains on LORD HOWE ISLAND, a 7 million year old volcano in the Tasman Sea. The caldera margin is marked by a volcanic breccia on the southern side, which indicates a violent explosive phase. Both the breccia and older lavas of the island are cut by many basalt intrusions, indicating a voluminous extraction of lava from the underlying chamber. The explosion and lava withdrawal caused a caldera collapse. This depression filled with lavas which now form Mt Lidgbird and Mt Gower, the two highest peaks on the island. The caldera was over 4 km across and the massive basalts resisted erosion. It is a dissected caldera and leads us into descriptions of eroded Australian volcanoes.

△ Lord Howe Island caldera lava flows, May 1994: resistant lava horizons forming Mt Lidgbird (centre, front) and Mt Gower (right, back). Foreshore rocks are young dune limestones.

▽ The Nut, a 12.5 million year old neck overlooking Stanley township and Bass Strait, northwest Tasmania, November 1978. Looking east from Green Hills peninsula.

THE WORN VENTS

As volcanic fields become extinct, erosion and changes in environmental settings take over. MT BURR RANGE near Millicent in South Australia shows a volcanic field in the initial stages of erosion. Volcanic features are more subdued than in the younger Mt Gambier and Mt Schank vents to the southeast. A recent uplift of the coast changed the setting from sea-lapped to coastal inland. Mt Muirhead and THE BLUFF VOLCANOES have asymmetrical cross-sections due to erosion and the action of onshore winds, and they are now 20 km inland. These volcanoes, which erupted 20,000 years to 2 million years ago, are now partly buried by old dune sands. In contrast, old volcanic shields near Portland, western Victoria, have tuffs and breccias caused by airfall and groundwater steam explosion underlying ground surface lavas, and they are now exposed by the sea at Cape Bridgewater, Cape Grant and Cape Nelson.

VOLCANIC OUTLIERS AND NECKS

As erosion advances, volcanoes retain some outer trappings, but reveal more and more of their inner structures: first peepholes, then windows, and finally doorways open into the volcano interior. After 1–3 million years of erosion, Australia's inland volcanic fields retain some vent and flow features, but older fields become more segmented losing their original profiles.

Progressive erosion appears among basalt exposures in the Bundaberg-Gin Gin region, Queensland. The 1 million year old HUMMOCK BASALT remains a fairly complete entity. The 3 million year old TARARAN CRATER AND FLOWS are largely intact, but the crater is eroded into a 20 m cliff on its east side. This reveals pyroclastic beds of Strombolian and Hawaiian type volcanism. The older STONY RANGE LAVA VOLCANO, 5 km SSW from Tararan vent, has lost its southern and westward flow extensions. The 5 million year old MT LANDSBOROUGH VENT 5 km WNW from Tararan vent shows a 40 m deep incision and only a small lava cap remains at the summit.

Basalt remnants become isolated outliers (younger rock within older rock). Such outliers descend the coastal escarpment west of Ingham, Queensland, in segments of a 1.5 million year old lava erupted from the STONE RIVER VENT. The advanced dissection for a flow of this age probably indicates that erosion caused by runoff from steep escarpment slopes has increased. An isolated outlier forms MT ST MARTIN VOLCANO, south of Bowen River, Queensland, where a 3 million year old lava caps coarse pyroclastic beds. Blocks of country sandstone up to 9 m across in the pyroclastics suggest the vent was once nearby. Only its flank protected by the flow remains, sitting on a small pedestal of sandstone isolated when the land surface eroded around the protected outlier.

Vents retaining parts of their pipes and pyroclastic deposits are called necks. Spectacular necks form coastal landmarks in northwest Tasmania at TABLE CAPE, near Wynyard, and THE NUT, near Stanley. These massive basalt headlands grade upwards into coarser-grained and lighter-coloured rocks as feldspar minerals increase and olivine decreases. The softer tuffs and breccias of the old crater are cut back by the sea, leaving the solid basalt as upstanding promontories. The necks were probably deep lava lakes 12 to 13 million years ago — long before Bass Strait seas rose to their present levels and attacked their thresholds.

BLINKING BILLY POINT VOLCANO forms an unusual remnant near Hobart in Tasmania. The 27 million year old lavas and pyroclastic beds overlie sedimentary clay beds, but all the layers are folded and dip down towards a vent under the Derwent waters. Magnetic studies of the lavas show that this flank of the volcano had collapsed back into the vent while the lavas were still hot and plastic and the underlying clays were wet and soft.

FL SUTHERLAND

◁ Mt Warning, New South Wales: central plug in erosion caldera, looking south from Numimbah valley, Queensland, June 1971.

▷ Breccia pipe being quarried for aggregate, Hornsby, northern Sydney, New South Wales, March 1979. Note steep contact of Jurassic breccia pipe with Triassic sandstones and Sydney Basin shales at top of quarry (centre left).

◁ Mt Wellington: jointed (background) and massive (foreground) dolerite of eroded sill cap overlooking Hobart, Tasmania, February 1980.

PLUGS, DOMES, DYKES, SILLS & SHEETS

These structures support a volcano's outer shape as a skeleton does. They are channels that fed lava into the volcano before the latter cooled into solid rock. As the volcano's outer layer becomes stripped away, these resistant structures protrude through its eroded flanks and depressions.

Denuded volcanic pipes that stand out as sharp or pot-bellied peaks are called 'plugs'. Another infamous type of plug was the towering spine of lava, 310 m high, pushed up by pressure below the MT PELÉE VENT after the May 1902 eruption. It stood like a memorial overlooking the devastated town of St Pierre on the coast below. Australia's most prominent plug is MT WARNING in New South Wales, a landmark for sailors and pilots. It rears up from the eroded heart of the TWEED SHIELD VOLCANO, the largest Australian central volcano. The main spine of this complex plug rises to 1,125 m and is trachyandesite, a rock resembling something between trachyte and andesite. The spine penetrated the volcano's surrounding rings of syenite as well as the gabbro intrusion that now forms the plug's outer slopes.

Domes are broader and more rounded than plugs. Contrasts between these structures are easily seen in the Warrumbungle National Park in New South Wales. Here BELOUGERY SPIRE forms an irregular plug-like feature and BELOUGERY SPLIT ROCK forms rounded domes of trachyte rock. The imposing BLUFF MOUNTAIN forms the largest lava dome in the WARRUMBUNGLE VOLCANO and is made of coarse-grained trachyte.

Dykes are long narrow bodies formed by lava filling volcanic fissures. If a dyke rock is more resistant than its surrounding rocks, it forms a wall-like body; if softer, it forms an erosional ditch. The best-known Australian dyke is THE BREADKNIFE in the WARRUMBUNGLE VOLCANO (photograph, p.98). This narrow trachyte wall 600 m long rises steeply from pyroclastic rocks around its base. Dykes, which may be very numerous, tend to radiate out from a volcanic centre (radial dykes), or form in series (dyke swarms), or run in multiple directions (cross-cutting dykes). Up to 2,800 rhyolite dykes intrude the MT GILLES and CAMPBELLS FOLLY vents in the Focal Peak central volcano in Queensland, which is overlapped by the TWEED VOLCANO to the east. Some intrude along circular fractures around volcanic cores, forming ring dykes. An elliptical ring dyke of syenite forms a narrow steep ridge encircling the MT WARNING PLUG.

A sill is a molten sheet of rock that invades between the layers of a bed, and often extends as a wide horizontal body. Heat escaping from a sill bakes the overlying as well as the underlying layers of the bed's 'sandwich' structure, whereas a lava flow only bakes beds beneath it. Tasmania owes its rugged plateau scenery to sills of hard dolerite rock that invaded the sedimentary strata in molten pulses. Without its Jurassic sills, Tasmania would look like the Sydney Basin in New South Wales. The sills reach 300–500 m thick and MT WELLINGTON sill dominates the mountain scenery above Hobart. Only in one place in Tasmania has erosion left the original Jurassic surface intact. Here LUNE RIVER LAVA, a 175 million year old flow from a fissure eruption, descended into a rift valley, burying vegetation of that age. The underlying sills spread from long dykes and funnel-shaped feeders called cone sheets. Similar sills cap the Transantarctic Mountains in Antarctica, but large stacks of lava that erupted from the sills still remain here (photograph, p.95).

In most other Australian volcanoes, sills, dykes and cone sheets are small local affairs, not grand disruptions as in southern Gondwana. Cone sheets are relatively rare compared to dykes and sills. Inward-dipping cone sheets of dolerite encircle Focal Peak central volcano in Queensland and irregular cone sheets invade the Lord Howe Island caldera.

TOMBSTONES AND BURIAL MOUNDS

With extinction and decay, only inner cores of volcanoes remain as headstones to mark their former life. Eventually, all remnants become buried under new geological deposits. First of all, erosion leaves only underlying pipes and dykes. An Australian volcano near this stage is the 20 million year old MT BELMORE CENTRAL VOLCANO in northeast New South Wales. It lies on the escarpment, where erosion has cut steep valleys through the heart of the volcano. This left plugs and dykes of trachyte and rhyolite on the eastern fall and a few small remnants of basalt and rhyolite flows on upland spurs.

With time, such erosion extends across whole volcanic fields, leaving tens or hundreds of volcanic gravesites across wide regions. In places where much lava was erupted, many dykes, sills or cone sheets rise from the underlying rocks, as in Tasmania. In contrast, where small amounts of lavas richer in water and dissolved gases erupted in maar-studded volcanic fields, breccia pipes now pockmark the eroded surface, as in the Sydney Basin. These pipes of explosive fragmented material usually extend down in narrowing fan-shaped bodies, and at greater depths they pass into a feeder pipe or dyke. Because of their fragmented and altered nature, tuff and breccia pipes usually show less surface relief than more solid intrusions do.

LARGELY DENUDED FIELDS

Several well-visited examples of such volcanic fields are found in central north Queensland around the Rockhampton, Mackay and Anakie regions. Between Rockhampton and Yeppoon scenic plugs of trachyte and rhyolite rise from the coastal lowlands and catch the eye of passing travellers. These are headstones of denuded central volcanoes over 70 million years old. MT WHEELER, the largest plug, forms a broad-topped elongated hill. Other plugs, such as JIM CROW MOUNTAIN, form sharp peaks with little vegetation on their steep slopes, or a ridge of multiple peaks, such as PINE MOUNTAIN. Several plugs form rounded, smooth-sided hills such as MT HEADLOW and IRONPOT MOUNTAIN. A few small peaks are crowned with heavy vegetation, such as MT MUNGAWOPPA. Erosion has removed all the higher lavas, leaving a few flows of basalt and rare trachyte as low skirtings around some plugs. On the coast, the sea has exposed the internal structure of plugs and inclined sheets south of Rosslyn Bay.

Cape Hillsborough, a scenic coastal park north of Mackay, presents some dramatic topographic contrasts. The eroded CAPE HILLSBOROUGH CENTRAL VOLCANO lies in an old rift valley, but MT JUKES and MT BLACKWOOD VOLCANOES have been reduced to subvolcanic chambers forming mountainous humps on the uplifted rift margin. The sea has carved through most of Cape Hillsborough volcano and the main vent now lies offshore. Coastal sections, combining flows, coarse and fine pyroclastic beds and old avalanche slides, are exposed in a 300 m thick flank of the volcano, and survive

◁ *Jim Crow Mountain, a trachyte plug (back) rising above eroded basalts (front), looking southeast from Mt Headlow property, near Rockhampton, Queensland, August 1987.*

▽ *Eroded flank of Cape Hillsborough volcano, looking south to Wedge Island, Queensland, August 1978.*

▷ *Large block of coarse volcanic fragments fallen from cliff face (back) Cape Hillsborough, Queensland, August 1978.*

▽ *Mt Jukes, an eroded volcanic chamber of fine-grained granite rock, looking east from Mt Blackwood, Queensland, July 1986.*

as shore platforms, headlands and seacliffs. Rhyolite flows, dykes and plugs are common rocks, but some lower flows and plugs are basalts. Pinnacle Rock is an impressive rhyolite plug 4 km west of Cape Hillsborough.

Inland, the massive MT JUKES and MT BLACKWOOD INTRUSIONS are made of fine-grained syenite and granite, surrounded by margins of gabbro and dykes of gabbro and syenite. Long rhyolite dykes and plugs to the east form THE LEAP, an abrupt hill overlooking the main highway north of Mackay. The difference in erosion between the Cape Hillsborough volcano on the one hand and Mt Jukes and Mt Blackwood on the other is partly due to the fact that the rift volcano is 33 million years old, whereas the intrusions are 42 million years old, but is also due to enhanced erosion on the uplifted rift margin.

The Anakie area, 300 km east of Rockhampton, is renowned for its rich sapphire deposits and is popular with visiting fossickers. It has 70 basalt plugs, large and small, in an area 50 km across. Prominent plugs include Mt Leura, MT HOY, MT BALL, MT DUMBELL and MT PLEASANT. A few hills, such as MT SCHOLFIELD, are the remains of sills. Some of the volcanoes that once crowned these plugs erupted sapphire-rich explosive deposits. From these pyroclastics and the basalts themselves, sapphires were washed into the alluvial drainages. Many became washed over and over again as streams cut deeper into the land surface. The basalt plugs range in age from 18 to 56 million years, but on the outskirts of the field some younger flows remain. ANAKIE HILL volcano is the youngest and its 15 million year old vent retains a crater-like infill.

COMPLETELY DENUDED FIELDS

The spectacular Glass House Mountains rise from the coastal region north of Brisbane. These old spines of central volcanoes formed 25 to 27 million years ago. Mt Beerwah, a trachyte, forms the highest peak at 556 m. Mt Beerburrum and Mt Miketeebumulgrai form dome-like intrusions of trachyte, whereas Mt Coonowrin is a steep spire of rhyolite. Mt Ngun Ngun and Mt Tiberogargan are made of coarser rhyolite rock. Mts Tunbubudla, Tibberowuccum and Coochin and the more conventionally named Wildhorse Mountain are all rhyolite intrusions. Further north similar plugs, domes and dykes of trachyte and rhyolite form prominent peaks at Mt Coolum, Mt Peregian, Mt Cooroy, Mt Tinbeerwah, Mt Cooroora and Mt Cooran, so that these denuded central volcanoes extend over 80 km.

A widely spaced chain of volcanoes once hugged the great rift line along the Tasman Sea. These volcanoes, now 90–100 million years old, are stripped down to small intrusive complexes. Their diverse alkaline rocks have fascinated petrologists since 1900. Besides the main rocks in coarse-grained intrusions a great variety of dyke rocks penetrate their margins. The Cygnet and Cape Portland intrusions in Tasmania, Mt Dromedary complex in southern New South Wales, and Mt Ridler complex near Gladstone in Queensland are best known, but centres were recently discovered on the downfaulted rift blocks under the Tasman Sea. Some dyke rocks look exotic as they contain large crystals of feldspar, mica, pyroxene, amphibole or garnet in a fine-grained matrix. A 'biscuit' rock of large, pale pink feldspar crystals in a dense, dark green matrix at Port Cygnet (photograph, top of p.155) was popular with collectors, but now needs preservation. These centres consist of 'halfway' rocks — that is, their characteristics lie between those of calcalkaline subduction volcanoes and alkaline within-plate volcanoes, but only rare remnants of their lavas remain.

A vast number of small dykes, plugs, sills and explosive pipes riddle the sandstones, shales and coal measures in the Sydney Basin in New South Wales. Many are the denuded remains of Jurassic volcanoes that flared into activity during the stretching which led to dolerite invasions in Tasmania. Over 600 dykes are known. They reach up to 8 km long, often form in swarms and are best exposed in coastal platforms and cliffs. Most dykes trend NW–SE, occupying the main fracture direction as the basin was stretched. The Bondi dyke and pipe exposed in sea cliffs near Sydney has heated adjacent sandstones, thereby creating prismatic columns where they cracked on cooling.

△ Sinuous dyke of mica-lamprophyre rock cutting older Jurassic dolerite rock, south of Cape Portland, northeast Tasmania, April 1968.

▷ Glass House Mountains: eroded volcanic plugs, looking east from Maleny plateau, Queensland, July 1979.

The chambers of several central volcanoes are exposed. Mt Gibraltar, north of Bowral, NSW, is a syenite mass uplifted along its western bluff by faulting. It supplied much sturdy stone for Sydney buildings. Not all intrusions broke the surface to erupt lava. The Prospect dolerite intrusion, west of Sydney, formed a laccolith, a blister-like body. This arched up the overlying shales, but rose no higher. Another intrusion drilled near Scone, revealed layers of gabbroic rocks that had crystallised in pulses as the body cooled.

About 150 explosive pipes, called diatremes, indicate that widespread maar and cinder cone activity took place during Jurassic volcanism. They provide many opportunities to investigate the mechanism behind diatreme emplacements. The pipes range up to 0.5 km across, but can extend up to 3 km in length. They are filled with tuffs, lapilli tuffs and breccias containing blocks that were torn off vent walls. Pipes that penetrate massive sandstones are usually hollowed out by erosion, unlike those in softer shaley rocks which usually form low hills. Diatreme development was aided by molten basalt transgressing into newly formed soft, wet sediments.

Different erosional levels in the pipes are exposed in quarries and coastal sections. The HORNSBY DIATREME, north of Sydney, reveals basin-like layers. These form at less than 500 m deep under maar volcanoes, as continuing air fall and base surge deposits collapse into subsiding vents. The MINCHINBURY and ERSKINE PARK DIATREMES show vertical bands and patches of chaotic broken blocks, typical of deeper explosions and the subsidence of material 500–1,000 m below the surface. Basalt breccia in the MINCHINBURY PIPE suggests a deep level just above a basalt feeder. The BONDI DIATREME tapers down to a narrow base, contains sandstone blocks blasted from its walls and includes small sills of basalt. It marks a deep throat, perhaps 1,200 m below the original surface vent. Deeper down, only basalt will remain as a small plug or dyke. The ST MICHAELS CAVE DYKE and BELANGLO COAL MINE DYKE may have been diatreme feeders.

The mixed bag of broken fragments in diatremes attracts quarry operations, as the fragmented rocks prove ideal for aggregate. Some fragments come from strata 300 m or more below the level of exposed wall rocks. In contrast, pieces of charred wood and plant spores of Jurassic age found in diatremes probably descended from the surface in the continuous turnover from explosions. Coal fragments from wall rocks show that temperatures during the explosive collapses were below 150–250°C. As well as basalt fragments, the diatremes also contain deeper fragments brought up by ascending basalt. These include rocks from the granite-metamorphic basement under the Sydney Basin, granulites from the lower crust and peridotites and pyroxenites from the mantle.

ANCIENT BURIAL GROUNDS

As volcanic fields are buried under sediments or folded into mountain ranges, their original nature becomes altered. Their minerals break down, and eventually their entire fabric is extensively recrystallised. The main episodes of geological deformation in Australia mostly affect rocks over 300 million years old, although limited zones of deformation continued up to 100 million years ago. Old volcanic sequences caught up in deformations are found in numerous places in Australia and only a few are selected here to illustrate some transformations.

BURIED AND BARELY FOLDED

In the Illawarra region, inland from Wollongong and south through Gerringong, volcanic rocks appear in the lower Sydney Basin sediments. Around Kiama

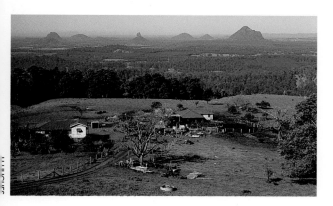

△ *Basin-like layers of tuffs and breccias filling the Hornsby diatreme pipe, northeast Sydney, New South Wales, March 1979.*

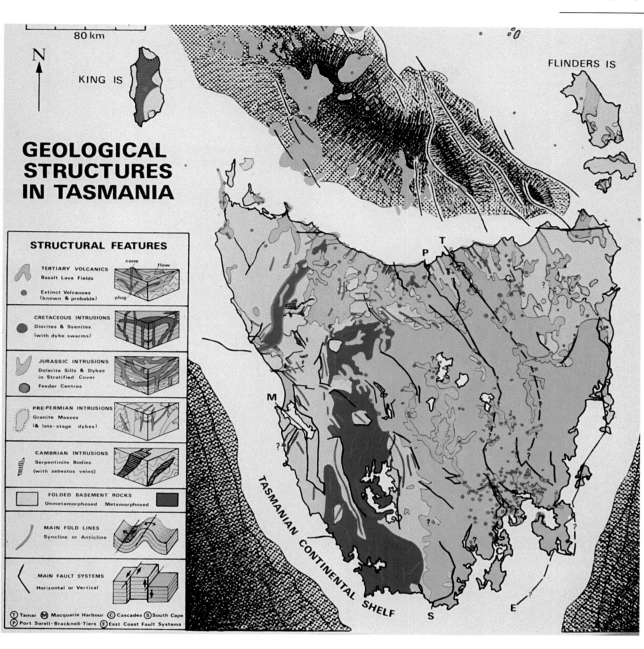

△ Bass Basin and Tasmania: later faults, volcanic basalt eruptives, earlier intrusion, and fold structures. 1980

◁ Altered Permian lavas at The Blow Hole, Kiama, New South Wales, April 1988. The fissure extending seawards has eroded along a narrow basalt dyke. The sea cave enters here under the blowhole, which has eroded into alteration zones in the lava.

these rocks form scenic attractions such as the Blow Hole and are highly visible in highway bypasses and public access quarries. The volcanic horizons are late Permian, between 230 and 260 million years old. The erosion exposing them has removed over 1 km of thickness from the overlying Sydney Basin strata. Nine flows and interbedded tuff horizons were erupted from volcanoes in the southern part of the field. The lavas include distinctive potassium-rich basalts, an andesite and some basaltic andesites, and they often contain large feldspar crystals. The early eruptions entered shallow seas, but later eruptions covered freshwater sediments.

◁ *Gloucester Buckets, looking west over Gloucester, New South Wales, September 1993. Note near-vertical dipping volcanic beds.*

▽ *Sheared, altered volcanic rocks exposed at The Blow, the original site of the Mt Lyell mine, west Tasmania, February 1980.*

The volcanic horizons are folded into broad warps with the enclosing Sydney Basin beds, but remain relatively undisturbed. The burial has partly altered the lavas – olivine crystals and some ground mass minerals are replaced by greenish minerals, and secondary cavity and vein fillings of carbonate and zeolite minerals are common.

A similar burial overtook the Garawilla volcanic field in the Gunnedah Basin, northern New South Wales. This basin is an extension of Sydney Basin, but was covered by further Jurassic and Cretaceous beds laid down in Surat Basin. The Garawilla volcanic field, which is younger than the Kiama-Gerringong volcanics, erupted in early Jurassic time 180–195 million years ago. The lavas have suffered similar burial, however, and were uplifted along a broad arch before being exposed by erosion.

BURIED AND DISMEMBERED

Between Victoria and Tasmania large rift basins developed along the Otway, Wonthaggi-Gippsland and Bass Strait regions. Stretching generated these rifts between late Jurassic and early Cretaceous times. Huge influxes of volcanic debris were washed from nearby volcanoes, active between 140 and 100 million years ago. As Australia separated from Antarctica and New Zealand, the volcanic horizons became buried under huge thicknesses of incoming sediments. These became built up in large deltas between 100 and 45 million years ago, before marine beds were accumulated by seas entering into Bass Strait.

Further rifting and folding associated with fault movements disrupted these beds in series of step faults. On shore, the Cretaceous volcanic-rich sedimentary beds now either outcrop above sea level or lie less than 2 km deep. Offshore, in the deeper basin, these volcanic horizons are down-faulted and buried under sediments as deep as 8 km. At these depths of burial, and with normal increases in temperatures with depth, the volcanic grains have partially broken down into secondary silica, clay, zeolite and carbonate minerals.

BURIED AND STRONGLY FOLDED

The scenic Gloucester Buckets form a ridge overlooking Gloucester township in eastern New South Wales. They are made of Carboniferous volcanic rocks 300–320 million years old. Rhyolite flows and breccias dip steeply east under Gloucester, but reappear on the opposite side. The rocks lie in the limbs of a steep downfold or syncline.

Similar volcanic formations appear again 40 km southeast of Gloucester in the Alum Mountains, preserved in another syncline. Only narrow strips of the volcanic beds are evident at the surface because their intervening fold crests, called anticlines, have eroded away and the keels of the synclines lie deeply buried. In these old folded terrains, the volcanic fields become more difficult to reconstruct and interpret.

A 'ghost' town is all that remains of Yerranderie, a silver mining settlement 100 km WSW of Sydney at the edge of the Sydney Basin. The old mines worked in volcanic rocks — rhyolites and dacites belonging to the Bindook porphyry complex of Devonian age, 380–400 million years old. They form part of the Lachlan fold belt. Detailed studies of volcanic structures around Yerranderie showed that the silver-lead veins were linked to explosive rocks near the old crater rim of a collapsed caldera.

Further west, similar rocks are found in the Lachlan fold belt near Mt Hope in central New

S ROBINSON

South Wales. These volcanics lie in segments of a very large down-fold, complicated by smaller folds, called a synclinorium. Reconstructions of the superimposed folding effects show how a huge pile of rhyolites and dacites, up to 3.5 km thick, originally accumulated in the region with their granite invasions. Many eruptions were explosive and submarine in origin. A major rift and undersea cauldron structure probably formed in a large depression, before the rocks became uplifted and folded into a continental landmass.

BURIED, STEWED AND FOLDED

Western Tasmania is famous for its minerals and ore deposits. They lie in the Tasman fold belt, which extends through Queenstown and Zeehan and into northwest Tasmania. Large ore bodies are found in the MT READ VOLCANICS, which erupted in Cambrian time 515–550 million years ago. The volcanics include extensive explosive deposits and flows of rhyolites, dacites and some andesites.

The structures are complicated by several folding and faulting events and many rocks are sheared, highly altered and difficult to diagnose. Decades of close study were devoted to unravelling their geological sequence because of its economic importance. This showed that the main ore bodies are held in the earlier volcanic units formed on sea floors, not in the overlying volcanic sequences. Gas-charged and metal-enriched thermal solutions streamed up fracture systems lying above hot chambers. Reactions with seawater precipitated large lead-zinc sulphide ore deposits, found at Rosebery, Hercules, Que River and Hellyer River. Some fluids precipitated minerals in the volcanic edifice to form copper-rich sulphide deposits, once mined at Mt Lyell. Deciphering the old volcanic structures will help to target new 'blind' ore deposits.

THE LAVA FLOWS

As lava pours forth it adopts many guises as it cools. Its final appearance depends on a host of factors — the amount and rate of flow, its chemical nature and volatile content, and the environment and terrains it encounters. In Australia only the youngest flows still preserve their outer surfaces.

Contrasting forms of lava develop in flows. Pahoehoe lava shows flowage best of all (photographs, p.5). This expressive Hawaiian term describes smooth, lobes and ropy drapes of lava. Entrail and toothpaste lavas are easily visualised varieties. A'a lava is a mass of broken-up clinkers of lava, the sharp surface being exactly described by its Hawaiian term (photograph, bottom p.85). Pahoehoe lava can change into a'a lava, as the rate of flow and shearing forces in the moving lava increase. Where lava flows into water, or burrows into wet sediments, the chilling accentuates rounded lobe-like tubes called flow feet. These split and

△ *Ropy pahoehoe lava formed 2–3 km from Kupaianaha vent, Hawaii, June 1990.*

bud into more lobes to form pillows. These may break up in steam explosions and cascade down as fragmented rock, forming pillow or flow foot breccias. Lava flowing completely underwater often keeps its pillows intact, forming pillow lava (photographs, pp.24 and 40). These forms develop in more fluid lavas, like basalt, but thicker parts may cool into massive lava.

More viscous lavas, such as basaltic andesites, largely develop a'a lava, while andesites commonly form block lava, in large, smooth-surfaced blocks, often metres across. A special block lava is called cleft lava. Here a cleft develops along a large flow lobe and continued pressure inside the flow forces blocks upwards and outwards along a wide furrow. The blocks show 'chisel marks' made on their sides as they break in jerks.

In very viscous lavas, such as trachytes, rhyolites and dacites, the lava often just domes up or descends in large, sluggish flow lobes called coulées. These slow-moving lavas commonly fragment as they flow, forming auto-breccias, and sometimes develop giant surface wrinkles called ogives.

△ *Duck Point Bay, Great Lake, central Tasmania, April 1968: Ropy flow (pahoehoe) developed inside a broken flow foot lava tube (pillow lava); exterior quenched in water and embedded in flow foot breccia.*

▷ *Partly preserved steam vent in basalt, resembling an eroded hornito, west slope of Mt Kooroocheang volcano, western Victoria, January 1992.*

SURFACE FEATURES

New spills of lava may overwhelm local vegetation, animals, vehicles, roads and buildings. They may register as hollows left by decayed trees or even elephants, cut-off roads, 'fossilised' car wrecks and isolated walls and compounds. As flows establish their courses they may build into lava levees — side ridges formed by solidified lava, the accretion of clinker, or the piling up of avalanching a'a lava. Channels are left as lava drains away. Side ridges remain on Harman Valley flow in western Victoria and channels have grooved wide lava sheets beside Lake Surprise crater at Mt Eccles volcano. The main canal here is 40 m wide and 6 m deep and canals radiate out to 3 km distance.

Pressure ridges form across lava flows. They often curve downstream, following the faster-moving central lava zone. Tumuli are domes pushed up by ponded lava below their crust, which in some cases has sagged or collapsed. Pressure ridges and tumuli remain on the HARMAN VALLEY FLOW, with numerous tumuli forming pock-marked landscapes. The typical 'stony rise' country near Mt Porndon volcano is a hybrid surface formed by pressure ridges, lava lobes and broken-up and collapsed lava. Hornitos are small spatter mounds formed over holes above surging lava or steam vents and an example remains on Mt Kooroocheang volcano.

INTERNAL CHANNELS

Lava surfaces sometimes collapse, forming 'skylights' into an inner world of lava tubes and tunnels. Extensive tubes may develop in larger flows, draining lava to its advancing frontlines. Old tubes can be entered through collapsed 'windows', to follow subterranean tunnels over many kilometres. Hawaii is famous for its lava caves. THURSTONS LAVA TUBE in Hawaii Volcanoes National Park at KILAUEA VOLCANO is easily accessible from a roadside walking track.

Simple lava tubes form when roofs build over streams of lava flowing in channel walls. More complex tunnels form at deeper levels where layered structures allow access to liquid lava. When the lava supply stops, this leaves empty or partly solidified tubes. A floor of pahoehoe lava commonly fills the bottom and may show gutters and a central rope-like ridge.

Many tunnel interiors retain the remains of former lava surges. For example, horizontal ledges

△ *Entrail pahoehoe lava, 2–3 km from Kupaianaha vent, Hawaii, June 1990*

and pavements indicate old lava levels. Lava stalactites (lavicles) and lava driblets from dripping lava may decorate walls where hot gas escaped. Lined cylindrical openings in roofs mark places where lava escaped to form hornitos or lava ponds.

Lava caves were discovered before 1880 in Australia. The Skipton Cave near Ballarat, Victoria, became noted for rare phosphate minerals crystallised in deposits of bat guano. A new mineral newberyite was named after a Melbourne identity, J. Cosmos Newbery, in 1879. The BYDUK CAVES are 19 caves formed by collapses along a tunnel in a 20 m thick flow from Mt Napier volcano. These are deeper, layered roof tunnels, in contrast to the simpler channel roofing in GOTHIC CAVE on Mt Eccles volcano. When lava tubes in the McBride basalt field were recognised in northern Queensland in 1960, it opened up one of the great lava tube systems in the world.

The UNDARA LAVA TUBES lie in a 160 km long, 190,000 year old flow from Undara volcano, in the McBride field, near Mt Surprise. One section, marked by surface depressions, includes over 60 underground arches extending over 6 km. This section lies upstream from 'The Wall', a narrow ridge 35 km long which may be an elevated extension of the tube system. The Undara caves join narrow depressions 30–50 m wide, probably old tunnel collapses. Some cave entrances are screened by rainforest and vines. Wider oval depressions up to 100 m across with raised rims may indicate former lava ponds similar to the once-famous HALEMAUMAU LAVA LAKE in Hawaii.

BARKERS CAVE, the largest tunnel in the Undara tubes, is over 560 m long and 13 m high. Rafted blocks remain jammed there by the final flush of lava. THE ARCH is the widest cave, 28 m across. TAYLORS CAVE shows prominent lava platforms above its floor and lines of old lava levels up to roof level. It divides near its termination, where boats are needed to negotiate cave waters. The structures suggest that lava was deviated around a collapse during tube activity. Most caves show a simple lava lining, but in PINWILL CAVE 17 layers plaster the walls. Where roofs have collapsed, as in PETERSEN CAVE, horizontal or arched flow layers are left exposed. These horizons display vesicles as gas bubbles escaped and followed the flow of lava.

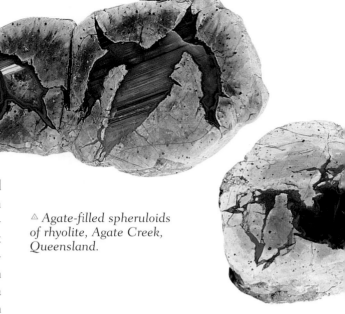

△ *Agate-filled spheruloids of rhyolite, Agate Creek, Queensland.*

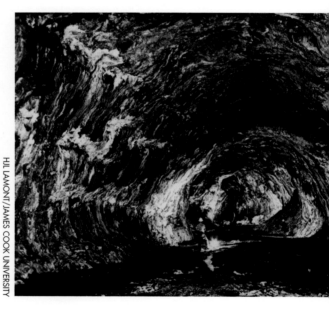

△ *Barkers Cave, Undara lava tube, McBride lava field, north Queensland, 1973.*

▷ *Taylors Cave, Undara lava tube, McBride lava field, northern Queensland, 1973. Note prominent lava platforms.*

GAS AND FLOW LINES

Flows often show more frothy lava towards their outer parts, particularly in flow tops where most gas escapes. The remaining gas holes are called vesicles. Vertical trains of gas bubbles in pipes mark the upward escape of volatiles, which can form by lava passing over swampy or watery patches. In older lavas the cavities and cracks become lined with secondary minerals deposited from percolating groundwaters. Cavity fillings, called amygdales, can carry glistening, beautifully crystallised minerals. Zeolite, carbonate and clay minerals are common in amygdales and silica-rich amygdales provide a great source of agates.

Flow structures develop in some lavas. These include partings forming platey layers, bands of more chilled and more crystalline material, or alignments of large crystals. In viscous lavas flow banding is conspicuous and is often folded by the drag effects of moving layers. Breccia zones may develop as material fragments during a flow. Some lavas have bands of spherulites and spheruloids. Spherulites are small spheres, formed where gel-like lava accumulated around a nucleus and crystallised in radiating clusters of minerals. Larger spherulites build up gases internally and distend into spheruloids with hollow or partly agate-filled interiors. They can rupture and draw in some host lava. Extended spheruloid growth forms large ball-shaped bodies called thunder eggs. Mineral-filled thunder eggs and rhyolites containing mixtures of spherulites and spheruloids in a hardened silicified matrix provide attractive lapidary materials. Fine thunder eggs are found in Queensland at Mt Hay, southwest of Rockhampton; at Mt Tamborine, southwest of Brisbane; and at Agate Creek in northern Queensland.

JOINTS

Widespread and often spectacular joints develop in lavas, especially in thick sections. Joints form as lava cools, solidifies, shrinks and cracks. Closely spaced, platy or irregular joints fracture lavas into thin slabs or uneven pieces. Widely spaced cross-joints give large square or rectangular blocks. The most impressive joints form several-sided columns resembling organ pipes. Some columnar jointing is so striking it provokes legends and draws tourists; for example, the Giants Causeway in Antrim, N. Ireland, and Devil's Post Pile in California, USA.

New studies suggest that cooling columns evolve in stages. The initial cracks propagate inwards, advancing in small steps. This breaks up lava into large irregular immature columns, since the cracks usually meet in T-shaped and X-shaped junctions. The cracking is then driven by thermal stresses into the largest, hottest columns, often forming Y-shaped junctions. This gives mature joint patterns as the columns develop more uniform, commonly six-sided cross-sections. Joints also break across the columns to produce stacked segments. If the transverse joints are curved, they form ball and socket jointing.

Columns propagate inwards from the tops and bottoms of flows, generating two regular sets of columns called the UPPER AND LOWER COLONNADE. These meet in some flows, but in thicker lavas more irregular cooling in the central part produces fans and rosettes of curved columns called the ENTABLATURE.

Cooling columns are ubiquitous in flows throughout eastern Australia. Ball and socket columns are displayed near Burnie wharf in Tasmania. Graceful curving columns and hexagonal cross-sections in mosaic-like pavements feature in The Organ Pipes National Park, near Melbourne. Curved fans in entablatures overlie regular columns in colonnades along Campaspe River in Victoria.

Huge, upstanding columns dwarf visitors to old quarry reserves at Bombo and Kiama, and separate sets of columns appear at the upper and lower Ebor Falls in New South Wales. Prominent vertical columns cap Anvils Peak and Lords Table Mountain east of Clermont in Queensland. It is easier to view them through binoculars than to climb them.

EROSIONAL LAND FORMS

Erosion of flows produce landforms bearing little resemblance to the original topography. Solid lava filling a valley often offers more resistance than the valley walls do. Over time, the flow becomes a ridge or plateau and former lowland becomes highland. This inverted volcanic relief is clearly seen at MT CAMERON WEST FLOW in Tasmania. The 170 m high ridge was once a single thick lava flow and its base sits on Tertiary marine limestones near sea level. The lava filled a steep gorge, but because the limestone walls of the gorge were very susceptible to erosion, they have vanished.

Where resistant lavas overlie soft clays and sands, strong undercutting occurs and the slopes are extremely prone to landslips. Great cliffs may slide down, sometimes with trees still attached. Extensive talus (large blocks) or scree (small fragments) cover underlying slopes. Slips often occur in rotation. The land drops downwards and outwards in concave curves, thus tilting backwards. Such slips abound in the Tamar Valley in Tasmania, where the Tamar River cuts through old basalt flows at Beauty Point, Bell Bay and Rosevears. Where jointing in basalt inclines outwards, blocks of basalt topple forward, adding to the extensive basalt waste. Basalt slopes in eastern Burnie in northwestern Tasmania were so slip-prone they were declared restricted building zones.

In flow sequences, the more vesicular tops and bases alter more easily and tend to erode along these planes. This also applies to softer pyroclastic and sedimentary horizons between flows. Stepped terraces are formed. They are well developed in basalts exposed along the Nive and Ouse Rivers and around Great Lake in central Tasmania. The resistant dolerite bedrock of the plateau assists this process by confining river downcutting close to the lava fills. Terraces in thick flows lie behind many scenic waterfalls in east Australia. Ebor Falls in New South Wales, Milla Milla Falls in Queensland and Trentham Falls in Victoria are well known.

◁ *Cooling columns and tessellated surfaces in eroded basalt, Bald Knobs, south of Armidale, New South Wales, August 1985.*

▷ *Large blocks of basalt toppled from Bradys Lookout, west Tamar River, Tasmania, 1965.*

▽ *Inverted topography with 16 million year old basalt flow (top darker horizon) overlying old deposits of Ringarooma River (lower lighter horizon, partly obscured by basalt slips), Tasmania, February 1980. The present Ringarooma River lies between this bank and Derby township (foreground). Breises tin mine (right) worked the gutter of the old river channel.*

In regions of uplift with strong downcutting by rivers, as in Grand Canyon, USA, lava flows that dam the river become preserved at successively higher levels. The ages measured in these lavas make it possible to estimate the rates of downcutting, and uplift can then be calculated. Such estimates are made for basalt-filled drainages in eastern Australia. In Tasmania, 50 million year old basalts filled former valleys around Weldborough at 500–800 m elevation, and 16 million year old basalts filled old courses of the Ringarooma River at 160–300 m elevation. The present river now flows 20–70 m below the basalts, a drop between 360 m and 570 m in 50 million years.

As erosion breaks through flows, segments become isolated. Long, flat-topped remnants form mesas, a typical land form where harder horizons overlie softer horizons. Mesas degrade into small rounded caps called buttes. These forms are very common in central Queensland, especially at Lords Table Mountain and Anvil Peak near Clermont and around Biloela and Monto. Even after flows have completely eroded, their former course may be still evident, especially when the underlying river sands and gravels have been cemented into hard silcrete rock and remain as a last vestige.

In some areas, as seen in the next chapter flows become so extensive that they spill out of the confining valleys and form broad lava plains, completely obscuring the previous topography. As erosion works down through this volcanic screen, the later river systems often divert from the previous drainage courses. The new valleys then expose the former interfluves and valley lava fills. Reconstructions of these old drainage lines become important when alluvial gold, tin or gemstone deposits are involved. The Ballarat region of Victoria, the Weldborough–Ringarooma region of Tasmania and New England's diamond fields are intensively worked examples.

CHAPTER

R HEWER COLLECTION/J FRAZIER/M E WHITE

9

VOLCANIC DISRUPTIONS AROUND AUSTRALIA

ERUPTING volcanoes dislocate local drainage, flora and fauna. Continued eruptions can cause the radical redistribution of river systems. Lava flows descending into valley systems often seal off river beds and preserve their sediment load. These buried river courses or deep leads may include both main channel and tributary leads.

Pyroclastic deposits and lava flows sometimes overwhelm local forest, swamp, pond or lake communities, preserving fossil forests, leaf beds, fish beds, diatomite deposits or bone beds. Such preservation gives us fragmentary glimpses of past terrestrial life. Lavas flowing into coastal bays seal off estuary, beach and shallow sea deposits, preserving past shoreline life.

Although eruptions disrupt local life and environments, the volcanic cover can provide excellent markers for dating buried fossils. In this way, eucalypt leaves buried under basalt lavas at Berwick, Victoria, were dated at over 22 million years.

◁ *Polished slice of fossil Casuarina tree trunk from a fossil forest, Bushy Park, Derwent Valley, Tasmania. The Tertiary wood shows a swirling pattern of wide medullary rays preserved in silica.*

EARTH
VOLCANIC

DRAINAGE DISRUPTIONS

Rivers become displaced by volcanic obstructions. Some rivers make a simple sideways shift to a new lateral drainage, and some divide around the blockage to form twinned lateral drainage. A severe blockage may dam the river, creating an upstream lake which accumulates sediments until the barrier is breached. In extreme cases upstream drainage diverts to an entirely different route. In rare circumstances, lavas even enter underground cave systems and can indicate past water-tables.

Sites of major eruptions build up large shield volcanoes, central volcanoes and volcanic divides. These create new drainage systems that descend from the elevated heights. Extensive lava fields and lava plains may obliterate previous drainage systems and their land forms, changing the character of the region.

New Zealand has some additional examples of drainage disruptions by volcanic features, other than those exhibited in Australia. The North Island includes plateaus formed by the eruption of massive young ignimbrite deposits. Large ignimbrites can blanket wide areas of the previous drainage systems, leaving a plateau-like surface. Rivers will cut down into the plateau, but their courses follow the gentle plateau gradients and adopt a partly radial drainage from the highest part. Good examples are the MAMAKU IGNIMBRITE plateau, extending west and north of Lake Rotorua, and the MATAHINA IGNIMBRITE plateau which covers 130 km² in the Rangitaiki Valley, both near the Bay of Plenty. As rivers cut down into the ignimbrites and meet the hard welded parts, they carve steep sides and gorges. An unusual eruption of rhyolite dykes from concentric fissures in the Waikato River has caused the river to fall in a series of rapids along a narrow race. This energetic water flow over a volcanic blockage at Aratiatia Rapids was used to commission a hydro-electric station.

DISPLACED DRAINAGES

Shifts in river positions after lava infill show many variations. In western Victoria, flows from MT HAMILTON VOLCANO diverted the south-flowing Mt Emu Creek 3–10 km eastwards. Similarly, the PLENTY FLOW near Melbourne moved the river eastwards from its valley until it eventually broke through and drained into a neighbouring valley. Around Ballarat, overflowing basalts forced the drainage to skirt the eastern edge of the Ballarat lava plain. Extensive alluvium has built up along

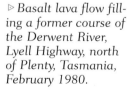

▷ *Basalt lava flow filling a former course of the Derwent River, Lyell Highway, north of Plenty, Tasmania, February 1980.*

◁ *Basalt lava plains, looking towards Mt Hicks from Lower Mt Hicks, northwest Tasmania, March 1978. Mt Hicks (centre right horizon) is one of the few recognised volcanic centres in this area. The smoke column is coming from a bush fire but helps one to imagine past eruptions.*

take place around wide lava fields. The Wannon River drains southwards from the Grampian highlands, but is diverted as it meets the lava plains and turns eastwards for 50 km to join the Glenelg River.

Lateral displacements are much in evidence among confined valley flows in the Southern Highlands of New South Wales. Here drainage flows both westward and eastward from the highland divide. Since basalt eruption, erosion has cut into the west-flowing basalt fills and has stranded east-flowing valley fills. West of the divide, the Lachlan River now runs east of its old source, which was filled by the 21 million year old BEVENDALE FLOW, and Wheeo Creek skirts west of the 19 million year old WHEEO FLOW. East of the divide, Wollondilly River has diverted west around the 25 million year old POMEROY FLOW. In contrast, the growth of the large WARRUMBUNGLE VOLCANO across the Castlereagh River caused its major diversion to the

the eastern tributaries flowing in towards the basalt-bounded drainage. The new drainage watershed now lies 30 km north of the old watershed before the lava eruptions. Large diversions may

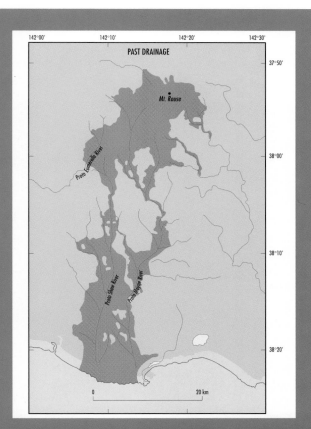
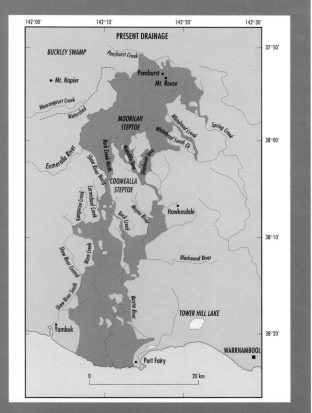

Mt Rouse lava flows, western Victoria. On the left is the former deep lead drainage system. The lava flows (dark areas, both diagrams) erupted from the vent during the last 2 million years and travelled 55 km down to the coast west of Port Fairy causing several lateral, twin lateral and diverted drainages, shown on the right. Modified from 1985 maps by C D Ollier, Royal Society of Victoria, Melbourne.

south. After this, the river resumes the same northerly flow direction taken by its companion Macquarie River as they flow to the Barwon River.

Basalt-filled gorges 300 m thick are exposed in the Johnstone River system, north Queensland. These gorges were cut by drainage flowing off the Atherton Tablelands onto the coastal plains. The thick lava fills, less than 3 million years old, forced the drainage to cut new gorges for the present river. Gorge-filling basalt, over 27 million years old, was discovered when the Expedition Range was drilled near Bauhinia Downs in central Queensland. This basalt descended east through a narrow defile cut in resistant massive sandstones, and it spread out beyond the barrier. The blockage banked up later flows coming off the Expedition Range to the south and the Staircase Range to the northwest.

Twinned drainage is demonstrated by the Mersey and Forth Rivers, which drain the northwest interior of Tasmania. They were born when thick lavas blocked an ancestral river course. The original river descended from 770 m at Maggs Mountain to 440 m under Gads Hill and Lorinna. Volcanism starting 27 million years ago filled the valley with basalts over 400 m thick. Afterwards the Mersey drained the east side of the valley and swept eastward to the coast at Devonport. The Forth drained the west side and reached the coast near Forth township.

A less imposing example of lateral twinning is shown by PADDY'S PLAIN FLOW, north of Dorrigo in New South Wales. Here the western Little Murray River is the dominant lateral and the 19 million year old flow remains relatively intact. Twin lateral streams formed by a young lava flow appear southwest of MT FOX, a 100,000–200,000 year old volcano west of Ingham in north Queensland. Four Mile Creek follows the southeast side of the flow, then cuts across the end of the flow to join its lateral stream counterpart.

Former underground drainages are preserved by lava in limestones near Timor in New South Wales. The present Timor limestone cave system exposes the basalts that entered an older cave system. These 73 million year old basalts were cooled by air, not water, which means that the underground drainage previously ran beneath this filling site and the topographic relief of the area in Cretaceous time was similar to the relief present today.

BURIED COURSES AND DAMMED DRAINAGES

Old drainages under lavas may preserve river conditions that differ from those in present drainages. Gravel beds underlying 40–60 million year old basalts in the Walcha-Nundle district of New South Wales contain a wide assortment of pebbles and boulders, many with percussion marks, indicating that swift tumbling streams once ran there, probably fed by wetter climates. In the Bathurst–Oberon area some buried river beds are full of resistant quartz fragments. They lack basalt pieces common in the present streams and depict river catchments devoid of basalts. Old river deposits under basalts capping Airly Mountain near Capertee were mined for diamonds and gold. The diamonds appear in some courses, but not in others. Diamonds are scarce in the present drainage, now at much lower levels, which probably means that the local diamond source was exposed before basalts sealed off the drainage 41 million years ago.

Where volcanic dams stall rivers, waters back up into substantial lakes. Damming totally altered the ancestral South Esk River that fed into the Tamar River at Launceston in Tasmania. The thick basalt

F.L. SUTHERLAND

under Launceston Airport was erupted from the COCKED HAT HILL DYKE and blocked the old river for 15 km, as far west as Longford. Clays and gravels with basalt pebbles were deposited over the basalt. The South Esk, forced into a wide detour, cut cataract gorges through the resistant dolerite hills at Launceston. Rose Rivulet, a tributary of the North Esk, also joins the Tamar at Launceston, but is cutting into the basalt barrier. A breach would capture the South Esk and re-establish its original source.

Continued eruptions into dammed rivers acquire aquatic characters. This happened repeatedly in large river systems in Tasmania, as volcanoes discharged at many places. In the ancestral Mersey-Forth drainage, flows blocked valleys at different stages at Paloona, Sheffield, Moina, Gads Hill and Borrodaile Plains. Extensive hyalotuffs and flow foot breccias formed until lakes were overrun by massive lavas. Hyalotuffs and breccias in the old Cam River course and in drill holes through the basalt-filled Guildford courses mark the occurrence of further explosive disruptions made by lavas entering northwest Tasmanian river systems.

The Derwent River System, Tasmania's major drainage, shows 18 sites where volcanism interacted

△ Coarse gravel beds in diamond-bearing deep lead, Monte Christo Mine, Bingara volcanic field, New South Wales, October 1990.

◁ Cross-section of a lava tube (pillow) in flow foot breccia, Duck Point Bay, Great Lake, Tasmania, April 1968. Note multiple cooling crusts where lava was chilled and cracked in water before breaking up in steam eruptions.

▽ Diamond (bright glassy grain) in water-worn gravel containing quartz pebble 2.5 cm across, near Inverell, New South Wales.

with the rivers. The thick SKITTLEBALL PLAINS FLOW in Ouse River probably dammed ancestral Great Lake. Extensive flow foot breccias flowed into Great Lake, the lake around LIAWENEE and REYNOLDS ISLAND VOLCANOES. Lesser flow foot breccias and hyalotuffs appear at the Nive-Derwent River confluence near Wayatinah, in Dee River around Ouse township, and in Ouse River near Waddamana. They reappear at the Styx River's junction with the Derwent at Glenora, downstream near Plenty, and in the Derwent estuary at Bridgewater, Old Beach and Claremont near Hobart.

Mainland Australia has some aquagene volcanics, but nothing approaching the Tasmanian scene. In the Monaro field in New South Wales the main basalts blocked streams flowing from the south and west. Lakes were formed deep enough to accumulate sediment up to 150 m thick. Later lavas flowed in, exploding into a chaotic mixture of glassy tuffs and lumps of lake bed clay. Pillowy lava, formed as lake waters and wet clays, squeezed up into the flows. The Southern Highlands, the New England Tablelands in New South Wales and the Atherton Tablelands in Queensland all have scattered examples of flow foot breccias indicating where lavas entered local water bodies. At Inverell lava flowed into the old MacIntyre River, and near Atherton extensive pillowed basalt marks a blockage of a northern tributary of Wild River.

RADIAL DRAINAGES

When large volcanic peaks grow and streams radiate off their slopes they change the local drainage pattern. Such drainages typify many eroded basaltic shields and central volcanoes in eastern Australia. Some radial patterns become modified even further by later erosion, faulting and tilting.

Deep radial valleys are eroded into the flanks of the MALANDA BASALT SHIELD, 20 km in diameter, on the Atherton Tablelands, Queensland. The older BUCKLAND VOLCANO in central Queensland developed a radial drainage which became modified by the underlying divide to form east and west flowing streams. Over 25 million years the drainage cut through the volcano into the underlying, massive sandstones and created spectacular gorges such as Carnarvon Gorge. Other volcanoes with clear radial drainage include the CANOBOLAS, WARRUMBUNGLE, NANDEWAR, COMBOYNE and

partly removed EBOR VOLCANOES in New South Wales. Radial drainages point into smaller, largely eroded basaltic centres at MT CORICUDGY and THE BOX HILL in the Singleton–Dubbo region. The MONARO BASALT PLAIN also shows a radial focus, but there is dispute over whether this drainage came from a large volcano or developed by coincidence.

In several volcanoes radial drainages became modified by adjacent volcanic drainages. BARRINGTON VOLCANO in New South Wales shows radial stream incisions, which to the west join a major southerly flowing drainage — the upper Hunter River. This river developed in a saddle between two shield volcanoes of similar age, the 54–60 million year old BARRINGTON and MT ROYAL RANGE VOLCANOES. The MAYBOLE VOLCANO, a large 33–38 million year old basalt shield in New England, has a partial radial drainage. The western drainage was re-routed by lavas from the younger 19–24 million year old INVERELL VOLCANO. This lava plain diverted the ancestral Gwydir River in a loop around its southern side. Radial drainage from the TWEED CENTRAL VOLCANO is obscured by its overlap with the FOCAL PEAK VOLCANO to the west and by an extensive headward erosion of its east side by the Tweed River.

VOLCANIC TOMBS

◁ Drainage descending east off the Buckland basalt shield volcano (left), cutting deep gorges in the massive sandstone horizons of Carnarvon Gorge (centre right), Queensland, November 1990.

Volcanic eruptions often bury the remains of local organisms. Such buried time capsules may catch life in catastrophic circumstances — Pompeii's city life is a human example. They also seal off records of previous environments and help to date burials older than historic records. Volcanic entombments augment the fossil record and our knowledge of past ecosystems and the evolution of life.

Australia with its fauna and flora has gradually separated in stages from its former continental neighbours — India, Antarctica and New Zealand. This has taken 130 million years. The continent has approached the island chains of southeast Asia in the last 130 million years, moving away from higher southern latitudes and closer to equatorial regions, and passing through many climatic and sea level changes. About 45 million years ago the gap between Australia and Antarctica widened, creating strong circum-polar ocean currents. These

◁ Fossil trunk of a Jurassic cyathealean tree-fern Oguracaulis from Lune River, Tasmania. The trunk is replaced by silica, which preserves the woody vascular core; inner leaf traces remain as protuberances and detail of mucilage canal details remain on the stem surfaces.

events have helped shape the evolution of Australia's flora and fauna. The regular volcanic eruptions along Australia's east coast have helped preserve fragmentary slices of this evolution.

BURIED FLORA

Tasmania has splendid plant remains preserved under volcanic rocks. These include petrified forests in growth position, and woody materials replaced by silica, often yielding spectacular lapidary material. These forest fragments, woods, leaf fossils and plant spores preserved in sediments under basalts allow scientists to follow vegetation changes in Tasmania. This analysis is ahead of studies for comparable regions in mainland Australia. The oldest extensive flora in Tasmania occurs under a Jurassic flow at Lune River and dates back over 170 million years, when Australia was still integrated with the giant southern continent, Gondwana. The next oldest floras are buried in the channel and lake deposits of large river systems such as the ancestral Derwent, Tamar, Mersey-Forth and Helleyer Rivers. These date back to early Tertiary times, between 25 and 60 million years ago and include petrified forests in the Derwent Valley. The last sub-basaltic floras are found in smaller drainage systems such as the ancestral Ringarooma course and are younger than 25 million years.

LUNE LANDSCAPE

The Lune River fossil flora contains silicified replacements of woody parts of Gymnosperms (non-flowering plants) such as conifers, cycads, ginkgophytes, seed ferns and tree ferns (photograph, previous page). The material comes from sediments, soils and volcanic ash under basalts which flowed into a rift valley. The rift broke up dolerite intrusions 170 million years old. It is uncertain whether plants were silicified before burial or afterwards by siliceous fluids invading the basalt. The Lune lava cavities and joints are filled with agate and quartz (sometimes replacing calcite crystals). These add to the lapidary lure of the site.

The conifer woods show rich colours and even evidence of drying, splitting and fungal degradation, before silica filled the voids. Fragments of tree ferns have been reassembled into trunks a few metres long, suggesting that little transport occurred before burial. Rare leaves of seed ferns and cycadophytes form impressions in a mudstone lens. However, it is the silicified materials, particularly the 'man' ferns, which provide the most glorious petrifications. *Osmund-acaulis* was an abundant fern trunk and two new species were described, *O. jonesii* and *O. nerii*. *Oguracaulis banksii*, another common Cyathealean fern, was found as well as a fourth new species, a related *Cibotium* fern which was named *C. tasmanense*. Petrified root mantle, fronds on trunks and frond bases in the trunk apex are also preserved. The prize find was a complete polycarp (cone-like fruiting body) of a Bennettitalean cycad, probably the most perfectly preserved known one from this extinct group.

TASMANIA'S TERTIARY WOODS

Petrified trunks were found standing 2 to 3 metres high in growth position in volcanic tuffs near Macquarie Plains in the Derwent Valley. Forests were exposed in a section of seven flows and intervening beds of layered pyroclastics and sediments. The lavas overlie lake deposits at Glenora and some pillow lava suggests water interactions. One forest overlies mudstone with leaf impressions of *Nothofagus,* a southern beech. The tree trunks were silicified by solutions infiltrating from the enclosing tuffs. Wide, wavy medullary rays identify some beautifully preserved woods as *Casuarina* (photograph, p.130). Leaves and wood of flowering plants (Angiosperms) indicates a late Cretaceous or Tertiary age, but pollen in the underlying lake beds refines the dating to between 40 and 20 million years ago.

A coniferous flora overwhelmed by tuff was described in 1874 from a railway cutting under the Cocked Hat Hill flow near Launceston. The early Tasmanian geologist R.M. Johnstone also recorded silicified logs in tuff near Badger Corner on Flinders Island, while silicified *Melaleuca* wood is found nearby at Petrifaction Bay. Fossil tree trunks, branches, leaves, fruit and even log jams were found in tin workings along the old Ringarooma River course below the 16 million year old basalts. Some remains were partly transmuted into coal or replaced by marcasite, an iron sulphide mineral.

TASMANIA'S FLORISTIC EVOLUTION

The changes in vegetation preserved below Tasmania's basalts illustrate a progressive process of adaptation. The Lune River flora depict a

Jurassic forest, a woodland scene of gymnosperm plants and tree ferns, with ground ferns on the open plains. Tree ferns were abundant and in moister parts horsetails, club mosses, mosses, liverworts and shady ground ferns flourished. Tasmania's higher latitude at that time (about 65° S) invokes long dark winters with short days and a short mid-summer with longer days. The climate was typical of overall Jurassic conditions: wet and warm. The forest scene then would resemble the remnant forests of kauri pine, podocarp conifers, cycad, tree-fern and fern communities surviving today in the Atherton Tablelands of north Queensland. Since Cretaceous flora have been found elsewhere in which Angiosperms, or the flowering plants, are present, these forms presumably entered Tasmania at that time. None are recorded from the scanty Tasmanian Cretaceous yet, although Angiosperm pollens are found in Cretaceous sediments in the adjacent Gippsland Basin in Bass Strait.

◁ *Fossil leaves from Vegetable Creek, Emmaville, New South Wales. Reproduced from drawings by C. Baron von Ettingshausen, 1888.*

The oldest Tasmanian Tertiary beds, from 40 to 65 million years ago, show that the vegetation supported sub-tropical Casuarinaceae, Araucariaceae and broad-leafed podocarps. More temperate plants, such as broad-leafed forms of *Nothofagus*, appear below the basalts some 40 million years ago when Tasmanian rainforests resembled present-day rainforests in northern New South Wales. Cool temperate rainforest species appeared about 25 million years ago, and smaller-leafed *Nothofagus* species and sub-alpine flora came into vogue. Eucalypts were probably present, as a silicified log 12 m long was found near Hobart, but these plants did not become prominent until the burning practices of Aborigines were introduced in the last 100,000 years.

AUSTRALIAN MAINLAND VEGETATION

Many scattered fossil flora have been discovered under volcanic horizons in mainland Australia. For instance, the Vegetable Creek tinfields near Emmaville in New South Wales produced an impressive array of plants. Samples assembled by Professor T.W.E. David from Sydney University were described by C. Baron von Ettingshausen in 1888. They underlie flows and tuff beds now dated between 20 and 45 million years old. The flora under the older basalts are preserved in clays and carbonaceous bands and suggest tropical to sub-tropical assemblages of complex evergreen rainforests. The younger flora in 'ironstone' matrix evoke drier, more open rainforests that still retain tropical to sub-tropical elements. Fossil fruits of uncertain age also came to light, preserved without distortion and retaining original organic matter. Ettingshausen's descriptions of the flora generated considerable controversy, as he interpreted them as cosmopolitan species rather than as more local derived forms.

Ettingshausen also examined flora from sedimentary beds under basalts in the Ipswich-Brisbane area in Queensland, now dated at 46–55 million years. A younger range of flora was recovered below 30 million year old basalts near Capella, central Queensland. It included fruits of *Sarcopetalum,* plant types similar to those in present northern Australian and Papua Niugini rainforests, and fossil wood showing termite attack. Floras from below 18–22 million year old basalts were recovered from Kiandra, New South Wales. They contain abundant Lauracean and Myrtacean leaves, *Nothofagus* pollen and fungi and mosses, and these indicate humid warm conditions.

Further south in Victoria, *Eucalyptus* is found below a 22 million year old flow at Berwick. Its presence with rainforest species suggests a rainforest in a drying environment and represents early evidence of seasonality in the evolution of Australia's flora. Among younger Victorian basalts, *Banksia* comes from Ballarat, *Eucalyptus* and *Banksia* from Redruth, and *Eucalyptus* from Daylesford, all in floras less than 5 million years old. Fruits include types resembling those of existing taxa with rainforest species, such as *Owenia, Flindersia* and *Elaeocarpus*. As average eruption time in western Victoria is about 13,000 years, modern vegetation may well become buried by a subsequent eruption.

BURIED FAUNA

The monotreme and marsupial members of Australia's terrestrial fauna are of particular interest, and volcanic horizons overlying marsupial finds help date their evolution. The oldest known mammal fossils in Australia include jaw fragments from early Cretaceous sedimentary beds in the Lightning Ridge opal field in New South Wales. An opalised jaw comes from a primitive-toothed platypus. There is an enormous gap of 65 million years between the age of this fossil and the next marsupial fragments, found under basalt at Murgon, Queensland, and another 30 million years between them and more complete remains found under basalts in Tasmania. The most complete mammal faunas, discovered in younger basalt terrains in western Victoria and northern Queensland, help to decipher the evolution of Australia's unique mammals over the last 5 million years.

Many other vertebrate remains are found under volcanic caps. They include a range of crocodiles, lizards, turtles, birds, freshwater fish and other miscellaneous strays. Invertebrates such as snails, freshwater molluscs, water fleas and insects have also been preserved.

EARLY TERTIARY BURIALS

There are early marsupial remains in greenish silt at Tingammara, near Murgon in central Queensland. The fossil bed overlies rhyolite tuffs and underlies a 29 million year old basalt from the BOAT MOUNTAIN VOLCANO. A clay mineral in these deposits was dated at 54 million years. Marsupial remains there consist of only a few teeth, and these are quite unlike those of other known marsupial teeth. Crocodile, turtle and bird bones give us some idea of the fauna existing at the time.

AUSTRALIAN MUSEUM F42459/C BENTO

△ *Fossil fish impression (trout cod* Maccullochella macquariensis), *Bugaldi, New South Wales, Australian Museum specimen*

Intriguing remains have been recovered from sediments below 46–55 million year old basalts around Ipswich and Brisbane. An avian foot, reptile parts, chelid turtles, three different fish, three types of water fleas, lacewing, plant-hopping insects and grasshoppers, and a host of beetles, bugs, waterbugs, scorpion flies and a termite suggest a busy early life.

△ *Fossil Bluff, Wynyard, northwest Tasmania, December 1969. The fossil marine sandstones (pale beds) are capped by basalt (vegetated upper slope) and overlie old glacial beds (darker cliff base and foreshore).*

Similar remains are found below the 33 million year old CAPE HILLSBOROUGH VOLCANO in Queensland. Fossils found at Wedge Island and Donna Bay are largely those of aquatic lake dwellers whose remains became broken up by water action. The spines and vertebrae of teleost fish, crocodile scutes, coprolites, turtle carapace and plastron pieces and water flea shells provide a fragmented vision of this lake life.

MIDDLE TERTIARY BURIALS

Tasmania boasts two unusual marsupial burials. A travertine spring deposit under basalt at Geilston Bay near Hobart yielded fragments of bone, snails and insect larvae. They were dispatched to England in the 1860s and sat in the British Museum of Natural History for a century. A visiting Australian scientist recognised their primitive nature and the basalt was dated at 23 million years, confirming their old affinities. The remains represent phalangerids, burramyids, dasyurids and diprotodontids, all types allied to later marsupials.

An old sea bed below basalt forms a weird marsupial burial at Fossil Bluff near Wynyard. The teeming tropical marine life was suddenly sealed by a basalt flow, so the discovery of a partly articulated marsupial skeleton, 40 cm long, embedded in marine fossils was unexpected. The animal, called

Wynyardia bassiana, was thought to be related to possums and was explained as a bloated carcass that had floated out to sea. The enshrining marine fossils suggest that this unique preservation is 24 million years old.

The Fossil Bluff sea-life was prolific in a shallow sea lapping the edges of Bass Strait. A coarse sandstone deposited by the rising sea is stuffed with over 300 species of marine animals. Bivalve molluscs such as *Eucrasatella* are conspicuous. Gastropods, brachiopods, echinoids, corals, algae and sharks' teeth are present. The overlying sea beds in the cliff consist of finer sandstones. Spire-shaped gastropod shells of *Turritella* are common and there are heart urchins, including *Lovenia*. Some unusual 'swimmers' appear besides the unfortunate marsupial *Wynyardia*. A skull of the now extinct toothed whale was named *Prosqualadon davidi*. The skull structure suggests that this whale used sonar for finding prey. A vertebral spine and rib section shows that other whales also swam off Wynyard. A well-preserved flathead fish skull became the first described fossil form, and shows that the modern sand flathead, *Platycephalus bassensis,* has changed very little over 24 million years.

Other marine fossil beds of this age appear under basalts of the MARRAWAH and REDPA volcanoes in Tasmania. Mollusc and gastropod shells, bryozoans and other past marine life are seen in outcrops along Marrawah beach. The BRITTONS SWAMP VOLCANIC NECK erupted through such beds, digesting pieces of fossiliferous limestone. This produced a startling specimen of a fossil pecten shell sitting in basalt — a so-called 'geological impossibility'.

LATER TERTIARY BURIALS

In the WARRUMBUNGLE VOLCANO in New South Wales some lakes in the volcanic caldera teemed with diatoms and deposited diatomite beds. Animal and leaf remains are found in diatomites at Bugaldi under a 13.5 million year old basalt flow. Codfish skeletons prove that fish inhabited the lake. The species named *Maccullochella macquariensis* resembled the modern Murray cod. The imprint of a primitive owlet-nightjar bird was named *Quipollornis koniberi*. Further south, near Gulgong, old river deposits overlain by basalts of similar age have delivered up pieces of turtles, such as

FL SUTHERLAND

Emydura and the oldest known *Meiolania*.

The young basalt fields in eastern Australia have preserved many fossil sites. The Hamilton fauna in Victoria below a 4.5 million year old basalt is remarkable for preserving only mammals. Out of 26 mammal species, the majority were small or intermediate animals typical of a temperate rainforest environment. The few big animals consisted of a sloth-like diprotodont, *Palorchestes,* and larger kangaroos such as *Macropus* and *Protemnodon*. Smaller marsupials included the brush tailed possum, *Trichosurus,* and the mountain pygmy possum, *Burramys*. The fauna found near Bacchus Marsh under 3.5 million year old tuff and basalt is called the Coimadai fauna. It includes diprotodontids, wombats and kangaroos.

△ Fossil shells embedded in marine sandstones, Fossil Bluff, Wynyard, northwest Tasmania, December 1969.

The Dog Rock fauna near Geelong comes from fissures sealed by a 2 million year old basalt and the host sediments suggest an age between 2 and 3.5 million years. The site demonstrates that the grey kangaroo, Macropus giganteus, and rodents like Pseudomys existed even then.

Northern Queensland lavas overlie several vertebrate fossil finds. The Bluff Downs fauna of Allingham Creek near Charters Towers were found in a channel fill that has become partly converted to limestone. Nearby basalt flows are 4–4.5 million years old, but basalt over the fossil site could be younger. The fauna are linked to modern communities — saltwater crocodiles, predatory varanid lizards and a variety of marsupials including grey kangaroos, red kangaroos, wallabies, euros, other wallaroos, rock wallabies and pademelons. The marsupial lion, Thylacoleo, wombats and diprotodontids were in attendance then. The large diprotodontids include Zygomaturus and Euryzygoma, the latter easily recognised by flared cheekbones in the males. Other fossils at Blaggard Creek and Tara Creek add to the record of Australia's developing fauna. The Blaggard Creek fauna yielded a crocodile, Pallimnarchus, with a broad flattened snout and upward-looking eyes.

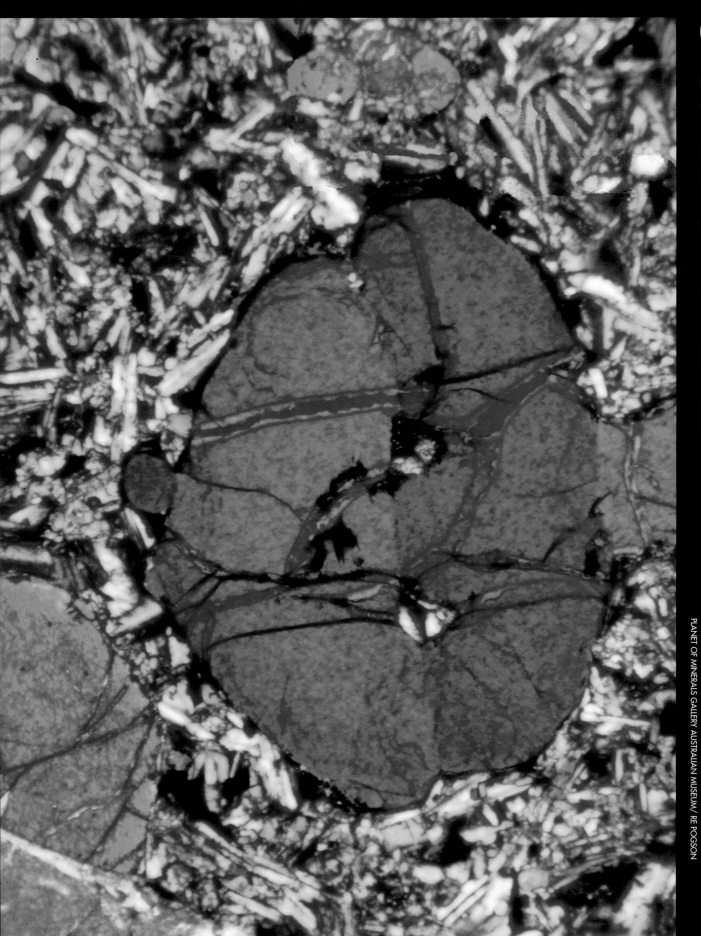

CHAPTER

PLANET OF MINERALS GALLERY AUSTRALIAN MUSEUM/ RE POGSON

10
VOLCANIC MINERALS AND ROCKS AROUND AUSTRALIA

◁ *Basalt rock, Lord Howe Island, 1984. Large early-formed olivine crystals (phenocrysts) in finer-grained groundmass of small, later-formed minerals, mainly feldspar crystals (pale thin laths). Microscope view: the central large olivine is about 2 mm. The olivine's bright colours are revealed by polarised light on the transparent rock slice.*

AUSTRALIAN volcanoes rarely retain the yellow stains of sulphur deposits once left by their hot gaseous breath — most of the superficial sublimates have worn away. Minerals in the lavas are mostly fine-grained and indistinct, although they come into focus when the rocks are slabbed and polished into thin transparent slices for viewing under a microscope. Volcanic minerals form at different stages, so most rocks show some variety in mineral size and texture. The wide range of volcanic rocks in Australia provides a correspondingly wide spectrum of minerals, including those below.

PHENOCRYSTS	crystals grown before or during lava eruption
GROUNDMASS MINERALS	crystals grown during lava flow and solidification
SECONDARY MINERALS	minerals formed in cavities after lava solidifies
XENOCRYSTS	minerals carried up from disintegrated wall-rocks
XENOLITHS	mineral aggregates carried up from dislodged wall-rocks

Some lavas chill so quickly that minerals hardly crystallise and volcanic glass is produced. These natural glasses are not clear and transparent like most commercial glass, but resemble industrial siliceous slags. Low silica lavas, such as basalt, form dark glass called tachylyte (scoria when frothy). High silica lavas, such as rhyolite, form glass that ranges from light to dark in colour and is called obsidian (pumice when frothy). The embryonic crystallisation of minerals produces less glassy rocks called pitchstones. There is every kind of variation between volcanic glasses and completely crystallised volcanic rocks. Many rocks retain a percentage of glass in their groundmass on cooling.

Certain minerals crystallise soon after the molten liquid stage, and others crystallise just before the rock solidifies. Which minerals form depends on the temperature, pressure and composition of the melt. An original melt, known as a primary magma in contrast to a modified magma, may rise directly to erupt at the surface. This is called a primitive lava. Most melts, however, lose some minerals as they rise or rest in chambers. The original melt composition changes and on eruption forms a fractionated or evolved lava. In this way, a wide spectrum of mineral associations and rock types can appear in volcanic sequences.

The percentage of each mineral in a rock is called its 'mode'. This can be used to name rocks, but partly crystallised and glassy rocks are hard to classify. More accurate schemes use the chemical composition of the rock. One, for example, uses the Total Alkali and Silica content of the rock (TAS classification). The mineral suite that would normally crystallise from a rock's chemical analysis can also be calculated. This is called a 'norm' and is used in some classifications. A classification using the norms of quartz, alkali feldspar, feldspathoid and plagioclase minerals in a rock has been used to estimate accurate rock names in Australian suites.

Lavas that solidify slowly in thick flows or intrusions usually show more complete crystallisation and a coarser grain size. Fluids may become trapped in the interior parts, where they assist the crystallisation of larger crystals in patches and veins in the rock that was formed last.

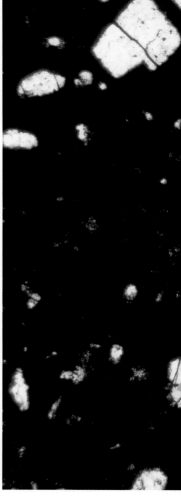

▷ *Quickly chilled volcanic basalt rock. Microscope view of early-formed olivine crystals (phenocrysts) up to 1 mm long, in a black glass base. Basalt flow, Murrumbidgee River, New South Wales.*

H HENLEY

THE MINERALS

Some minerals are regulars in volcanic rocks, others appear in certain rocks, and odd minerals are found in rare rocks. The primary minerals are covered in the box, next page. Silicate minerals dominate; oxides including quartz are the next most common. Only the main minerals are discussed, but a wider range and their chemical compositions are listed in Appendix 2. Details of these minerals can be found in the *Encyclopedia of Minerals* listed in the bibliography.

THE SECONDARY MINERALS

These minerals form when lava degasses and expels hot liquids, or else they crystallise when later groundwaters infiltrate cavities and cracks in lavas. In the first process, olivine often changes into clay and iron oxides and shows up as red or orange replacements instead of its more common form —

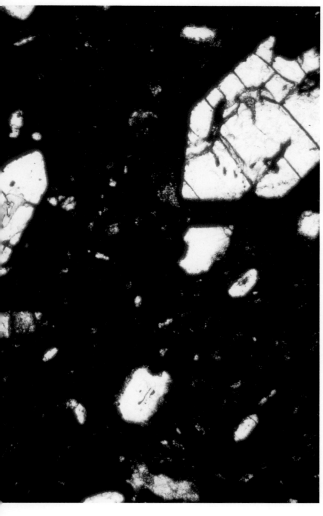

and filled cavities left by the original steam explosions. Calcite formed globules and scattered single crystals in the cavities. The minerals were recent groundwater depositions, as the aragonite had not converted into calcite.

Another aragonite–calcite association found in a highway cut in basalt near Ben Lomond was notable for varied aragonite and calcite crystal forms. The aragonite showed unusual bladed habits. The calcite had eight different forms including sharp needles, rhombs, elongate bipyramids, globules, and powdery coatings. Zeolite minerals mingled with the carbonates.

The diverse zeolite minerals in volcanic rocks deserve detailed treatment (see box, p.152). Their aluminium silicate frameworks are able to accommodate exchangeable alkali and alkaline earth elements as well as water molecules. This structural and chemical flexibility yields minerals in several crystal systems. Australian volcanic rocks feature some world class zeolite associations.

▽ *Basalt cavities and fissures filled with secondary zeolite and carbonate minerals, New England Highway quarry, near Ben Lomond, New South Wales, March 1988.*

glassy green crystals. Alkali feldspars turn into microscopic shreds of mica or clay minerals; pyroxenes often become amphiboles; amphiboles, in turn, break down to green hydrous minerals called chlorites; and dark micas may change into golden vermiculite. Plagioclase feldspars decompose to a range of zeolite minerals. Iron oxides change into the constituents of rust, forming highly oxidised and hydrous iron oxides.

In later fillings, in voids and veins, the minerals remain stable at lower temperatures and hydrous conditions. The elements of the minerals often come from weathered host rock. Carbonate minerals are common, particularly the calcium carbonate species, calcite. Aragonite, a related mineral, has a different crystal structure. It is less stable under surface pressures and eventually transforms into calcite. Both minerals were found in the Kulnura pipe in New South Wales. Aragonite crystals formed in fine radiating groups

PRIMARY VOLCANIC MINERALS

△ Minute zircon grain (0.1 mm long), Windera, Queensland, used for dating a 240 million year old rhyolite plug.

RE POGSON

THE REGULARS

OLIVINE is a common mineral in many low-silica rocks. It occurs in some intermediate rocks, but is absent from silica-rich rocks. Quartz and olivine are incompatible chemical partners. They may overlap in the same rock, if the olivine and quartz crystallised early and late respectively. Quartz is typical of high-silica rocks.

Minerals of the clinopyroxene family are usually present but the exact mineral depends on the lava composition. Feldspar minerals of the plagioclase family prevail in all except the more extreme lava compositions. Iron-titanium oxide minerals are ubiquitous subordinate minerals, occurring either alone or in combination. Apatite is a widespread subordinate mineral and is an example of a phosphate mineral.

THE PART-TIMERS

Minerals of the alkali feldspar family feature in silica-rich and alkali-rich rocks. They differ from the plagioclase feldspars because they contain more potassium and are useful for dating volcanic rocks.

Minerals of the orthopyroxene family differ from the clinopyroxene family in crystal structure, but the two sometimes occur together. The orthopyroxenes usually appear in rocks that tend to be silica-rich.

Minerals of the amphibole group are an extremely complex group of hydrous silicates found in hydrous volcanic rocks. Micas, another group of hydrous silicate minerals, are rich in alkalis and appear in hydrous volcanic rocks that are similar to amphiboles. Feldspathoid minerals are alumino-silicate minerals found in low-silica and alkali-rich rocks.

THE ODD ONES

Zircon, a zirconium silicate mineral, usually appears in rocks very rich in silica and alkalis. It is useful for dating volcanic rocks because it bears traces of uranium. Many other rare subordinate minerals are encountered in Australian volcanic rocks, including titanium silicates, titanium oxides, sulphide minerals and rare earth element minerals.

Carbonate minerals characterise the unusual carbonatite lavas. These rocks are rare compared to the overwhelmingly abundant silicate lavas. Only old carbonatite pipes are known in Australia.

EARTH VOLCANIC

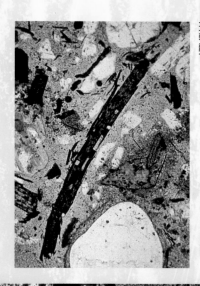

◁ Larger pyroxene crystal (phenocryst) in basalt rock, Barrington Tops, New South Wales. Microscope view: polarised light shows up the growth zones in the 8-sided crystal (darker core and lighter rims).

▷ Large rounded quartz and bladed dark mica crystals in rhyolite rock, Good Hope area, Murrumbidgee River, New South Wales. Microscope view: long curved mica flake is about 4 mm long.

▽ Amphibole (orange brown to yellow) and apatite (cloudy white to pale green) crystals from an inclusion in volcanic tuffs, Tararan, Queensland. The black opaque grains are the iron oxide mineral, magnetite. Microscope view, polarised light.

149

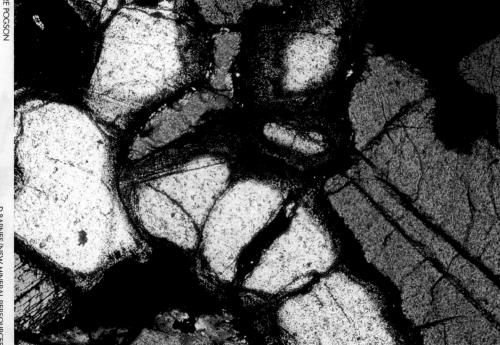

△ Large plagioclase feldspar crystal in basalt rock, Mt Bruce, Eungella, Queensland. Microscope view: polarised light shows up the repeated twin zones (dark and light bands) of the 2 cm-sized crystal. Note small feldspar laths in the lava around the crystal.

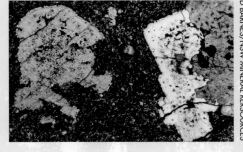

△ Large alkali feldspar crystals (light and grey) in trachyte rock, Mt Landsborough, near Nebo, Queensland. Microscope view: polarised light shows the intergrown grains and partly eaten appearance due to enclosure in hot lava. Large composite crystal is about 4 mm across.

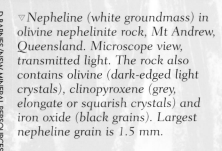

▽ Nepheline (white groundmass) in olivine nephelinite rock, Mt Andrew, Queensland. Microscope view, transmitted light. The rock also contains olivine (dark-edged light crystals), clinopyroxene (grey, elongate or squarish crystals) and iron oxide (black grains). Largest nepheline grain is 1.5 mm.

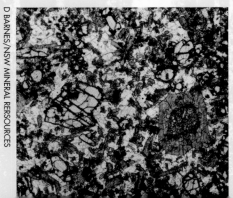

CHAPTER 10

THE ROCKS

This account concentrates on the volcanic rocks that decorate Australia's last 230 million years. They are less affected by folding and metamorphic events and less eroded and obscured by later sedimentary covers than the older volcanic series. Only the main rock groups are mentioned, and those who wish to delve into the thorny details of rock nomenclature and classification can find texts listed in the Bibliography. A simple classification scheme for the main Australian volcanic rock names is outlined in the table below.

COMMON OLD BASALTS

Australia has its share of the common, fine-grained, dark rock that makes up 90% of the Earth's volcanic inventory. Basalt rocks are not monotonous to geologists who specialise in rocks (petrologists); they include many different types. It takes a detailed microscope and chemical examination to yield their inner secrets, but experienced geologists can spot differences in the field. Basalts containing feldspathoid minerals often have a spotty and cracked appearance caused by their breakdown.

The presence of feldspathoid minerals, either in the form of visible crystals or inherent in the rock's chemistry, splits basalts into silica-unsatisfied and silica-satisfied series. The silica-satisfied series contains silicon-rich minerals such as orthopyroxene and quartz. All basalts, however, contain plagioclase feldspar as a main mineral. The silica-unsatisfied basalts are called alkali basalts and the silica-satisfied basalts are called tholeiitic basalts. In-between types are called transitional basalts.

For petrologists peering into basalts, the division into alkali and tholeiitic basalts is not enough. They use the type of plagioclase feldspar to name evolutionary types. As the plagioclase becomes more rich in sodium than in calcium, the alkali basalts divide into: alkali basalt, hawaiite, mugearite and benmoreite.

From there, petrologists distinguish basalts even further by using feldspathoid minerals to subdivide silica-unsatisfied rocks. Those with over 5% of feldspathoid minerals, such as nepheline, form basanite and its evolutionary types correspond with nepheline hawaiites, mugearites and benmoreites. Basalts rich in large olivine crystals are called picrites, and those rich in olivine, pyroxene and sometimes feldspar crystals are called ankaramites.

The silica-satisfied (tholeiitic) basalts are also divided, but no feldspathoids are involved. This

SIMPLIFIED CLASSIFICATION OF AUSTRALIAN VOLCANIC ROCKS, BASED ON SILICA (QUARTZ), FELDSPATHOID AND FELDSPAR CONTENTS (PLAGIOCLASE/ALKALI FELDSPAR)					
Quartz content	Largely calcic plagioclase	Largely sodic plagioclase	Sodic Plagioclase & alkali feldspar	Largely alkali feldspar	Largely feldspathoid & alkali feldspar
common	dacite	sodic dacite	rhyodacite	rhyolite	
sporadic	andesite	sodic andesite	trachyandesite	trachyte	
minor	tholeiite	sodic tholeiite	icelandite	alkali icelandite	
absent	ol tholeiite	sodic ol tholeiite	ol icelandite	alkali ol icelandite	
Feldspathoid content					
minor	alkali basalt	hawaiite	mugearite	benmoreite	
sporadic	basanite	ne hawaiite	ne mugearite	ne benmoreite	phonolite
					nephelinite
					analcimite
					melilitite
					leucitite
					lamproite
(ol: olivine, ne: nepheline)					kimberlite

▷ Alkali basalt from north Bowen Basin, Queensland. Microscope view, polarised light. The rock mostly consists of plagioclase feldspar crystals (long white dusty laths) intergrown with clinopyroxene crystals (dark to medium grey zones). The small scattered grains (pale grey) are olivine and the opaque grains (black) are iron oxide minerals. The dusty alteration of the feldspar marks the feldspathoidal mineral analcime. The feldspar and pyroxene grains are up to 3 mm long.

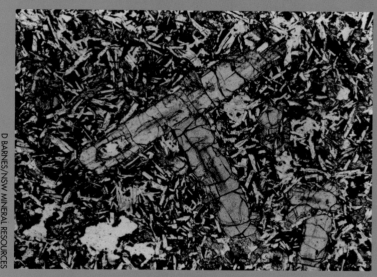

▷ Tholeiitic basalt from lava flows near Goonyella, Queensland. Microscope view, transmitted light. The large, elongated crystals with transverse cracks are orthopyroxene and the absence of olivine indicates that the basalt is composed of quartz tholeiite. The other smaller minerals are plagioclase feldspar (white laths), clinopyroxene (dark grey grains), and there is some iron oxide charged glass (black) and irregular gas cavities (white). The longest orthopyroxene crystal is 5 mm long.

series includes tholeiite, basaltic icelandite and icelandite, as the plagioclase feldspar composition increases in sodium. The tholeiites include those with olivine (olivine tholeiites) and those with quartz (quartz tholeiites).

THE FULL FELDSPATHOID SERIES

Since rocks rich in feldspathoid minerals are much rarer than basalts they are more treasured by petrologists. They are named after the dominant feldspathoid mineral and include the nephelinites, melilitites and leucitites. Olivine is common and often included in the name, e.g. olivine nephelinite. Nephelinites rich in large olivine crystals are called ankaratite.

THE LIGHT FELSIC SERIES

Petrologists distinguish the rocks that no longer look basaltic because they have become so rich in lighter-coloured, lighter-weight minerals. 'Felsic' is shorthand for feldspar, feldspathoid and silica. These minerals form over 75% of the rock and any plagioclase that joins in is usually sodium-rich.

As felsic rocks evolve from alkali basalt parents, they gain in silicon content and felsic character. Eventually they crystallise quartz, even though they have originated from a silica-unsatisfied stock. The main rocks include nepheline trachyte (up to 10% nepheline), quartz trachyte (up to 10% quartz), phonolite (over 10% nepheline) and rhyolite (over 10% quartz). The felsic rocks that evolved from

THE ZEOLITES

OLDER volcanic rocks develop zeolites during burial, folding and regional heating. In a flow of Permian age at Unanderra, near Wollongong in New South Wales, laumontite crystallised first in long crystals which then became embedded in calcite crystals. This zeolite forms at higher temperatures (about 200° C) than most other zeolites do. Next came ferrierite, a very rare zeolite, growing in bladed radiating crystals as fine as any ever found. Heulandite formed last, in between episodes of calcite crystallisation. These minerals formed when a high heat flow from the opening of the Tasman Sea altered the coastal rocks about 90 million years ago.

The Garawilla volcanic rocks of Jurassic age near Coonabarabran in New South Wales yield exceptional zeolites — pink, orange, red and brown fans of heulandite, and a rare zeolite, stellerite. This is the most abundant zeolite at Garawilla, although it is scarce in most other places. A more common counterpart known as stilbite has a slightly different crystal structure. Crystals of analcime, laumontite, glistening spheres of quartz and finally calcite were formed. The minerals were probably formed by hot spring waters as volcanism waned.

Basalts of the Tertiary Period throughout eastern Australia carry many fine zeolites. At Ben Lomond they formed as the basalt cooled, and afterwards during hot spring activity. Phillipsite crystallised first, with twinned, blocky and slightly cruciform-shaped crystals, then natrolite developed in typical radiating needles. The hot spring waters crystallised more phillipsite, but this time in distorted, strongly cruciform twinned crystals, and then produced chabazite in many crystal habits. The simplest habit formed rhombs, which have a more or less cubic shape. More complex crystals formed twins. Some of these grew as

◁ *Crystals of the zeolite mineral gmelinite (orange pink), with natrolite (white needles) and other secondary minerals in a basalt cavity, Flinders, Victoria.*

interpenetrating crystals, others as repeated twins forming almond-shaped crystals. Greater twinning usually indicates high deposition temperatures.

In the nearby Liverpool Range, at Ardglen, basalt contains natrolite in spectacular radiating needles, that line cavities and grow on analcime. Very rare zeolites were discovered at Merriwah, south of the Liverpool Range. Offretite forms fibrous sheaves and overgrowths on thin tablets of levyne. Erionite forms rare groups of radiating crystals.

Zeolites have featured in Victoria's mineralogy since 1861. Among the older basalts, Flinders and Phillip Islands have become celebrated zeolite collecting sites, so much so that Flinders has been declared a reserve. Flinders is a renowned site for gmelinite, a rare zeolite that forms bright red or orange hexagonal crystals. Thomsonite, a more common type of zeolite, assumes several different habits. Other rare zeolites in the region include ferrierite, offretite, erionite and cowlesite. Melbourne basalts are widely known sources for chabazite, especially the twinned, almond-shaped variety called phacolite, and for cruciform twins of phillipsite. The chabazites presented unfamiliar crystal forms and Australian and European mineralogists vigorously debated whether these were new species. More recently, a fibrous radiating zeolite, once called mesolite, proved to be intergrowths of natrolite and another rare relative gonnardite.

Tasmania is well endowed with zeolites, particularly in the aquagene pillow breccias in coastal and inland river sequences. Chabazite varies in crystal habits and grows in combination with white gmelinite near Redpa. Phillipsite is common and near Scottsdale it accompanies gonnardite crowned with bristles of natrolite. Some associated minerals also have special interest. Apophyllite, named from its fish-eye lustre, grows with chabazite. Also found there is a surprisingly abundant snowy white mineral, tacharanite, which is scarce elsewhere in the world. Its name means 'changeling', as the original mineral from Scotland soon broke down when exposed to air.

New Zealand boasts a wide range of zeolites, as alterations in the volcanic rocks and from the action of hot groundwaters in the thermal areas of the North Island. One of them is wairakite, a calcium zeolite related to analcime, reported from the Wairaki geothermal field in 1956.

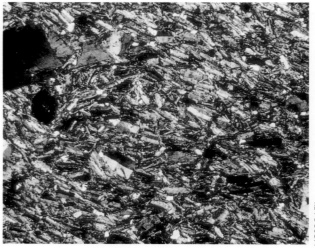

◁ *Trachyte, from Pinnacle Rock, near Rockhampton, Queensland. Microscope veiw, polarised light. The rock is mostly potassium-rich alkali feldspar (grey to white laths) lined up along the lava flow direction. The large dark grey grain of anorthoclase feldspar, is about 1 mm across.*

tholeiitic basalts include dacite (75–85% alkali feldspar and quartz), quartz trachyte and rhyolite.

Although quartz trachytes and rhyolites can evolve from either alkali or tholeiitic parents, the tholeiitic members usually show more aluminium than alkali elements in their makeup. This helps petrologists pin down their origin.

INTRUSIVE AND EXTRA NAMES

Volcanic rock names were developed for lavas with fine-grained crystals (less than 1 mm across). However, the pipes, dykes and chambers that spawn lavas are intrusive rocks and they usually cool more slowly and develop coarser grain sizes. These rocks are given separate names, but have minerals and chemistry similar to those of the equivalent lavas. In the simplest scheme basalt becomes dolerite (crystals 1–5 mm) and gabbro (crystals over 5 mm), trachyte becomes micro-syenite and syenite, and rhyolite becomes micro-granite and granite.

The texture of many lavas and their intrusive counterparts consists of two grain sizes: coarser phenocrysts in a finer groundmass. Such rocks are called porphyritic and the main phenocryst minerals can be designated by the rock name — e.g. porphyritic olivine basalt, porphyritic quartz-feldspar rhyolite and so on.

THE SUBDUCTIVE ANDESITE SERIES

These angry discharges from subduction settings form rocks that are similar to the tholeiitic rocks. They range from basaltic andesite through andesite to dacite, trachyte and rhyolite. A porphyritic olivine-rich andesite is called oceanite.

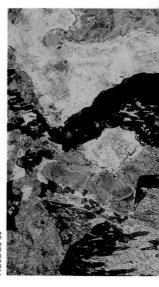

△ *Rhyolite, from near Mt Sebastapol, Queensland. Microscope view, polarised light. Pointed quartz crystals (white grains) and alkali feldspar (squarer grey grains) lie in a finer matrix mostly comprising alkali feldspar laths and quartz grains. The larger grains are over 1 mm across.*

The subductive andesite series has more porphyritic textures and greater amounts of hydrous aluminium-rich minerals (e.g. amphibole and mica) than the within-plate theoleiitic series does. This tendency reflects higher contents of silicon, aluminium and magnesium in relation to the alkali and iron elements. Where rocks of the two series overlap, petrologists turn to trace elements to fine-tune

◁ Porphyritic dyke containing conspicuous crystals of the alkali potassium-rich feldspar mineral sanidine in chilled alkaline rock, Port Cygnet, Tasmania, November 1978. This is a shoshonitic rock called 'biscuit rock'.

▽ Rhyolite crystal tuff, Good Hope area, Murrumbidgee River, New South Wales. Crystals include quartz (larger light grains), altered alkali feldspar (cloudy pale brown grains), amphibole (small elongate green-brown crystals), mica (dark flakes) and iron oxides (opaque grains). Microscope view, transmitted light. The largest quartz grain is about 1 mm across.

◁ Nepheline syenite, a coarse equivalent of phonolite, Mt Ramsay, near Baralaba, Queensland. Microscope view, transmitted light. Large nepheline crystals (white) with alkali feldspar crystals (white partly cloudy grey), pyroxene and amphibole minerals rich in sodium (elongated dark green to black grains and intergrowths). Cavities are filled with secondary carbonate and clay minerals (pale brown patches). Large nepheline is over 3 mm across.

their assignments. For instance, if the amount of the element niobium is low in relation to uranium and thorium, this usually signifies a subduction suite.

The coarser intrusive equivalents to andesite are micro-diorite and diorite. The coarse equivalents to dacites and rhyolites are granodiorite and granite rocks, usually found in large, deep, subvolcanic chambers.

POTASSIUM-RICH SHOSHONITIC SERIES

These rocks form a bridge between the andesite and alkali basalt series. They usually carry some potassium-rich minerals such as alkali feldspar, dark mica or leucite, and occur in the outer regions of subduction zones. Since they are transitional rocks, their mineralogy and chemistry tend to be ambiguous. Their distinction and terminology in Australia was championed by Germaine Joplin, a petrologist who worked on them in the 1960s and 1970s.

The lavas are commonly porphyritic types. Olivine is prominent in shoshonite and alkali feldspar is prominent in latite. Intrusive equivalents to shoshonites include potassium-rich diorites and intrusive equivalents to latites include monzonites. The shoshonitic series develop swarms of strongly porphyritic dyke rocks, including complex types called lamprophyres which are full of different phenocryst minerals.

AUSTRALIAN SYNTHESIS

Basalts and their related rocks dominate eastern Australian regions. Many erupted from within-plate volcanoes over the last 95 million years. Similar basalts also erupted earlier when Australia was joined to Gondwana, 95–230 million years ago. Tasmania formed a flood basalt province at this stage. Only in northeastern Australia did andesites and other unstable margin volcanic rocks develop in this interval. For the major part of Australia, volcanic offerings of these periods are sparse. A few basalts fringe the southern and western rift margins and low silica alkaline rocks of kimberlite and lamproite affinities are scattered in southern and northwestern Australia. Some offshore volcanoes erupted on the northwest shelf beyond the Australian margin.

Andesites and accompanying basalts, trachytes, dacites and rhyolites of unstable margin affinities make up some significant exposures in fold belt regions formed 230–320 million years ago. The volcanoes follow old subduction zones along the New England–Hodgkinson Basin fold belts, extending from New South Wales into northern Queensland.

Further tracts of unstable margin volcanic suites appear in older eastern Australian fold belts, formed 320–570 million years ago. They lie in the Tasman and Lachlan fold belts and extend into central and southern New South Wales, Victoria and western Tasmania. Some altered sea floor basalts appear in older parts of the New England fold belt. When unstable margin volcanism in eastern Australia began it was accompanied by extensive within-plate

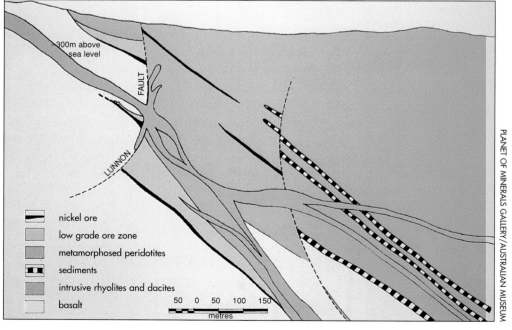

◁ Section depicting metamorphosed basalts and peridotites of the Archaean greenstones, Kambalda, Western Australia, showing their relationship to the nickel sulphide ore-bodies.

PASSENGER MINERALS AND ROCKS

As molten rocks, charged with volatiles, rush to eruption, they may dislodge and break up minerals (xenocrysts) and rocks (xenoliths) from the channel walls. Such passengers come from many depths, even below 300 km in exceptional circumstances. The exotic travellers, trapped in the solidified hosts, tell scientists what really lies below. Their volcanoes deliver goods of great scientific and sometimes commercial wealth. Australia has prospered from such volcanoes to the extent of becoming a world supplier of diamonds and sapphires.

XENOCRYSTS

Many common minerals break off the wall rocks. Those from coarse deep rocks grew as large crystals. The hot melt corrodes and etches their surfaces, modifying or even obliterating the original crystal forms. Octahedral diamond crystals, with 8 triangular shaped faces, may react to become more rounded dodecahedral crystals with 12 rhomb-shaped faces.

Once the rocks reach the surface, xenocrysts become released by erosion, particularly from loose, broken-up explosive deposits. Minerals susceptible to weathering, such as olivine, feldspars and micas, break down to clay and carbonate minerals. Moderately resistant minerals such as pyroxenes and garnets may linger awhile. Highly durable oxide and silicate minerals, such as quartz, spinel, ilmenite, corundum and zircon, will wash and rewash in alluvial deposits. Heavy minerals like sapphire, zircon and diamond become concentrated by gravity into alluvial traps. Some heavy alluvial minerals, such as garnets of particular compositions, are used to indicate parent volcanic sources.

DIAMONDS

Australian diamonds come from three general regions. In the western region, the main area — Kimberley, in northwestern Australia — contains lamproite and kimberlite pipes. Some carry diamonds brought up from depths over 150 km. The prolific Argyle pipe is the main producer (see over).

△ Colour range of cut sapphires from New England sapphire fields, New South Wales. Each sapphire is about 0.5 cm across.

▷ Cut zircon, 2.3 cm across, weighing 40.5 carats, from Inverell area, New South Wales.

flood basalts in central-northern Australia. These Antrim plateau basalts probably covered 500,000 km² and leads into the volcanic history of the western two-thirds of Australia.

Beyond this lie the vast episodes of Precambrian volcanism in Australia. These extend back to 4 billion years and are obscured by many metamorphic events. The extensive Proterozoic silica-rich volcanics in the Gawler Range, South Australia, and the extraordinary silica-poor lavas of the Archaean 'greenstones' in Western Australia were formed in this era. The magnesium-rich Archaean lavas, which were a few hundred degrees hotter than present basalt lavas, sometimes melted and excavated the rocks they flowed over. These rocks are called komatiites and a 2,700 million year old example from Kambalda contains 32% magnesium oxide and is associated with rich iron, nickel and copper sulphide ore deposits. These high temperature lavas mark a time when the whole underlying mantle was in a hotter state.

VOLCANIC EARTH

DIAMOND MINES

An explosive pipe carried diamonds to the surface before they become washed into alluvial drainage deposits. Theoretical model.

1. Diamonds from 200km depth blasted up pipe by gas
2. Diamonds & boulders fall back into pipe
3. Diamonds & ash fall around crater
4. Diamonds washed into crater
5. River erodes diamond bearing ash

CLUFF RESOURCES PACIFIC P/L SYDNEY

THE 1,180 million year old ARGYLE PIPE became buried with its flared throat and exposed crater lake still preserved. Molten olivine lamproite, which contained diamonds, rose in dykes and exploded on contact with groundwaters. Tuffs full of sandy materials from collapsing country rocks were blasted out along with the diamonds. The diamonds are mostly brown, rounded, partly dissolved dodecahedral crystals, up to 40 carats in size. White, cognac and champagne colours are cut and the rare fancy pinks and reds, produced by strong crystal strains, fetch premium prices. Minute garnet and diopside crystals included in the diamonds date their birth to a 1,580 million year old eclogite metamorphic parent rock. A 400 million year confinement took place before the diamonds were whisked up from the base of the Australian plate into the Argyle eruptions.

The ELLENDALE PIPES, in contrast, form much younger lamproite volcanoes — about 20 million years old. Their diamond grades are lower than Argyle grades. As well as diamonds of eclogite parentage, Ellendale pipes contain just as many octahedral crystals grown from a parent peridotite rock. The peridotite diamonds possibly formed over 3 billion years ago — a long time before seeing the surface.

Small diamonds appear in the central Australian region. Some come from 170 million year old ORROROO KIMBERLITE DYKES in South Australia, but are too sparse to mine. Although outwardly typical, these diamonds contain some high-pressure minerals — notably magnesio-wüstite, a magnesium iron oxide. This suggests that the Orroroo diamonds grew in the mantle under the Australian plate well below 300 km. Sparse, larger diamonds found around Adelaide are suspected to be escapees from local lamprophyre dykes. Even stranger are small cubic diamonds recovered from 2050 million year old beds near Coanjula in the Northern Territory. These oddities concentrate around a trachyandesite plug, a rock not normally associated with diamonds. However, the diamonds are unlike any of those recovered from kimberlites or lamproites.

Eastern Australian diamonds from the third region are equally mysterious. Dribs and drabs and minable pockets of diamonds extend along the eastern fold belt country. Despite intense exploration, mining companies failed to find the ancient basement rock, kimberlites or lamproites usually favoured by diamonds.

Diamond crystal 25 mm across in dolerite rock from Oaky Creek diamond mine, Copeton, New South Wales.

JF LOVERING

Indeed, rare diamonds appear in basalts, which usually generate from melting well above the diamond zone. Diamonds found around Copeton in New England, a district which supported early mining, are strange characters; tougher than normal, their internal mineral inclusions differ from

CHAPTER 10

EARTH VOLCANIC

◁ Contact of Argyle diamond pipe (foreground workings) and dipping beds of Proterozoic quartzites (background), Argyle, Western Australia, August 1986

▽ Cut diamonds from Argyle, Western Australia. These were the first stones cut and polished and show the cognac, champagne and white colours.

those usually found in diamonds. For instance, they contain coesite, which is a high-pressure silicon oxide, and a calcium-rich garnet, and they are made of a heavy form of carbon. A possible birth age of about 300 million years makes them abnormally young compared to most diamonds born at least 1–3 billion years ago. There are unsound theories that mighty rivers or glacial meltwaters brought them large distances from old areas to the west or south.

Brand-new ideas of diamond origin were proposed to explain the Copeton stones. These 'subduction diamond' views were based on the assumption that diamonds were created in relatively cool seafloor slabs as the slabs were forced under the New South Wales fold belts. These slabs may rest at higher levels than is usual for diamond-bearing slabs. Later basalt or lamprophyre eruptions could then pluck up the diamonds, without any of the special deep generation of kimberlite or lamproite melts. In this theoretical model, Eastern Australia probably erupted small parcels of diamonds more often than western Australia, which had long waits between its diamond bounties.

▷ Inside adit, Oaky Creek diamond mine, Copeton, New South Wales, May 1991. An altered dolerite dyke (yellow-brown rock, centre left) intrudes granite (speckled grey rock, right) and is overlain by diamond-bearing sediments (top left).

SAPPHIRES AND ZIRCONS

Eastern Australia boasts a wide spread of sapphires and some famous gemfields. Productive fields lie around Inverell and Glen Innes in New England, New South Wales, and in the Anakie area, central Queensland. Smaller fields are found at Lava Plains and Lakeland Downs in northern Queensland, around Oberon in New South Wales, and in scattered finds in Victoria and Tasmania.

Sapphires shed from basalt sequences become concentrated in local drainages. Crystals in barrel, dogtooth or flat hexagonal shapes reach several centimetres in size. Many have corroded surfaces and some are found still embedded in basalt. Mostly blue, green and yellow stones are cut, some as parti-coloured zoned stones or star stones. In the USA a few large black and dark blue more opaque stones were carved in the likeness of the heads of famous American presidents. Very rich sapphire deposits are still being explored and mined at Kings Plain, in the New England field, and east of Rubyvale in central Queensland.

Gem zircons abound with the sapphires and share an intimate origin. Large coloured zircons are cut locally into fine stones (photograph, centre p.157). Pale stones bear low traces of uranium, deep orange, brown or red stones bear higher traces of uranium. Zircons, unlike sapphires, can be dated and help solve the xenocryst story. Boat Harbour zircons, from northwestern Tasmania, grew at the time their evolved basalt hosts erupted, 15 million years ago. Minute quartz grains in these zircons suggest that they grew against quartzites in the underlying Precambrian crust. The crystal forms indicate that their temperature growth was lower than that of most gem zircons. Many gem zircons elsewhere in Australia grew from hotter, deeper alkaline melts, linked to greater sapphire formation.

Breakthroughs in understanding sapphire genesis came from several studies:

1. Fluid inclusions — these show that some sapphires grew 30–40 km down, from melts charged with carbon dioxide gas.
2. Silks, where rutile needles lie along crystal directions — these show that sapphires grew at higher temperatures which dissolved this titanium oxide mineral.
3. Mineral inclusions, such as alkali feldspar, uranopyrochlore, columbite, thorite and zircon —

△ Sapphire-bearing gravels overlying weathered granite bedrock, underground in Bobby Dazzler Mine, Rubyvale, Queensland, May 1991.

these show that sapphires grew from highly evolved alkaline materials enriched by traces of large metal elements like zirconium, niobium, uranium and thorium.
4. Zircon inclusions, which date the sapphires — these show that sapphires grew under large basalt volcanoes.
5. Zircon fission track ages, which date the eruptions — these show that sapphires were probably erupted several times.

Shuffling eruptions may explain Australia's sapphire story. The main New England gem volcanoes grew one by one between 43 and 28 million years ago. Each volcano erupted for a few million years. When one died, the next grew up alongside, first on one side, then the other. The volcanoes match the times and positions in which the Australian plate is estimated to have passed over some small hotspots which sprang up beneath the Coral and Tasman sea floors. Each hotspot melted the mantle rocks in the overlying Australian plate, just enough to make alkaline melts. These crystallised in deep, coarse mineral veins. As melting increased, basalt melts broke through incorporating minerals from the veins. These erupted as refractory residues such as sapphire and zircon. As Australia continued northwards, the underlying hotspots would then have

△ Ruby crystals found as alluvial grains, Barrington volcano, New South Wales. Grains range up to several millimetres in size.

produced more gem volcanoes further south. This accounts for the sapphire fields around Oberon, where zircons make it possible to date several gem eruptions. The age range here is a few million years later than the New England age range, and is consistent with the estimated drift of the area over the same hotspot system.

RUBIES

Rare rubies grace a few basalt fields in eastern Australia, especially the Barrington Tops and Tumbarumba fields in New South Wales. The Barrington rubies grade from pink sapphires into true rubies, which have a characteristic colour due to the trace element chromium. A few stones are cut locally and exploration companies sometimes express interest in potential ruby deposits. A large blue sapphire with a zoned core of ruby was found near Glen Innes in New England. Some fine blue and red stones cut from this stone were displayed at gem shows.

Barrington rubies are strongly etched by their immersion in hot melt. A clue to the ruby source came from small samples of the parent rock. Its peculiar combination of ruby, sapphires of all colours, a sapphire look-alike mineral (sapphirine), and a spinel made this a unique scientific treasure. All these minerals are oxides rich in aluminium and crystallised from an aluminium-rich parent of mysterious origin. Zircon found with these crystal clusters is of a special type, very high in uranium. The zircon age suggests that the crystals and rubies burst onto the Barrington scene 4–5 million years ago. The pipe that discharged them lies hidden in heavily vegetated, rugged terrain, but its fairly young age is a clue that may lead to its discovery.

OTHER GEMSTONES

A galaxy of lesser gem minerals come from basalts and are cut locally. Many of these xenocrysts are silicate minerals, stripped from the mantle and crustal rocks encountered by the rising basalts. Some crystals are large phenocrysts grown in melts in deep chambers.

Olivine and its gem form, peridot, come from mantle peridotites. The basalt soil around Chudleigh, northern Queensland is a happy hunting ground for loose peridot crystals. Garnets freed from mantle garnet pyroxenites include gem pyrope-almandine from the aptly named Garnet Gully near Brigooda, southern Queensland. Almandine garnets, like those from Mt Wyangapini in the Queensland Darling Downs, probably come from crustal rocks.

Anorthoclase feldspars abound at basalt centres such as Mt Anakie in Victoria and Brigooda in Queensland. Transparent crystals are cut as moonstone. Large, limpid yellow plagioclase feldspars from some basalts are also cut into fine stones, like those from the Hogarth Range in New South Wales and the Bunya Ranges and Springsure area in Queensland.

XENOLITHS

Eastern Australia is a leading world source of xenolith finds. These transported rock fragments assist scientists to reconstruct the foundations of the continent, and help seismologists discover how earthquake waves behave in rocks of different densities. Xenolith rocks also reveal the temperatures and pressures under which rocks exist at depth. This checks the continent's physical state — rather like the thermometer and blood pressure readings taken from patients. While rock temperatures and strain measurements can be measured down boreholes, these probe only the shallower layers. To test the deeper underlying condition, petrologists submit samples for chemical analysis.

Natural xenolith minerals are compared with synthetic rocks grown in the laboratory under precisely controlled conditions of temperature, pressure and volatility. By matching xenolith minerals against duplicated minerals, the appropriate conditions can be estimated. In certain xenoliths the minerals allow only an estimate of temperature, but others contain minerals that make it possible to estimate the pressure. This is important because the pressure gives the depth of the rock — atmospheric pressure increases about 300 times for every km in depth.

The xenoliths ideal for prescribing temperatures and pressures contain garnet, orthopyroxene and clinopyroxene minerals. By analysing several xenoliths, a temperature-depth gradient can be measured for the section under a volcano. This geothermal gradient measures how much heat flowed in the region when the volcano erupted, and thermal gradients under Australia can be measured over hundreds of million of years by this method. Once a geothermal gradient is estimated, the xenoliths that yield only temperature estimates can be matched against the gradient to indicate their pressure and its equivalent depth. This builds up a much more detailed section of the crust and mantle.

◁ *Cut feldspar (labradorite) of 33 carats in size, Hogarth Range, New South Wales and cut olivine (peridot) of 9 carats, Chudleigh, Queensland.*

DOWN UNDER AUSTRALIA

Australia's xenolith volcanoes give us glimpses of the underlying rocks. Combining these glimpses creates an overall picture of the Australian plate and its thermal variations. Four main layers of increasingly dense and pressurised rocks are found:

1. COVER ROCKS — ranging from unconsolidated to consolidated; largely unfolded horizons covering older basement rocks. These horizons include sedimentary beds, lava flows and intrusive igneous sheets. They extend from shallow surface layers to layers several kilometres deep in the sedimentary basins. These layers are usually well known from drilling, so that xenoliths merely provide extra details.

2. UNDERCOVER CRUST — folded, metamorphosed and intruded basement rocks. These

encompass folded metamorphic rocks such as slates and schists, large igneous intrusions such as granites and gabbros, and highly metamorphosed crystalline rocks such as gneisses. These upper crust rocks usually reach 10–25 km in depth. Their xenoliths help to indicate structures hidden under cover rocks.

3 LOWER CRUST — strongly banded, high-pressure, recrystallised metamorphic and igneous rocks. These include banded feldspar and pyroxene-rich rocks called granulites. Some are garnet granulites and others are metapyroxenites and garnet metapyroxenites. They extend 20–50 km deep and xenoliths are common.

4 UPPER MANTLE — dense, metamorphosed, high-pressure metaperidotite rocks, interleaved with some metapyroxenite and garnet metapyroxenite rocks. These rocks lie below the Moho, the boundary between the lighter feldspar-rich lower crust and denser olivine-rich mantle. In some sections the boundary is gradual. The mantle usually starts below 30–50 km and the xenoliths tap three zones of increasing pressure down to the zone of melting for volcanic eruptions.

The spinel peridotite mantle supplies abundant xenoliths. Beautiful examples wrapped in basalt bombs are ejected from young Victorian and north Queensland vents. Some show intricate pyroxenite veins and alterations to hydrous minerals such as amphibole and mica. Unusual xenoliths found in dykes near Kiama, New South Wales, are rich in amphibole and apatite formed by previous influxes of deep fluids. As pressure increases with depth, spinel (an oxide in the peridotite mantle) reacts with pyroxene to convert to the denser silicate mineral, garnet. This takes place 50–60 km down.

The garnet peridotite mantle is tapped sparingly. Pieces appear in a basalt flow at Bow Hill, Tasmania, but most finds come from breccia pipes. Rare garnet pyroxenite rocks from this zone eventually turn into eclogite, the high-pressure variety. Deeper in the garnet peridotite zone, diamonds crystallise at depths of over 150 km. Diamond peridotites found in the ARGYLE LAMPROITE PIPE probably came from a depth of 180 km.

▽ *Fragment of mantle peridotite (xenolith) wrapped in basalt, forming a volcanic bomb 10 cm across, from Mt Leura, Victoria.*

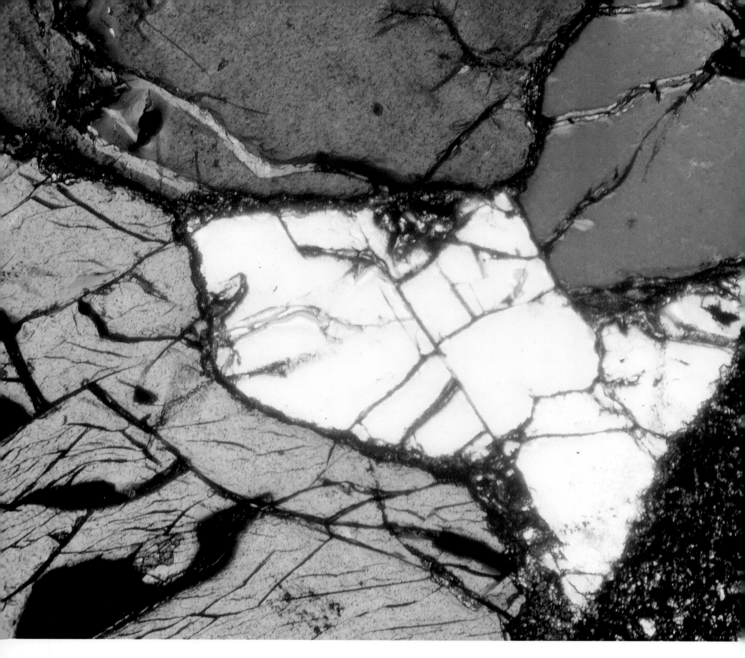

AUSTRALIA'S HEAT FLOW

The volcanic xenoliths record how Australia's continental heat flow has fluctuated over geological time. For example, from young Victorian and Queensland volcanoes xenoliths make it possible to measure relatively recent heat flows, and LAKES BULLENMERRI and GNOTUK CRATERS erupted many garnet pyroxenite xenoliths which define the heat flow for western Victoria as a very hot gradient — not surprising for a young volcanic field. It suggests that many molten injections were built up below the crust when the lava field erupted.

Heat flows measured on xenoliths from older east Australian volcanoes come from the following:

- 3 million year old Mt St Martin volcano, north Queensland;
- 13 million year old TABLE CAPE VOLCANO, northwest Tasmania;
- 25 million year old BOW HILL VOLCANO, central Tasmania;
- 36 million year old BRICK CLAY CREEK VOLCANO, New South Wales.

These sites all record hot gradients, and the Tasmanian and New South Wales gradients are as hot as any recorded around the world. Such long-lived high heat flows in eastern Australia probably reflect the continent's continued movement over a swathe of hotspots. These were inherited from the thermal pulse, rifting and seafloor spreading events in the Tasman and Coral sea regions 95–55 million years ago.

Heat flows measured on xenoliths from even

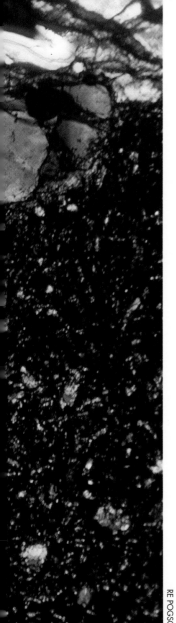

◁ *Coarse-grained mantle peridotite rock (xenolith) in chilled basalt host, Bow Hill, Tasmania. Microscope and polarised light show up the different mineral grains, up to 5 mm across. The four minerals, by which the original temperature and pressure of the peridotite rock are measured, are garnet (dark grey, top left), orthopyroxene (pale brown, bottom left) clinopyroxene (yellow, bottom centre) and olivine (purple red, top right). The host rock in contact with the xenolith is the dark brown rock speckled with minute crystals (lower right). The rock came from mantle about 70 km below at 1,200° C.*

▷ *Section of crust and upper mantle rocks below Western Victoria. Model based on fragments brought up in volcanic pipes.*

older volcanoes suggest that some cooler periods prevailed during volcanic events 100–300 million years ago. These xenoliths come from the DELEGATE, GLOUCESTER, BINGARA and KAYRUNNERA PIPES in New South Wales and kimberlite pipes in South Australia. The Kayrunnera, White Cliffs and South Australian xenoliths record the lowest heat flows, which accords with their location in thick, cool Precambrian crust. Eclogites appear in these xenolith suites because in cooler conditions granulites, which are stable under higher temperatures, convert to eclogites which are denser.

Diamonds are sensitive to heat flows. Their chances of eruption in small pockets of melting are increased by cooler conditions because they can remain stable at higher levels in the mantle. A diamond-bearing peridotite from the Argyle pipe records a 1,200° C temperature at 180 km depth. This is typical of ancient diamond-producing areas such as South Africa, Russia, India and Brazil. Under higher heat flows diamonds convert more readily to graphite, and peridotite melting spreads to higher levels. The eruption of unconverted diamond in xenoliths or in released xenocrysts becomes more unlikely. This is why a special origin is proposed for eastern Australian diamonds, where heat flows are high and odd diamonds come up in higher level melts. The insertion of thick cool slabs of oceanic peridotite and overlying carbon-rich sea floor sediments under pressure may make new diamonds at higher, shallower levels than normal.

CHAPTER

THE SILENT PICTURE SHOW/MULLUMBIMBY TECHNICAL AND FIELD SURVEYS/SYDNEY

11 DYNAMIC VOLCANOES AROUND AUSTRALIA

◁ *Satellite image of the Tweed shield, Australia's largest, eroded central volcano, formed when two migrating volcanic lines intersected. This produced Australia's largest concentration of central volcanoes: the Main Range, Focal Peak and Tweed volcanoes, extending from Queensland into New South Wales. Mt Warning plug lies in the circular ring dyke in the centre of the volcano. This is a false colour image, but the rich green denotes thick forest cover; yellow-pink shows cleared land; red and magenta patches indicate sugar cane and other crops; the dark blue channel is the Tweed River. Landsat 5 spacecraft image taken from 750 km, September 1989.*

THE rise and fall of thousands of volcanoes along eastern Australia enlivens the continent's last 200 million years. This prolific performance accompanied Australia's rifting from Gondwana, but only scatterings of small volcanic vents appear in the remaining continental vastness. The great imbalance in volcano distribution must have been caused by some dynamic mechanism. What explains Vulcan's generosity to the east? Let us start with the largest and most voluminous volcanoes — the central volcanoes. What makes them come and go?

The TWEED VOLCANO typifies a central volcano. Its core of evolved lavas formed during extended growth of a basalt shield. The large volcanic chambers indicate much mantle melting. The central volcanoes form overlapping or spaced chains, which become younger along the chain. The vast melting and migration indicate a deep hotspot origin (p.91).

EARTH
VOLCANIC

THE GREAT MIGRATION

About 30 central volcanoes, with scenic peaks of rhyolite and trachyte rise from their basaltic aprons. They extend from CAPE HILLSBOROUGH VOLCANO in northern Queensland down through the eastern hinterlands to the MACEDON VOLCANO in Victoria. Apart from the older central volcanoes around Rockhampton, all show a common trend: they become younger to the south, decreasing from 33–42 million years old around Hillsborough to 6–8 million years old in central Victoria. This migration was noticed by scientists from Australian National University after they had dated many central volcanoes. They explained this southward sweep by a simple plate tectonic model. The Australian plate (lithosphere), drifting slowly northwards from where the Southern Ocean opens, passes over deep upwellings (hotspots) in the underlying mantle (asthenosphere). In other words, each volcano was formed over a hotspot, then became carried away with Australia's movement. This created an ever-lengthening chain of extinct volcanoes (see p.91 onwards).

The rare leucitite volcanoes, which pass through central New South Wales into northern Victoria, also join the great migration. From the 17 million year old BYROCK VOLCANO, the chain passes through the 14 million year old CONDOBOLIN and CARGELLICO VOLCANOES to the 7 million year old COSGROVE VOLCANO.

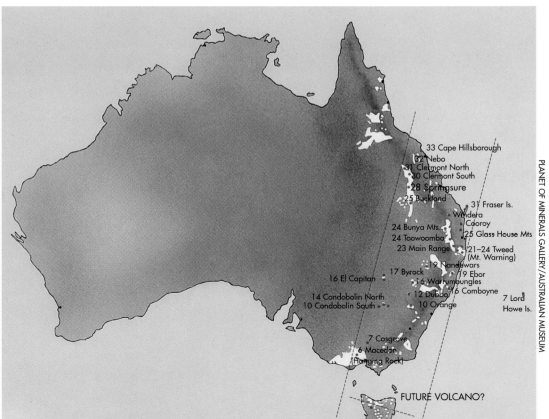

◁ Australia's eastern basalt fields (white areas) and hotspot central volcanoes, which erupted during a span of 35 million years. The numbers denote age in millions of years.

CHAPTER 11

TASMAN SEA MIGRATION

The hotspot model for Australia's central and leucitite volcanoes extends to the adjacent Tasmantid seamounts on the Tasman sea floor. This chain shows the same rate of drift. It marks another rising plume under the Tasman floor, which left an oceanic hotspot trail like the Hawaiian chain. LORD HOWE ISLAND VOLCANO and the seamounts to the north form another hotspot trail. Apart from Lord Howe Island none of the seamounts are dated, but they deepen away from the island as expected for aging hotspot volcanoes. However, at 7 million years Lord Howe Island is only half the age of its Tasmantid and Australian central volcano counterparts, so that a completely separate hotspot was involved.

VOLCANIC BOOMERANGS

The migratory central and leucitite volcanoes stand out from the ubiquitous basaltic volcanoes because of their distinctive alkaline rocks. However, many basalts, when dated, also fit the great migration. These include the 20–27 million year old Monto and Expedition Range fields in Queensland, the 18–24 million year old Inverell and 12–14 million year old Dubbo fields in New South Wales, and the 0–5 million year old western Victorian field. Such basalt fields can form over hotspots, as similar basalts appear in hotspot trails outside Australia.

When all the migratory central volcano, leucitite, tholeiitic basalt, alkali basalt and nephelinite fields are considered, a remarkable feature emerges. The migratory chains form a series of crescents, with the convex bulge of each one pointing westwards, shown on the map, next page. The types of volcanoes are not random; they increase in size and contents of silica-rich rocks as they approach the curved bulge in the chain. These features are termed volcanic boomerangs because:

1. Each shows a boomerang-like shape in plan, as volcanic fields taper in size towards each end.
2. They return, further along the migratory line.
3. They are a distinctive Australian volcanic feature.

△ St Peter's Dome, a large eroded rhyolite intrusion in the Springsure central volcano, Queensland, May 1986. Looking west from north of Springsure.

▷ Gravity map of east Australian coast and Tasman Sea based on satellite measurements. The Tasmantid seamounts (centre top to bottom) are shown as circles of increasing high gravity concentrations (bulls-eyes) on the sea floor. The seamounts become younger south, from 24 million years (Queensland seamount) to 6 million years (Gascoyne seamount).

ANATOMY OF VOLCANIC BOOMERANGS

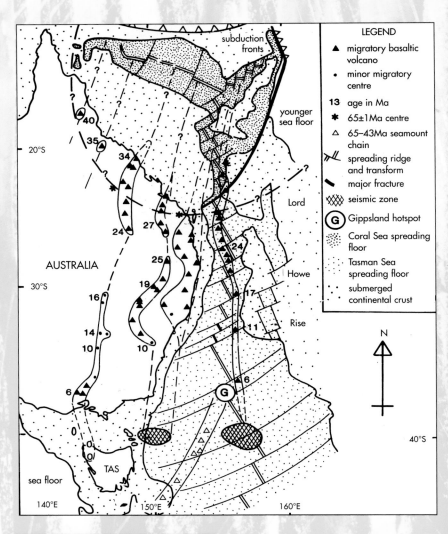

EXTENDING from southeastern Queensland near Brigooda to Oberon in southeastern New South Wales exists a well-defined migratory volcanic boomerang, shown opposite. It progresses southwards with decreasing age from:

1 A minor eruption of 28 million year old zircons (Brigooda).

2 A small 25 million year old alkali basalt field (Millmerrin).

3 An extensive 24–19 million year old tholeiitic and alkali basalt field (Inverell).

4 A prominent 21–18 million year old double central volcano (Nandewars).

5 A large 18–15 million year old triple central volcano (Warrumbungles).

6 A moderate 14–12 million year old tholeiitic or central volcano (Dubbo).

7 A small 12–11 million year old central volcano (Canobolas).

8 A minor eruption of 10–9 million year old zircons (Oberon).

The larger the volcano in each boomerang, the further west it lies (149° E for the Warrumbungles compared to 151.4° E for Brigooda). The increase in volume, size and silicic character of volcanoes approaching the western bulge suggests rising degrees of melting in the underlying mantle — in other words, a plume or hotspot surge. The migratory boomerangs range in size: the Brigooda-Oberon chain is 800 km long, while others like the Victorian chain are under 300 km long. Some of the large Queensland and Tasmantid boomerang chains which extend offshore may exceed 1,000 km when their length is extrapolated from undated seamounts.

The less active regions between successive boomerangs show small isolated or minor clusters of migratory volcanoes which erupted low-silica alkaline lavas. The leucitite line through central New South Wales is the clearest example. These small bursts trace Australia's movements over the slight stirrings of a slumbering plume. As the plume flares up into an active hotspot and mantle melting rises, volcanism increases and erupts more silicic lavas. The

△ *Section of Nandewar central volcano, New South Wales, showing its two successive volcanoes and the original form at the end of Nandewar volcanism 17 million years ago. (Ma: million years old).*

◁ *Eastern Australia and Tasman Sea floor: main hotspot volcano migrations, boomerang bends and positions of seismic zones over present hotspot positions. (Ma: million years old).*

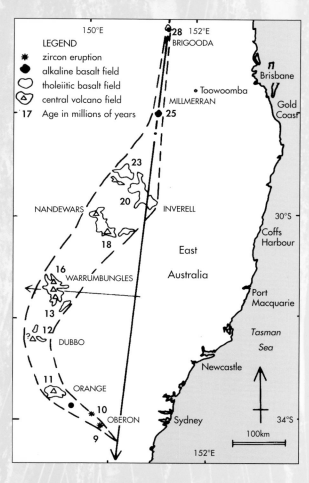

◁ *Volcanic boomerang, according to the Brigooda-Oberon line, eastern Australia. Numbers show ages of volcanoes, arrows show direction of volcanic migrations.*

eruptions step ever westwards, as they migrate south. The rising volumes of melt start to interact with the crust and central volcanoes appear. This is illustrated by the NANDEWAR VOLCANO, the first central volcano to form in the Brigooda–Oberon boomerang. The two Nandewar volcanoes show subtle differences. In the older KILLARNEY GAP VOLCANO the rhyolite rocks are rich in the element strontium, which yields isotope values related to crustal rocks. Thus crust was melted by underlying basalt invasion. In the younger southern KAPUTAR VOLCANO, strontium isotopes show that the rhyolites evolved from the basalts themselves — the basalt melts had evolved in large crustal chambers.

A central volcano usually climaxes at the westward bend of a boomerang, as in the case of triple-crowned WARRUMBUNGLE VOLCANO. After this, volcanism dies down and lavas decrease in silica as they tap deeper mantle sources. The eruptions shifts ever eastwards back to the original line. Even while the volcanism veered on its westerly excursion, rare small bursts of low silica lavas remained, marking the original line. The boomerangs resemble bows with a bowstring.

Dynamically, they represent three arrows of direction:

1 A southerly arrow (100–1,000 km long), in the direction of volcanic migration.

2 A westerly arrow (10–200 km long), in the direction of maximum volcanic deflection.

3 A vertical arrow (10–100 km long), in the direction of maximum plume rise.

The question is where do these volcanic arrows come from and where do they point?

AUSTRALIA'S VOLCANIC DIRECTIONS

The largest volcanic boomerangs, central volcano complexes and seamounts appear in Queensland and the north Tasman Sea. The structures diminish in size southwards — a hotspot system dying with time. The largest central volcano complex, the overlapping Tweed and Focal Peak volcanoes, does not sit at the bulge of a boomerang but marks the intersection and interplay of two boomerang lines — the Fraser Island-Barrington boomerang and the Main Range-Comboyne boomerang. These easterly lines seem to die out after the 16 million year old Comboyne volcano. All this speaks of older and larger parent plumes to the north. The volcanic chains all head towards one of Australia's large evolutionary events, the birth of the Coral Sea floor. This 65 million year old burst of rifting and melting, across a 1,500 km zone, has the right width and spacing in its rift zones to generate the Australian hotspot lines.

Where do the plume systems lie now? The migration arrows all point to Bass Strait and the Tasman Sea floor to the west at 40° S. The plumes are certainly there, triggering earthquakes over their predicted positions off Flinders Island and in the Tasman Sea, discharging deep volcanic gas under northwest Tasmania, and releasing strong heat flows under southern Victoria and Tasmanian. A hot zone over 400 km deep affects earthquake waves below northeastern Tasmania. The system is ready to awaken and activate a new boomerang somewhere along its length.

TASMAN LINE OF FIRE

The Tasman Sea, washing Australia's eastern shores, hides old spreading floors and submerged continental rises. The rises are dismembered strips of former continental lands that once bordered New Zealand. The Tasman margin marks a great north-south rift, formed 95 million years ago, when the Australian plate failed under intense stretching. Hot underlying mantle welled up into this gaping wound 4,000 km long. This mighty thermal pulse generated the Tasman Line of Fire.

The surge of heat cooked the crustal rocks along the continent's edge, warming them up to 250° C and disturbing their magnetic and mineral properties up to 150 km inland. Coal seams in the Sydney Basin had matured and the whole margin was uplifted into the eastern highlands. Massive melting in the Tasman mantle triggered seafloor spreading. Beginning in the South Tasman, 80–85 million years ago, the mid-ocean ridge volcanoes gradually propagated north to open the north Tasman, 65 million years ago.

The whole Tasman spreading zone, connecting with the Cato Trough and Coral Sea spreading zones, suddenly switched off 55 million years ago. The MORB volcanoes became extinct and the newly formed Lord Howe plate joined the Australian plate. Mantle melting dwindled and was channelled into separate upwellings. A line of spaced hotspot plumes began a new volcanic regime.

TASMAN—ANTARCTIC CONNECTIONS

A large plume lay under the triple junction joining the south and central Tasman spreading arms and the Gippsland Basin, a failed spreading arm. The plume built seamount chains on the newly formed sea floor that was being carried away from the spreading ridge. These included the 60–70 million year old Heemskirk and Zeehan seamounts on the Lord Howe side and perhaps the 50–60 million year old Janszoon seamount on the Australian side. After Tasman spreading stopped, the new Australian plate moved over the plume to form a seamount line migrating southwest to Tasmania. It finished at Soela seamount, a large submerged basaltic shield. From there, the seamounts took a new line south and vanished under the active Australian–Antarctic spreading ridge. This bend is usually matched with the 43 million year old bend in the Emperor–Hawaiian hotspot chain, a time when spreading directions in the Pacific changed.

The south trending seamounts reappear on the Antarctic plate under the young Balleny Islands volcanoes off Antarctica. Somewhere underneath lies the plume which once erupted off Gippsland Basin, but now erupts below the Ross Sea. In its great volcanic odyssey, the Gippsland plume records the activity of only one Tasman Sea plume. Its trace, however, can help to reconstruct the paths and present sites of other Tasman rift plumes.

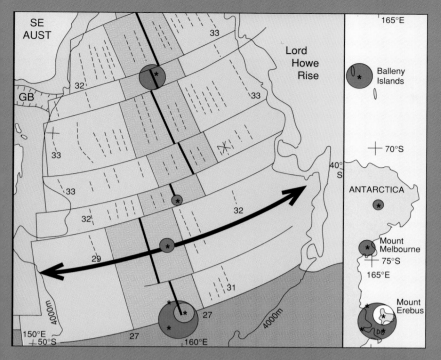

△ Australia–Antarctica: positions of original Tasman Line of Fire rift junctions compared with the present active volcanoes of the Ross Sea region.

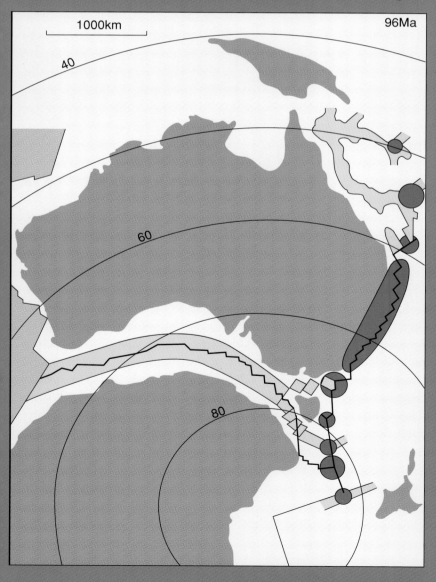

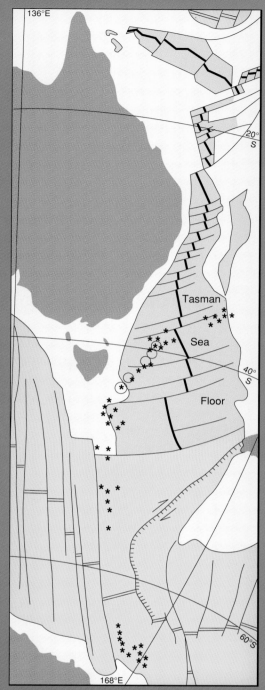

△ Line of seamounts (stars and circles) left by the Gippsland hotspot as Australia separated from Antarctica over the last 80 million years. The seamount chain bends SE of Tasmania, with a gap over the mid-ocean Australian–Antarctic spreading ridge, and its final position is at the young volcanic Balleny Islands off the Ross Sea.

◁ Tasman 'Line of Fire': main thermal upwellings (circles) along the Tasman–Antarctic rift zone at 96 million years ago. The fault lines and broad stippled zones between the reassembled continents and Lord Howe Rise (east side) are rift zones.

If we assume that each important Tasman rift junction overlay a plume, some astonishing volcanic correlations emerge. All the present active Antarctic volcanoes lie over former Tasman plume positions. MT EREBUS, the most active Ross Sea volcano, sits over a plume that once lay under the main Tasman Sea–Southern Ocean junction. MT MELBOURNE STRATOVOLCANO rises over a former rift junction of the South and East Tasman Rises. The ephemeral PLEIADES eruptions mark a lesser rift junction of East Tasman Rise. The rise of the BALLENY ISLANDS VOLCANOES over the old Gippsland plume position completes this Tasman–Antarctic connection.

TASMAN—EAST AUSTRALIAN CONNECTIONS

The South Tasman plume sites marooned by Australia's northern movement were crossed by the Antarctic plate. However, plumes below the central and northern Tasman spreading rift would have taken more northerly paths. Using the same movement of the Gippsland trace, these Tasman plumes would have been overrun by eastern Australia as it moved north. This could have initiated the vigorous older basaltic volcanism along Australia's margin.

Every time a part of eastern Australia moved over a plume, volcanism burst out, followed by another outburst over the next plume in line. This model explains the origin of many volcanic outbreaks in Tasmania over 50 million years as the island crossed over the line of Tasman plumes, north of the Gippsland plume. It also explains repeated eruptions of BARRINGTON and IPSWICH VOLCANOES in the 55–28 million year period, well before these areas would have encountered plumes from the Coral Sea rift zone. Much detail needs filling in, but the Tasman Line of Fire accounts for the earlier melting under eastern Australia.

The Tasman rift branches north of Cato Fracture, the northern limit of Tasman sea floor spreading. One branch formed the Capricornia Basin and other branches formed rift junctions off northeastern Australia. Any plumes here would have activated early volcanism in northeastern Australia as it slowly moved northeast from Southern Ocean spreading. Such a plume could generate the central volcanoes migrating south through Rockhampton to Baralaba, 80 to 70 million years ago. These volcanoes seem to define a west-pointing boomerang, suggesting mantle flows similar to those that occurred under later boomerangs.

PG QUILTY/AUSTRALIAN ANTARCTIC DIVISION

GONDWANA'S LAST VOLCANIC FLING

As supercontinent Gondwana fragmented after Triassic time, the highly stretched and fractured region gave ready access to basaltic melts. Extreme melting produced a sea of basaltic magma under Victoria Land, Antarctica, and Tasmania.

The uniform tholeiite composition of Gondwana's massive dolerite sills and lava outpourings indicates that voluminous mantle melting occurred 175–170 million years ago. The invasions going north stopped at Bass Strait, although further west volcanic outposts extended into southwest Victoria and Kangaroo Island, South Australia. It is not certain what caused this meltdown because the elongated zone does not suggest a starting plume area. Similar sills and lavas overwhelmed the Karoo region on the South African side. However, the Tasmanian–Antarctican rocks show subtle chemical

△ Antarctic hotspot volcanoes: active Mt Erebus (with steam plume, left back); flank of Mt Terror (right back); extinct cone of Crater Hill (centre front). Looking north from Hut Point Peninsula, Ross Island, 1965.

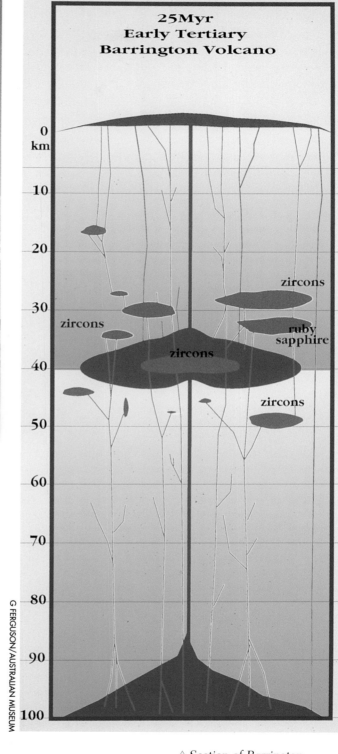

△ Section of Barrington volcano, New South Wales, at 25 million years ago, showing several chambers and zones of zircon crystallisation ranging in age from 27 to 60 million years.

differences which suggest that the melts digested some old subducted sediments under this region.

Elsewhere in eastern Australia melts were generated in smaller bursts over greater periods, from 210 to 150 million years ago. Central volcanoes of various ages, with cores of rhyolite, trachyte or phonolite intrusions, erupted around Gunnedah, Rylstone, Kandos, Dubbo, Bowral and Cooma in New South Wales. Basaltic lavas, dolerite intrusions and explosive alkaline pipes repeatedly broke through Sydney Basin beds and the adjacent Lachlan fold belt. The volcanic scenes resembled those of the younger eastern Australian fields. The central volcanoes may mark hotspot trails, but more dating is needed to confirm such migrations. Whether volcanic boomerangs can be recognised among the myriad Mesozoic volcanoes is a challenging point to follow up.

Australia's remaining Gondwanan scene was quiet, broken only by small fizzes of deeply formed melts. Kimberlites with rare diamonds and some carbonatites erupted about 170 million years ago in South Australia, around Port Augusta, Orroroo and Terowie. On the fringe of the activity unusual picritic alkali basalts containing the odd diamond pierced Carnarvon Basin in Western Australia 164 million years ago.

▷ *Big Ben volcano, Heard Island, Southern Ocean, 1954.*

◁ *Jurassic dolerite sill at top right invading Triassic sandstones and shales, Frankford Highway, west of Rubicon River, north Tasmania, February 1980.*

THE BUNBURY-KERGUELEN PLUME

Then came the Bunbury basalt, announcing the separation of Western Australia from India. Its eruption 128–135 million years ago may mark the early effects of a hotspot plume rising under the Australian–Indian–Antarctic rift junction. New sea floor pushed Australia and India aside as the volcanism built up the Naturalist Plateau basalts about 124 million years ago. Eastern India on the distant edge of the plume spewed out the Rajmahal basalts about 115–117 million years ago, while next door to Australia the plume built up the huge Kerguelen plateau 118–95 million years ago. The early Indian–West Australian spreading stopped about 95 million years ago, leaving the Kerguelen Plateau on the Antarctic plate. A new, more westerly ridge system, now near Christmas Island, rapidly pushed India northward. A prominent migratory seamount chain traced the Indian plate movement over the Kerguelen plume, forming the Ninety-East Ridge.

About 42 million years ago the old Kerguelen plateau passed over the plume again as it came under the Antarctic plate. Volcanism revisited the plateau as it slowly moved over the plume. This persisted over the last 40 million years and now extends to BIG BEN VOLCANO, on Heard Island. Only rapid Australian movement away from Antarctica and the Kerguelen Plateau prevented this extraordinary volcanic trace from circling back to its Bunbury beginnings.

The plume's present position is uncertain. Some place it west of Kerguelen Island, influencing MT ROSS VOLCANO and the southwest geothermal fields. Others think it could lie under Big Ben. It is complex enough for both to be correct.

THE PSEUDO-SUBDUCTION LINE

Since Tasman rifting, east Australian volcanism exhibits mostly rift and hotspot characters. However, early Cretaceous volcanism 95–145 million years ago has a calcalkaline character more typical of unstable subduction margins. The volcanoes are well developed in the Bowen–Brisbane region in north Queensland. More isolated volcanoes and intrusions extend from Townsville to Tasmania, so the zone extends over 3,000 km. Offshore volcanoes in the Whitsunday and Cumberland Islands were originally terrestrial volcanoes, before the sea encroached on the continent.

CALCALKALINE LINE-UP

Dacitic, rhyolitic and andesitic ignimbrite deposits dominate the volcanoes in the Whitsunday and Cumberland Islands. They suggest violent Plinian explosive activity and caldera formation. Granite chambers, flows of andesites and rhyolites are exposed on many islands. Dykes abound in a spectrum of rocks ranging from dolerites to rhyolites.

Onshore, around Proserpine, volcanoes commonly erupted basalts and andesites and no calderas are recognised. The volcanic scene evokes rift depressions filling up with large ignimbrite eruptions. Avalanche deposits are rare, suggesting that stratovolcanoes were low, rather than steep. Larger intrusions in the inland ranges mark subvolcanic chambers. Two main pulses of activity took place about 115 and 125 million years ago. Fallout from major explosive eruptions was carried west to the Eromanga Basin 300–400 km away.

Further south, rhyolitic to basaltic volcanoes erupted inland of Rockhampton and Gladstone. Rhyolite at Mt Hay is noted for its spherulites and thunder eggs. Early Cretaceous volcanic rocks, over 1,000 m thick in places, mark the western edge of Maryborough Basin, and are 135–145 million years old. In the Gall and Bucca Ranges lower tuffs, agglomerates and volcanic breccias underlie trachyte and rhyolite flows, followed by basaltic andesite and dacite flows. They are capped with trachyte, andesite and rhyolite tuffs and rhyolite agglomerates and breccias. The rhyolites often show spherulite and thunder egg formation, flowbands and flow fragmentation.

EARTH VOLCANIC

CALCALKALINE MISFITS

The ages of Cretaceous calcalkaline volcanoes along the central Queensland coast suggest that rifting and volcanism progressed from Maryborough north to the Whitsunday region. The volcano chemistry advertises a subduction event here, yet the vents lie in rift zones created in an extending continent. No obvious subduction zone or major folding event shows up in geological reconstructions, nor in the late Triassic volcanoes in central east and southeast Queensland that originated 210–230 million years ago. These volcanoes erupted around North Arm, Gympie and Arabanga and form granite cores offshore in the Shaw and Kent Islands. They include the large MUNGORE CALDERA, 50 x 35 km across, in which earlier basaltic lavas, ignimbrites and domes of rhyolites are intruded by rhyolites and granites after caldera collapse.

The calcalkaline chemistry in these pseudo-subduction volcanoes must have been acquired from mantle melted by older subduction events.

◁ Basalt dyke intruding granite, Yetholme, Lachlan foldbelt, New South Wales.

▷ Mt Stuart, Townsville, Queensland, September 1978, from James Cook University entrance, Douglas. The prominent jointed face is a granitic core, about 250 million years old, intruding earlier Permian pyroclastic deposits composed of andesite and trachyte (lower foothills).

◁ Triassic andesite and rhyolite volcanic beds (North Arm volcanics) exposed in Obi Obi gorge, west of Baroon Pocket Dam, near Maleny, Queensland, April 1991.

THE LAST SUBDUCTION

Old subduction volcanoes and chambers extend along eastern Australian fold belts north of the Sydney Basin. They pass through New England and along Queensland's hinterlands as far as Cape Weymouth and even beyond Cape York to the Thursday Islands.

AUSTRALIAN ANDES

Some deeply eroded volcanoes show that they were formed between the early Triassic and late Permian ages, 230 to 255 million years ago. They include the Nera volcanics in the Esk rift and extensive granodiorite, diorite, granite and gabbro intrusions in southeastern Queensland and northeastern New South Wales such as that shown on p.179. Large calderas in New England erupted the DUNDEE RHYOLITE IGNIMBRITE, a huge emplacement probably fed by several calderas like Indonesia's famous LAKE TOBA CALDERA.

These eruptions followed the climax in the Hunter–Bowen folding, 255–265 million years ago. Strong folding had compressed earlier rocks and twisted their original depositional basins into giant bends, along large thrust and tear faults. Further away, around Townsville and Cairns, volcanism and granite intrusions were more dominant. Eastern Australia resembled the present Andes Mountains fringing the Pacific subduction front along South America. Some sections slipped sideways, behaving like the San Andreas fault zone of western USA.

In the earlier stages of this subduction event, 270–310 million years ago, island arcs were prominent. However, the last great fold movements in eastern Australia have dismembered the volcanic details. Not all the activity of this zone was caused by subduction. Recent work shows that enormous volumes of Carboniferous and Permian volcanics and granite intrusions occupy a 1,000 km belt from Townsville to Mornington Island in the Gulf of Carpentaria (eg, Mt Stuart, previous page). This zone does not follow the old subduction margin, but is a major within-plate zone of melting and mineralisation. Before we can understand the secrets of this region's mineral potential we have to unravel the origin of its past volcanism.

NEW ZEALAND'S VOLCANIC CLIMAX

The exceedingly complex 'cooking pot' forming New Zealand's volcanoes developed over the last 30 million years. Several migrations of both subduction and within-plate volcanoes appear. Because of this complexity of age and movement factors, there is much debate by New Zealand geologists over the exact pattern and causes of North Island volcanism in particular.

In the North Island, old sea floor with its oceanic volcanic features was pushed up in sheets over Northland some 23–24 million years ago. Two parallel volcanic belts sprang up west of a subduction zone, where the sea floor descended below northeastern New Zealand, one being 150–200 km

▷ *Columnar latite flow from the last subduction in the Permian, Bombo quarry, New South Wales, December 1984.*

west and the other 200–300 km west of the trench. The nearest of these belts generated andesite volcanoes first, then in the last 10 million years it generated rhyolite volcanoes and hydrothermal activity, and finally basaltic volcanoes joined in. The farther belt began with andesitic volcanoes, then included some basaltic volcanoes and finally within-plate basalt volcanoes. There were shifts in alignments and migrations along these belts about 14 and 4 million years ago. Since 2–3 million years ago, the active Taupo volcanic zone of andesitic, rhyolitic and hydrothermal volcanic systems has grown over a subduction zone dipping inwards northwest from the Hikurangi trench. Short migration lines developed in the younger Taranaki volcanoes, to the southeast and in the within-plate basalt fields, which have migrated north along the east coast from the Okete to Auckland volcanoes. This last migration is not caused by the plate moving over a stationary hotspot because the volcanic migration is in the wrong direction for this.

In the South Island, in contrast to the North Island, the last 30 million years have produced a spate of within-plate volcanoes. These seem to show a migration sweeping southeast, forming the large central volcanoes at Banks Peninsula and Dunedin volcanoes between 12 and 8 million years ago and extending offshore to volcanoes rising from the Campbell Plateau and Chatham Rise. It includes the Auckland, Campbell and Chatham Islands volcanoes, finishing at the Antipodes Islands which were active 250,000–500,000 years ago. This migration is heading in the right direction for a hotspot movement, but the patchy pattern is still not fully understood.

FL SUTHERLAND

12
FUTURE VOLCANOES AROUND AUSTRALIA

◁ *Hypipamee crater, formed by a gas-rich volcanic blast, Atherton Tablelands, Queensland, March 1994.*

AUSTRALIA'S volcanic past points irresistibly to future eruptions. When, where and why remains to be seen, but some predictions can be made from the past volcanic trends. (The map on p.215 shows trends, while Chapter 6 covers prediction techniques). About 10 million years ago Australia's volcanism changed its character. Spreading increased in the Southern Ocean and Australia collided more strongly with its northern border. This effectively changed the stress fields imposed on the moving plate. Although Victoria predictably moved over the established hotspot system, a new rash of volcanoes broke out in Queensland. This left a largely inactive New South Wales between two volcanic ends. The new volcanic patterns set the stage for understanding Australia's potential for future activity.

Between 10 and 6 million years ago the new volcanic order was relatively subdued. A few small central volcanoes pimpled the highlands northwest of Melbourne, while some minor basalt and leucitite volcanoes erupted in northern Victoria. Lady Julia Percy Island rising off Victoria may have added a dash of excitement. Far away in northern Queensland small clusters of volcanoes spat and spread their wares to form limited basalt fields near Cooktown, Atherton, Conjuboy and Clarke Hills. The main volcanic show was held off New South Wales. Here Lord Howe Island volcano ruled the waves as the hotspot 'Hawaii' of the Tasman Sea.

The last 6 million years, in contrast, quickened the volcanic pace. Widespread basalt lavas poured from volcanoes, forming large lava fields in western Victoria and northern Queensland. Small basalt fields broke through areas previously calm for 15 million years or more. Western Victoria became smothered under basalt, but northern Queensland fields kept their separate identities.

Several mechanisms were in play, controlling the venues and volumes of lava production. The positions of underlying hotspots, the crimping effect of compressive stress fields and the releasing and uplifting effects of tensional stress fields, all affected the eruptive balance.

Hotspots below compressed regions break out only in small doses, but under stretched regions they can break through the floodgates. As hotspots surge, the rising plumes progressively catch any mantle flow and deflect towards the prevailing flow direction. Volcanic boomerangs form where plates move across the underlying mantle flow direction. All of these tectonic elements are taken into account when potential venues and styles of future eruptions are predicted.

THE SOUTHERN FORECASTS

A small volcanic front surged south through central Victoria, peaking 6 million years ago and reflecting a westerly mantle flow towards the Australian–Antarctic downwelling. This hotspot was tracked from earlier, more vigorous discharges under central Queensland before it took a deep, steady line down central New South Wales. The limited Victorian discharge suggests an area under compression. Victoria left the subsiding plume behind

▷ *Mt Schank volcano on skyline, looking south from Mt Gambier volcano, South Australia, March 1974.*

EARTH VOLCANIC

185

◁ *Southeast coast, east of Lady Barron, Flinders Island, and looking south to Cape Barren Island, February 1980. Earthquakes from a seismic zone offshore have damaged Launceston on the Tasmanian mainland 175 km to the southwest.*

about 5 million years ago. It receded further south and further east as it dwindled away, and probably now forms a deep dormant plume under northwestern Tasmania.

Another hotspot generated a prominent surge down southeast Queensland and through eastern New South Wales (the Brigooda-Oberon boomerang). Its line would have intersected far eastern Victoria 5–6 million years ago. It left no volcanic record, which implies that strong compression occurred at that time. Another hotspot lay some 500 km off the Victorian coast under the Tasman Sea. Again, it left no volcanic trace after building the 6–7 million year old Gascoyne Seamount off southern New South Wales.

THE TASMANIAN–TASMAN FORECAST

Potential volcanic plumes lie along northwestern Tasmania, Flinders Island in northeastern Tasmania and in the western Tasman Sea, about 155° E and 40° S. Carbon dioxide gas is escaping from deep

mantle sources below northwestern Tasmania and active earthquake zones lie off northeastern Tasmania and under the western Tasman floor. The gas, along with high heat flows measured in Tasmanian rocks, recent uplift tilting coastal dune lines and suspected young volcanic features offshore, make all these areas viable for future volcanism.

Nevertheless, stress measurements in Tasmanian rocks indicate that compression holds sway, and temperatures in boreholes through volcanic gas seeps have shown no source of high-level heat. This reduces the possibility of an imminent eruption. An eruption here would probably discharge alkaline lava from a small monogenetic vent. If this erupted in shallow coastal waters, a large steam explosion would raise a Surtsey-like island. An eruption from the western Tasman hotspot would emerge in deep water and have little surface effect.

THE VICTORIAN–SOUTH AUSTRALIAN FORECAST

Although they have moved away from the central Victorian hotspot, the western Victorian regions developed into a vigorous volcanic zone over the last 5 million years. This dramatic change needs some explaining. Although one hotspot had come and gone, its earlier record in northern Queensland suggests it was only one of several active over 8 million years. Therefore, further hotspots could be expected below Victoria. Indeed, there is evidence that several minor trachytic bursts took place in central Victoria from 8 to 2 million years ago. Such a hotspot train would supply continued melting under Victoria, but the great westward expansion of volcanism needs an extra factor — a change from compressive stress fields to those of tensional stretching.

It is possible to calculate the likely stress fields distorting the Australian plate as it moves north and collides with adjacent plates. The plate is mostly compressed, but a few regions are stretching under the slight buckling. This stretching includes western Victoria and northern Queensland, both areas of rejuvenated volcanism. It probably began about 5 million years ago when the Pacific plate slightly adjusted its motion and impact along its boundaries. This estimate is based on the slight deviations in the Hawaiian hotspot trace on the Pacific floor that probably occurred when the Pacific plate plunged more steeply and quickly below the Aleutian–Asian margin. Western Victoria, stretching apart over a hotspot train and lying within the stream lines of a deep westerly mantle flow, would become a highly elongated zone of volcanism.

The whole front has future eruptive potential. Electrical conductivity measurements suggest that partly molten material still lies under the southern Victoria-Bass Strait region. Here, heat flow readings remain high. Victoria may be just starting to cool because mantle xenoliths in the volcanoes indicate that heat flow has declined since volcanic activity peaked, 2 million years ago. Future eruptions would probably continue as small outflows of alkali basalts within maar and scoria cone activity. Exact positions for forthcoming eruptions are hard to predict because the activity over the warm front shows no clearcut trends. Eastern Victoria is not exempt from further volcanism, since odd eruptions have occurred in the last 2–3 million years around Uplands and Toombullup. However, the odds overwhelmingly favour eruptive showers on the western side.

The last volcanic storm activity happened at the South Australian end, where MT GAMBIER and MT SCHANK VOLCANOES erupted 5,000 years ago. Tremors still occur under these volcanoes, but no abnormal heat flows are detected. A shift in activity from the older volcanoes of the Barr Range means that future volcanism may move offshore. Noticeable earthquakes shook the coastal towns of Kingston in 1897 and Robe in 1948, and were registered 40 km offshore from Mt Gambier in 1975. Lava flows of young morphology are interpreted on the continental shelf off Robe, so that this shift in activity may have already started.

THE NEWCASTLE–SYDNEY BASIN FORECAST

Although New South Wales seems devoid of young volcanic vents, it has some zones of significant earthquakes such as the disastrous 1989 Newcastle earthquake. Recent studies of zircon grains washing from basalt areas show some unexpectedly young ages, which seem to decrease southward — from 7–12 million years in southeastern Queensland to 2–7 million years in northeastern New South Wales. The youngest ages (2–3 million years old) lie south of the Armidale–Walcha region.

In southeastern Queensland a few basalt flows show ages matching the dated zircon ages. This southern age decrease raises the spectre of a small hotspot now under New South Wales, blasting out zircons in volcanic explosions every 1–3 million years. The projected hotspot position now lies under the Newcastle–Blue Mountains–Wollongong region, within the earthquake zones.

▷ Newcastle, New South Wales, June 1990. Store reinforced by steel beams after damage by December 1989 Newcastle earthquake.

Chances of an eruption in the populated Newcastle–Sydney–Wollongong area are far more remote than the chance of a significant earthquake, which recurs between tens and hundreds of years. However, the southward volcanic trend and some erratic high heat flows in Sydney Basin boreholes give pause for thought. An eruption here would probably form small explosive flashes, causing minor damage in rural settings but a dangerous disturbance near large city precincts.

THE NORTHERN FORECASTS

The young volcanic fields in southeastern Queensland, northern Queensland and Torres Strait differ in size, duration and rate of activity, which means that they vary in eruptive potential. It is not clear what caused their activity. Evidence of hotspot sources further north may be submerged offshore or lie within tectonically disturbed margins. The region's volcanic history is similar to Victoria's, with limited initial volcanism 10–6 million years ago followed by a major expansion of activity.

Hints of volcanic migrations appear in the fields. The older 6–9 million year old Cooktown and Conjuboy fields appear to project south to 3 sites where helium gas of mantle origin discharges into local artesian waters. These discharges near Hughenden (21° S), Oakwood (25.5° S) and Juanobong (27° S) indicate deep basalt sources and it is possible that they mark the hotspot positions of former northern Queensland activity. The Hughenden discharge lies near 900,000 year old basalt flows and tracks back through the Sturgeon, Chudleigh and McBride fields to Cooktown activity that happened 6 million years ago. The Oakwood trace goes past suspected basalt intrusions less than 1 million years old between Blackall and Charleville, and can be traced through the Nulla, Walleroo, Atherton and Cooktown activity that took place 5–9 million years ago. The Juanobong trace shows that there was no activity for the last 7 million years, but it then tracks through the Sturgeon, Clarke Hills and Conjuboy activity that occurred 7–9 million years ago. From this basis it can be estimated that the Hughenden discharge has the greatest eruptive potential, the Oakwood discharge has slight potential, and the Juanobong discharge has little or no potential for a future eruption.

SOUTHEASTERN QUEENSLAND FORECAST

Small lava fields around Gin Gin and Bundaberg mark a region of stretching and jostled fault blocks. The area is seismic and a large offshore earthquake shook Bundaberg in 1918. The Gin Gin field erupted 6–3 million years ago (photograph, p.104), but became inactive. Massive lavas issued from BUNDABERG VOLCANO about 900,000–1 million years ago. Since then, volcanism migrated southwest through the GAYNDAH CRATERS AND FLOWS, about 600,000 years ago, to BRIGOODA CRATER about 300,000–400,000 years ago (photograph, next page). Earthquakes such as the 1935 Gayndah event are known on this line. The southward age trend implies a minor volcanic risk area between Bundaberg, Chinchilla and Brisbane.

▷ *Edge of crater rim, Garnet Gully, Brigooda, Queensland, September 1978. Explosions 300,000–400,000 years ago blew out the crater and brought up fragments of mantle rocks. These include gem quality garnets that are dug up in this fossicking site.*

NORTH QUEENSLAND AND TORRES STRAIT FORECAST

Considerable lava fields demonstrate that extensive volcanism occurred in north Queensland 5.5 million to 20,000 years ago. Unlike the Western Victoria–South Australia fields, these are not subject to any westerly mantle flow towards the Australian–Antarctic downwelling. This may account for the discrete regions of eruption, and for future outbreaks between these regions being unlikely. The youngest eruptions — Nulla field 13,000, McBride field 50,000–70,000, and Atherton field 10,000–200,000 years ago — are hot favourites for further eruptions.

In contrast to the Victoria–South Australia volcanic front, northern Queensland is not waning in activity. Voluminous, long flows appear among the young eruptions, particularly in the McBride field. The greatest likelihood of explosive maar eruptions lies in the Atherton–Cooktown fields because of their wetter climates and extensive groundwaters. Small fields like Mt Fox with a cone 100,000–200,000 years old have minor eruptive potential. A few small extrusions on Cape York Peninsula include the 3 million year old SILVER PLAINS PLUG.

Several volcanoes erupted in Torres Strait 1–3 million years ago (photographs, p.108 and next page). The outburst subsided and its advent remains mysterious. . Though currently stable, it is unwise to consider the area extinct.

FINAL VOLCANIC WARNING

Even accounting for all young lava fields in eruptive forecasts, unexpected surprises may appear in seemingly extinct areas. MT ST MARTIN VOLCANO, for example, a 3 million year old monogenetic volcano, rose from a north Queensland volcanic field that has been otherwise inactive for 18 million years. Moreover, it is a strange, hybrid volcano. Its alkaline, low-silica lava carries mantle fragments and is partially mixed with high levels of trachyte lava. It is a miniature 'one-off' central volcano. Again, high in the Buckland volcanic shield, on the central Queensland divide, MT RUGGED VOLCANO sits in an older, deeply eroded shield (photograph, page 90). Eruptions in the shield ceased over 20 million years ago, but Mt Rugged preserves a youthful crater rim. Such random young vents mean that no area in eastern Australia is entirely safe from future eruption.

Some seismic zones in eastern Australia defy ready tectonic explanation. These include the Newcastle–Blue Mountains–Wollongong seismic zones, for which a minor underlying hotspot was proposed. Does this mark a dormant hotspot? Only further study will tell. If it does mark one, no capital city in eastern Australia is entirely immune from a future eruption. Perhaps this future lies a great way off, but the tectonic and thermal flux below Australia holds the answer.

▷ Wyer Island, an emerged volcanic island in Torres Strait, August 1974. Outcrops of dipping explosive beds formed by steam explosions, as the island erupted from the sea.

OTHER FORECASTS

West of the eastern Australian highlands, the severe volcanic drought will probably continue with no relief in sight. The last volcanic showers were confined to the eastern Kimberley (WA) area, where leucitite pipes and lavas erupted through the Fitzroy trough 18–21 million years ago. Some diamond hailstones accompanied the eruptive clouds. A north to south migration in this volcanic front may indicate its' passage over an ephemeral mantle hotspot. Any further activity from this hot cell is unlikely, but it may have taken a dormant position under Eucla Basin.

Offshore, the last eruptions on Christmas Island took place 3–5 million years ago, probably during a local perturbation in the stress field. Big Ben volcano (p.176) on the Heard Island outpost is predicted to continue intermittent activity from its hotspot. Pulsing glows, vapour and ash plumes and small lava flows were observed during 1992–93 and pumice was erupted offshore.

Antarctica will remain active, particularly Mt Erebus. Others to watch along the Ross Sea coast include Mt Melbourne, the Pleiades and the southern Balleny Islands. Likely regions in Marie Byrd Land lie west of volcanic chains, where volcanism migrated inland before shifting west. Under ice eruptions are expected in odd places.

NEW ZEALAND FORECASTS

The North Island is expected to continue activity in the TAUPO VOLCANIC ZONE. Here MT NGAURUHOE shows nearly continuous activity, MT RUAPEHU'S outbursts are more irregular (photograph, p.87) and active spells of WHITE ISLAND regularly rise and fall (photograph, p.20). LAKE TAUPO VOLCANO remains a major, sleeping hazard. MAYOR ISLAND is on notice for a further outburst. The dormant EGMONT VOLCANO is due for a further outbreak and the Auckland field is a potential hazard. This last field shows little regularity in the spacing and size of its volcanic seizures, so any exact trouble spot is hard to predict. The youngest volcanic fields in Northland and Coromandel Peninsula have remote possibilities for future eruption, but since their patterns show no systematic trends, trouble spots are hard to predict. Mt Ruapehu presents a potential problem for the tourist industry — eruption would spoil the ski season with cinders, ash and meltstreams.

The South Island should remain calm on the volcanic front, but a coastal eruption between Christchurch and Dunedin cannot be completely dismissed.

THE RUMBLES seamounts will continue to live up to their name and erupt under the ocean. A new submarine volcano may emerge beyond the Antipodes Islands.

◁ *Mt Rugged, a young volcanic crater rim preserved on the older eroded Buckland shield volcano, Mt Moffat National Park, Queensland, July 1986.*

COMBATING FUTURE AUSTRALIAN ERUPTIONS

Suppose volcanoes erupted in Australia again, say, within the next century or millennium. How would this affect the populace and how would people deal with such events? For greatest effect, let us imagine an eruption near each of the two largest cities, Sydney and Melbourne. Our assumptions about the nature of these hypothetical eruptions will be based on the youngest volcanic features of these areas and their course will follow a realistic set of events.

These scenarios are set close to present time because a long way off in the future would involve too many unforeseen growth and technology factors. One scenario presents a minor eruption and the other presents a longer, larger one.

SYDNEY OUTSKIRTS ERUPTION

Sydney was preparing for a hot summer. People were worried by prospects of serious bush fires in the dry bushland parks and reserves around the city and in the Blue Mountains to the west (map, p.192).

The collective consciousness of citizens was jolted by an earthquake and several earth tremors overnight. These superficially damaged many buildings in the central city district, and caused minor subsidence, cracking, and warping in many suburban dwellings. Seismologists from the Australian Seismology Centre in Canberra studied the earthquake records, and located a centre for the earthquake on the Kurrajong fault line, 15 km west of Richmond township and 50 km westnorthwest of Sydney city centre. There was no undue alarm, as this fault was known to be active and had triggered an earthquake of magnitude 4 on the Richter Scale near Katoomba in 1919. This new earthquake, however, was of a greater magnitude and deeper focus. Sydney settled back into its summer preoccupations.

Two weeks later, residents at Kurrajong Heights were startled from their sleep by a shaking and a roaring. Running outside in various stages of dress, they felt gritty dust falling and glimpsed red pulsing glows and exploding red traces through the silhouettes of trees to the south. Early morning commuters travelling into Sydney were surprised to see a large dark plume rising rapidly into the still morning air above the Blue Mountains. Word spread rapidly: a volcano was erupting! By nightfall, thousands had driven up to the base of the escarpment to view the vivid fire fountains, as sprays of lava were flung up over 1,000 m into the dark sky. The scene resembled some of Sydney's celebrated firework displays and many thousands of onlookers caused huge traffic jams in the western suburbs. There seemed little immediate threat, as the display erupted from isolated bushland.

The next day, investigating helicopters reported that a cone of dark cinders had risen well above the trees, with small spills of lava slopping out under the fire fountains. A gusty wind was fanning fires ignited by the molten rain falling around the vent. Fire-fighting crews sprang into action, ready to deal with a better known foe. By afternoon, widespread bushfires shrouded the Blue Mountains and the lava fountain and its eruption plume were obscured by a haze of dense smoke. In the next few days, as fire-fighters fought for control of the spreading fires, Sydneysiders became accustomed to dark basalt ash and cinders falling over the northern suburbs. Minute bright red crystals in the ash were identified by geologists at the Australian Museum as the mineral zircon. Sometimes the fallout spread to the harbour and city centre, depending on the strength and direction of the prevailing winds. The fallout usually retreated as afternoon sea breezes sprang up and blew it inland.

Heavy falls of cinders around Kurrajong Heights and nearby Kurrajong forced the evacuation of residents to Richmond airbase. Aerial runs with water bombs were made in an effort to curb the bushfires, although the volcanic plume was too hazardous to approach. Volcanologists and media crews from Australia and overseas flew into Sydney to record and analyse the volcanic spectacle. The first lava flows began to descend the escarpment, streaming down into the Grose River in spectacular cascades. The molten streams spread downstream into the Hawkesbury River, encroaching on Richmond and Windsor townships. Residents from these towns were evacuated by road, while aircraft from the Richmond airbase were hastily flown out before lava lapped over the runways.

At the vent, the fire fountain dwindled, the bush fires were brought under control and Sydney was no longer troubled by volcanic fallout. However, lava continued to gush from the vent and began to bury large sections of the historic towns of Richmond and Windsor. Lava flowed north along the less populated sections of the Hawkesbury River, extending as far as the large bend at Wisemans Ferry. Other lava backed up along the Nepean River, threatening the city of Penrith. Emu Plains and the west side of Penrith were evacuated. Emergency squads desperately pumped water over the moving lava, trying to chill its advances along its north and south fronts. These efforts were hampered by a lack of water flowing into the Hawkesbury, as its tributary sources, the Grose and Nepean Rivers, were cut off.

After five weeks, the pressure of lava outflow diminished, gradually stopping the flow in its tracks. The Blue Mountains had a new monogenetic volcano, joining those which had erupted every few million years (or over greater spans) to form lava caps on Mt Wilson, Mt Irvine, Mt Tootie, Mt Tomah and Mt Banks. The main difference was that the new lava-filled valleys were much lower because rivers had cut further down since the last eruption 13 million years ago. This led to an unforeseen development.

The thick, solid lava choking the Grose River outlet had dammed the drainage, gradually flooding its upstream courses and tributaries. This created a many-fingered lake stretching 30 km inland. This water body provoked much debate among citizens and governing bodies. Some wanted the basalt barrier breached to restore the original lands and rivers, a restoration that required major engineering work. Others wanted the new waterways left alone as a new source of potable, irrigation or even industrial water for the Sydney region's future needs. Others pressed for the development of the lake waters as aquatic playgrounds, to serve the Blue Mountains tourist towns along the southern margin. Farsighted authorities advocated new emergency plans for the Sydney region. They realised that the volcanic event 50 km from the metropolis centre could easily have taken place even closer to it, with much more devastating results.

One new eruption had significantly changed the environment on the fringe of a major city and alerted its inhabitants to such geological hazards.

MELBOURNE HARBOUR ERUPTION

Melbourne had settled into winter. People engaged in indoor pursuits as southwesterly winds brought in cold air, accompanied by falls of hail and snow on surrounding highlands.

Strange stirrings in Port Phillip Bay between the ports of Melbourne and Geelong were observed by vessels using the shipping lanes. Unsettled, broken and 'boiling' water was reported off Portarlington on the Bellarine Peninsula. Incidents of more severe turbulence forced vessels to veer around this site and seemed to trigger tremors, which were registered at the seismology research centre in Melbourne. Seismologists traced the source of these shocks to the junction of two fault lines, where the Bellarine upland meets the Phillip sunkland. Earthquakes have formerly reached magnitudes of up to 5 on the Richter Scale at Port Phillip Bay, although most quakes in the region came from Selwyn's fault along

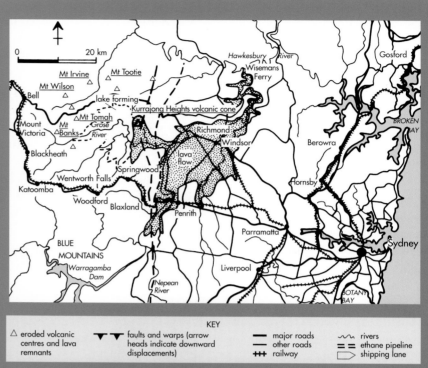

▷ *Sydney–Blue Mountains region, after imagined eruption of Kurrajong Heights volcanic cone and its lava flow. The lava forms a dam across Grose River, which has backed up into a lake. The Kurrajong Fault and associated faults and warps are indicated, as are and eroded volcanic centres and lava remnants (less than 20 million years old). Richmond, Windsor and Penrith are partly 'buried' by lava. Named volcanic and geological features are underlined.*

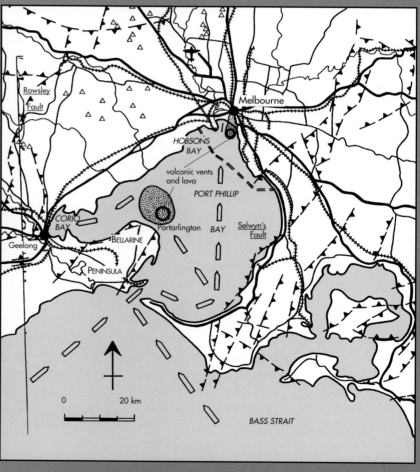

▷ *Melbourne–Geelong region, after imagined eruptions in Port Phillip Bay, off Portarlington and Hobsons Bay. The volcanic vents are shown within built up explosive deposits and lavas. Other volcanic centres on the surrounding basalt lava fields (less than 5 million years old) are indicated. Shipping lanes into Geelong and Melbourne ports and Tullamarine airport are marked, and, as with the Sydney map, concentrations of roads and railways are indicative of dense urban and industrial areas.*

the southeast margin of the Bay or Rowsley fault running north from Geelong. Melbourne and Geelong settled back into their winter preoccupations.

Three weeks later Portarlington residents were startled by muffled roars and huge spouts of water exploding from the bay. These spouts rose hundreds of metres, before crashing back into the bay with a loud smack. Word flashed from Geelong to Melbourne that a volcano was erupting! By late afternoon boats were circling the watery crescendos (at a respectful distance) and the shores near Geelong along Corio Bay were crowded with cars and onlookers. There seemed little immediate threat, as the display erupted offshore and vessels could detour into northern lanes along Corio Bay.

On the next day, a dark shoal was reported to be building up under the water fountains. Dark ash and cinders began to discolour the growing water spouts, making approach more hazardous. Large lava bombs, hurled out in the showers damaged roofs, gardens and cars at Portarlington, forcing residents to evacuate. A volcanic cone rose explosively from the bay. In larger explosions, base surges scudded across the water surface so that small boats had to keep well clear and ships could pass only during quieter periods. The surges damaged the nearby shores, which became uninhabited. When northwesterly winds blew, the eruptive fallout drifted across Mornington Peninsula. This caused inhabitants some discomfort, but did not force them to evacuate. Aircraft flying into Melbourne circled well clear of the eruption zone and only unusual wind shifts disrupted air schedules.

Four months later the vent had built up above the bay waters. Lava flows running down its flanks now entered the water with great hisses and billowing steam clouds, but the violent explosions and base surges were over. Clean up operations began along the damaged shores, but another danger loomed. The lavas from the emerged cone were building up underwater deltas as they encroached on the shores northeast of Geelong. This growth threatened to block off Corio Bay and the port of Geelong.

A massive operation began, combining high-density aerial water bombing with high-pressure hosing from fireships, to quench and chill as much lava as possible before it plunged underwater. Helped by dwindling lava flows from the vent, the project finally sealed off the new volcano, leaving a narrow navigation gap on its north side. Victoria had experienced another volcanic upheaval. Geelong and Melbourne could concentrate on the coming summer Christmas festivities, without a volcanic cloud in the sky.

It was too good to last. In the New Year holiday break, Port Phillip Bay once more trembled and erupted in turbulent gushes of exploding water, this time in Hobsons Bay off Port Melbourne and close to the city centre. A second vent along a volcanic fissure had opened up. The ashy fallout, directed by northwesterly and southwesterly summer winds, drifted mostly across the southeastern suburbs or the city and inner suburbs. The authorities closed the city to commuters and extended people's holiday leave to keep them out of it. The ethane gas pipeline across Port Phillip Bay and the nearby Coode Island chemical storage complex were shut down as a precaution.

This second vent, in shallower water, surfaced within days, and flows built up a delta that joined the shore north of the beach suburb of St Kilda. Lava descended inland, poured into the artificial lake in Albert Park, and solidified there. A more extensive lava flow diverted west around the earlier one and entered St Kilda Road. It lapped up against the Shrine of Remembrance, Government House and the Music Bowl before streaming through the Arts Centre complex and plunging into the Yarra River at the Princes Bridge crossing. Lava banking up here coated and crusted over the bridge, forming a lava tunnel through which a tongue of lava ran past Flinders Street railway station, along Swanston Walk and into the heart of the city. The lava buried the recently laid basalt blocks from quarries near Port Fairy in western Victoria.

Once again, emergency crews used great loads of water to slow the lava stream. Melbourne and the world media watched and waited with bated breath. The lava came to a standstill. Melbourne was spared the fate it was dealt in the Hollywood film of Nevil Shute's 1957 novel, *On the Beach*.

The government and citizens were uncertain about how to treat the new volcanic features on their doorstep. Some wished to quarry the basaltic rock for aggregate and building stones. Others talked about completely removing the new features in a protracted effort to restore the original conditions. In their wisdom, the authorities declared the volcanic edifices would remain as reserves. Untouched, they could be enjoyed as natural features to remind citizens and tourists of their vulnerability to the environment.

CHAPTER

FL SUTHERLAND

13

TRANS-TASMAN VOLCANO SPOTTER'S GUIDE

◁ *Lake Surprise in crater, Mt Eccles National Park, Victoria, looking west, February 1973.*

MANY volcanic areas described in this book are accessible to travellers driving or hiking through eastern Australian and New Zealand regions. A few special trips have been selected as a guide to volcanic features for students and those with a consuming interest in volcanism. Contrasting regions from each eastern Australian state and the main New Zealand islands have been selected to present a wide view of natural volcanic wonders. The maps section (pp.214–223) contains both regional and large scale maps of the districts covered in this chapter.

The tour starts with the young volcanic fields in Victoria and South Australia and moves on to some older basalt fields in Tasmania. These features are then contrasted with large eroded hotspot volcanoes in New South Wales, and offshore on Lord Howe Island. In Queensland, the tour peers into Australia's largest hotspot volcano, the Tweed shield, and follows the coastal highway from Brisbane to Cairns, to look at older eroded volcanoes. From Cairns the tour diverts into the Atherton Tablelands and its young volcanic features.

Crossing the Tasman Sea to New Zealand, the tour examines eroded hotspot volcanoes in the South Island. It then visits the North Island, looking at the dormant Auckland field and finishing at the active volcanic fields in the Rotorua–Tongariro region.

VICTORIA — SOUTH AUSTRALIA

(Map p.217) Places of volcanic interest abound west of Melbourne in the central uplands and western plains where 300 volcanoes built a vast basaltic field. This route only touches the volcanic centres and flow features, but the whole region presents volcano-spotters with an extensive hunting ground. The recent craters around Mt Gambier in South Australia mark an accessible extension of this field

NEWER VICTORIAN VOLCANOES: CENTRAL WEST TOUR

(Map p.217) The Organ Pipes National Park is a good starter, only 27 km northwest of Melbourne on the Calder Freeway. An impressive basalt outcrop shows elegant, slim, cooling columns in a thick flow which filled an old Maribyrnong drainage. The Calder Highway continues to Woodend, 66 km from Melbourne, and 5 km further north a turnoff leads 7 km east to the infamous Hanging Rock (photograph, p.30). This steep lava dome is fairly easily climbed and was featured in the Australian novel and film, *Picnic at Hanging Rock*. It consists of 6 million year old trachyte, like other nearby volcanoes at BROCKS MONUMENT and CAMELS HUMP. These small volcanoes are overshadowed by Mt Macedon, a remnant of huge outpourings of rhyodacite lavas 370 million years ago.

Returning to Woodend and turning west, a 44 km link road to Daylesford connects to the Midlands Highway. A further 46 km southwest lies Ballarat, a celebrated goldmining city built on a small lava plain with surrounding landmark

volcanoes such as Smeaton Hill and Mt Buninyong. A visit to the Gold Museum at Sovereign Hill recreates earlier mining days, when alluvial deposits below four separate basalt flows were mined to depths of 150 m, and over 60 tonnes of gold were recovered. From Ballarat, the Midlands Highway connects southeasterly 86 km to Geelong and the Princes Highway. Colac is reached 74 km west along Princes Highway, although there is a shorter 100 km connection along lesser roads from Ballarat through Rokewood and Cressy.

West of Colac, 3–9 km along the highway, roads branch north between Lake Colac and Lake Corangamite. These water bodies were formed by lavas blocking earlier drainages. The branch roads extend for 6 km into the Red Rock volcanic complex, which is a typical mix of young maars, tuff rings and scoria cones. Some may be only 8,000 years old. The highway continues south past Lake

◁ *Mt Gambier craters, South Australia, March 1974. Aerial view looking west.*

H HUGHES

Corangamite and curves north around and through the lava apron of MT PORNDON VOLCANO, 23 km from Colac. MT PORNDON LAVAS form 'stony rise' topography typical of the lava plains and are probably under 20,000 years old. From here, the highway reaches Camperdown, 45 km west of Colac.

Approaching Camperdown, the highway crests over the MT LEURA TUFF RING. Just to the south, MT LEURA CONE rises from this 2.5 km diameter structure. Thin-bedded pyroclastic surge deposits in the ring are exposed in cuttings and quarries beside the highway. Mt Leura's cone is made of scoria that contains basalt lapilli, blocks, bombs, and fragments of mantle peridotite rocks — some encased in basalt bombs. The Lookout gives fine summit views. LAKE PURRUMBETE TUFF RING and MT PORNDON VOLCANO lie 8 km and 17 km southeast respectively; Lake Corangamite lies 20 km east and the Red Rock volcanic complex lies 25 km east. The twin BULLENMERRI and GNOTUK CRATER lakes lie only 7–9 km west, MT NOORAT CONE (photograph, p.102) is 22 km westnorthwest, and MT ELEPHANT, Victoria's largest cone, is visible 33 km north (drawing, p.10).

Mortlake, another volcanic town, is reached after turning northwest off the highway, 17 km west of Camperdown. MT SHADWELL SCORIA CONE is 1.5 km north and visits to the Shire Quarry, which excavates the layered scoria, can be arranged through Mortlake Shire Office. Bands in the scoria yield basalt bombs which enclose pieces of mantle peridotite. This is a gem fossicking site for olivine, which is cut into peridot. From Mortlake, a 50 km southwesterly road connects back to the Princes Highway at Warrnambool.

TOWER HILL is a complex volcano besides the Highway, 12 km west of Warrnambool (painting, p.104). An elliptical maar contains multiple scoria cones within its crater lake. A Visitors Centre with a Tower Hill history display operates at the Wild Life Reserve entrance. Complex explosive deposits can be seen in the CRB Quarry near the Visitors Centre entrance. Steam explosion and airfall deposits intermingle their layers in at least 13 changes of activity.

At Port Fairy, a 47 km northwesterly road from the highway connects to Macarthur. From here a 7 km southwesterly road enters Mt Eccles National Park. This allows close views of LAKE SURPRISE CRATER LAKE (p.194), THE SHAFT, TUNNEL CAVE, NATURAL BRIDGE, THE PIT, THE ALCOVE, lava canals and MT ECCLES SCORIA CONE. This centre fed the TYRENDARRA LAVA FLOW, which reaches the coast 30 km to the southwest. From Macarthur, roads connect via the Henty Highway for 44 km to Heywood and the Princes Highway. Another 75 km takes the Highway into South Australia and Mt Gambier.

YOUNG SOUTH AUSTRALIAN CRATERS: SOUTHEAST TOUR

(Map p.217) Mt Gambier township is the doorway to MT GAMBIER VOLCANO. Volcanic ash beds are exposed in Crouch Street, just south of Gwendoline Street. Here some volcanic bombs have pierced the ash layers with their impact. South of Lake Terrace lie the lake-filled explosive craters of the multiple maar system. Lookouts on the crater rims overlook Browne's Lake, Valley Lake, Leg of Mutton Lake and Blue Lake from its north and south sides, and John Watson Drive completely encircles the BLUE LAKE CRATER RIM.

At Apex Lookout, white Gambier marine limestone and dune sands form cliffs above Blue Lake and are overlain by volcanic breccias and ash beds baked red on top by the heat from an overlying lava flow. Above this dark lava are tuffs and agglomerates from the explosions which blew out BLUE LAKE CRATER (photograph p.16). Boulders of limestone and basalt weighing up to several tonnes are seen in the tuff and agglomerate. At Potters Point Lookout on Hay Drive, the lava flow is visible in the Razorback, a ridge between Leg of Mutton Lake and Valley Lake. A buried scoria cone is seen at Bootlace Cave southeast of Browne's Lake. Northwest of the Lookout, Tenison College Oval lies in an early formed maar. At Marks Lookout you can examine the DEVIL'S PUNCHBOWL STEAM EXPLOSION VENT, layering in the tuffs and agglomerates, and a range of volcanic and country rock fragments. A track leads down from the Lookout into the western craters and the Public Pleasure Resort. Where it divides to encircle Browne's Lake, the south side exhibits glassy spatter from lava fountain activity. This resembles twisted rope and fresh cow dung and gives the crater floor an irregular, hummocky appearance.

MT SCHANK VOLCANO (photographs pp.16 and 184) is visited by a 15 km drive from Mt Gambier along the Port MacDonnell road and lies on the east side. A turnoff just south of quarries in a basalt flow leads 1 km east to a north branch and a carpark. A small subsidiary explosion crater south of the carpark exposes white Gambier limestone, buff dune sands, a dark lava flow and explosive deposits from later crater eruptions. From here, Mt Schank can be climbed for a view into the main crater. Beyond the volcano's north flank a fissure line with small cones is visible. These can be reached by continuing north along the carpark track and then turning east. A road cut exposes red and black scoria that is used as decorative stones in Mt Gambier gardens.

For further exploration of slightly older maars, cones, crater lakes and lavas, the Mount Burr Range volcanic field beckons along roads between Mt Gambier, Millicent and Kalangadoo.

TASMANIA

Tasmania's volcanic rocks form fertile fields or forested areas in the northwest, and along the coast fine volcanic sections can be seen by following the Bass Highway west of Devonport. Small vents and lava flows in the Derwent estuary are easily viewed around Hobart, the capital city.

BASS STRAIT: NORTHWEST TOUR

(Map p.218) Typical basalt lava flows form coastal cliffs and rock platforms west of Devonport. At Don Heads, west of Don River and 5 km northwest of Devonport city centre, two flows of massive and columnar jointed olivine basalt can be seen if you walk along the shore towards Lillicos Beach. They filled an old course of Don River, before the Bass Strait sea level rose 25 million years ago. About 1 km from Don Heads a younger flow appears high in the cliffs, filling a valley cut in the older flows. This lava flowed northeast to Don Heads from near Forth Township on Forth River. This distinctly different rock is easily examined in fallen boulders littering the shore below. It is dense fine-grained olivine nephelinite rock, packed with pieces of coarse mantle peridotite rock.

At Burnie, 50 km west of Devonport, another olivine basalt flow shows classic vertical cooling columns, broken across by 'Dutch cheese' ball and socket joints. It appears in an old quarry on the road bend at Burnie Wharves.

Wynyard boasts two notable features related to volcanic activity. At Fossil Bluff, north of Inglis River, visitors can inspect fossil sea life in cliffs preserved under a basalt cap (photograph pp.141 and 142). This has more appeal for fossil enthusiasts, as the basalt is poorly exposed. Older glacial beds underlying the fossil beds are a bonus and show a mixture of rocks dumped by melting ice 290 million years ago.

Excellent views of Table Cape, 4 km north, mark a resistant, coarse-grained volcanic core that erupted 13 million years ago. Sheer sides rise 175 m above sea level and coarse-grained rock is found along the road to the lighthouse on top. The volcano's contact between crater tuffs and overlying lava is exposed at sea level and can be approached by walking down tracks to the shore south of Table Cape. Property owners' permission is required first, and the walk involves some clambering along a

◁ *The Nut, Circular Head, Stanley, Tasmania, November 1978. The coarse dolerite of the neck forms the steep faces (background) where it intrudes through tuff beds (foreground). The lower slopes are mostly covered with blocks shed from the dolerite, but dipping beds of tuff are exposed (bottom right).*

boulder-strewn shoreline. Beds of fused, glassy, explosive tuff beds abut against north-dipping massive jointed basanite rock extending below sea level. The basanite carries conspicuous fragments of mantle peridotite rocks and grades up into coarse olivine dolerite rock infused with patches and veins of coarser rock. It probably filled the bowels of a deep lava lake in the original TABLE CAPE CRATER.

THE NUT, another large volcanic core, forms Circular Head. It forms a dramatic, flat knob about 60 km west of Wynyard and 7 km from the Bass Highway turnoff to Stanley. The track from this charming fishing village climbs steeply to the top; it provides a chain to help you up, and a chairlift (at a cost). The Nut consists of dolerite rock similar to that of Table Cape, but at 12 million years age it is a slightly younger centre. From the northwest side you can see explosive crater beds bounding the volcano and extending out in platforms that were cut by wave action. The pleasant adjoining peninsula stretching from Stanley to North Point is made of lava flows that outcrop along the shorelines. Dated at 8–9 million years, these lavas are the youngest found in Tasmania.

DERWENT ESTUARY: SOUTHERN TOUR
(Map p.218) The Derwent River at Hobart reveals several volcanic fillings in its former channels. BLINKING BILLY POINT VOLCANO at Lower Sandy Bay, 4 km south of the city, exposes 27 million year old flows and tuffs overlying clay beds in road cuts at a sharp bend on Taroona Road. They slope east down to the shore below Blinking Billy Point, where they are reached by pathways. Here the lavas and tuffs form folds pointing to a submerged collapsed vent. The lavas are unusual, evolved alkaline rocks related to nepheline benmoreite. On the opposite Derwent shore, 4 km east, small hills cap the Rokeby Hills near Droughty Point. These mark small plug and flow remnants of related uncollapsed alkaline lavas. Another alkaline volcano is MARGATE VOLCANO, exposed on the Channel Highway entering Margate township, 18 km southeast of Blinking Billy Point. It features an outcrop of coarse explosive deposits with geological explanatory signs and shows good examples of spindle-tailed basalt bombs.

A different style of lava appears north of Hobart, in old infillings of the Derwent River. Selfs Point, 4 km north of the city, exhibits weathered columnar olivine-basalts. At Claremont, 8 km north, three flows of tholeiitic olivine basalts are exposed in cliffs below Cadbury's chocolate factory. The lower two flows are interbedded with glassy tuffs and breccias. Pillowy lava with glassy cooling crusts and current bedding in the tuffs suggest that some lava erupted into flowing water.

NEW SOUTH WALES

Excellent views of volcanic significance abound in reserves, parks and lookouts throughout New South Wales. Many involve central volcanoes around Mt Warning, Nandewars, Warrumbungles, Ebor-Dorrigo, Comboyne and Mt Canobolas. Offshore lies beguiling Lord Howe Island. Basalt fields are easily reached near Sydney in the Blue Mountains at Mt Tomah and Mt Wilson. The New England basalt field caters for gemstone fossickers around Inverell and Glen Innes.

WARRUMBUNGLE MOUNTAINS: GUIDED DRIVES AND WALKS

(Map p.219) The Warrumbungle National Park in this central volcano lies east of Coonabarabran, 475 km from Sydney. Coonabarabran is reached either by the Oxley Highway from Tamworth via the New England Highway or by the Newell Highway from Dubbo via the Great Western Highway. From here, John Renshaw Drive leads to a visitors' centre in the park, 34 km from Coonabarabran.

The volcanic rocks are 13–17 million year old trachytes and basalts, exposed in a hotspot shield volcano 50 km across. Its earlier eruptions were largely trachytes and its later coverings were largely basalts. The viscous trachytes formed many intrusions and explosive deposits, compared to the more fluid basalts that provided extensive flows.

For a quick 'armchair' observation of WARRUMBUNGLE VOLCANO, a motorist can travel along John Renshaw Drive for 50 km from Coonabarabran and make a 79 km circuit to Tooraweenah, before returning via Oxley Highway. TIMOR ROCK TRACHYTE NECK rises behind a picnic area 13 km west of Coonabarabran and a road cut at 15 km shows a spectacular chaotic deposit of trachyte boulders in a volcanic mudflow or lahar. MOPRA ROCK TRACHYTE DOME rises on the north at 21 km. A dark basalt flow overlies creamy-coloured tuffs in a quarry near 24 km, just before the Siding Springs turnoff branches off the main drive. A diversion to the Anglo-Australian telescope and visitors' centre on Siding Springs mountain can be made here. From 25–26 km sandstone beds of Jurassic age under the volcano are seen, just before a view of the eroded volcano from White Gum Lookout Track a few kilometres further on. A trachyte dyke cuts the horizontal sandstones.

The visitors' centre, 34 km from Coonabarabran, provides park information and an easy walk, with wheelchair access, along Gurranawa Track. From the carpark, another 2 km from the centre, there is a short 1 km walk along Canyon Camp Nature Walk to a gorge, columnar jointing

◁ The Breadknife trachyte dyke, from near Lughs Throne, Warrumbungle National Park, November 1987.

and rock pavement in trachyte. Views of the volcanic shield appear at 42 km and 45 km from Coonabarabran, before a south turnoff goes to Tooraweenah at the 49 km mark. Bouldery trachyte is crossed at 58 km and near 96 km along the circuit from Coonabarabran, Wallumburrawang Creek forms Hickey Falls, which drop over a 15–20 m cliff of trachyte lava.

LORD HOWE ISLAND SEASCAPE TOUR

Lord Howe Island, Tasman Sea, looking south.

Balls Pyramid, Tasman Sea, March 1987.

LORD HOWE

ISLAND VOLCANO (map p.220) is a basaltic bastion in the Tasman Sea. Probably only one-fortieth of the shield has survived since its eruption 6–7 million years ago. Its listing as a World Heritage Site of great natural beauty was partly due to its volcanic foundation. The island lies 700 km northeast of Sydney from where it is connected by air and shipping. Car access is limited, but many volcanic features can be reached along cycling or walking tracks, or by boat.

The lower, older part of this emerged hotspot volcano lies north of the airport. The rocks are exposed along the shore lines, reached by tracks to Transit Hill, Malabar Hill and Mount Eliza. The lowest tuff beds outcrop at Searles and Stevens Points, under Malabar Hill and on the Admiralty Islands to the north. Most tuffs are yellow, but red tuffs at Stevens Point probably mark the vent's explosive emergence from the sea. Dykes cut the tuffs and cross-cutting dykes are seen north of Neds Beach. Overlying lavas make up North Ridge and Transit Hill and most have a weathered appearance. The contact between tuffs and lavas is seen below Malabar Hill. Vertical dykes cut the lavas, and a striking example above a sea cave at Old Gulch is accessible at low tide. The dykes tend to line up towards Mt Lidgbird and the younger, southern sections of the volcano.

South of the airport, hills of volcanic breccia mark a violent explosive phase that formed a crater rim. Breccias containing blocks torn from vent walls outcrop at Boat Harbour, many of them cut by dykes. The inner crater margin dips southward at Red Point, but the contact is best seen by boat or by air. The youngest, most prominent parts of the volcano are lavas filling the old caldera crater. These form Mt Lidgbird at 777 m and Mt Gower at 875 m. Mt Lidgbird lavas are approached from Smoking Tree Ridge Track up to the 420 m level, or can be viewed on Grey Face above the Mount Gower Track. Columnar jointing is clearly visible among the flow horizons. Mt Gower lavas can be traversed all the way to the summit; this takes a full day.

BALLS PYRAMID VOLCANO, 23 km south of Lord Howe Island, is one of the most dramatic spires of eroded lavas imaginable. A circumnavigation by boat or air is a perfect finale for the seascape tour.

EARTH

VOLCANIC

202

▷ *Belougery Spire, a trachyte plug, Warrumbungle National Park, New South Wales, November 1987.*

▽ *White Cave, Binna Burra National Park, Queensland, June 1971. Sculpted beds of rhyolite tuff.*

FL SUTHERLAND

CHAPTER 13

Walking tracks allow a more leisurely, informative and breath-taking acquaintance of the volcano. Grand High Tops Track from Camp Pincham, a 14 km steep circuit, is a highlight that offers a view of the heart of the volcano and spectacular features such as THE BREADKNIFE TRACHYTE DYKE and BELOUGERY SPIRE TRACHYTE NECK. Uninterrupted views of the whole volcano are seen from Lughs Throne.

Side walks in this circuit include Goulds Circuit, a 6 km detour around MACHA TOR VOLCANIC NECK. Bridget and Bress Peaks Track is a 7 km return with a steep climb to BRIDGET PEAK DOLERITE DOME and the contrasting hydrothermally altered rock in BRESS

▷ *Mt Tibrogargan rhyolite plug, Glass House Mountains, Queensland, April 1991.*

PEAK TRACHYTE DOME. Buff Mountain Track is a steep 3 km ascent up BLUFF MOUNTAIN TRACHYTE DOME, the largest lava dome in the Warrumbungles. Fans Horizon Track, a 4 km return from Camp Pincham, follows the underlying sandstones and shows how much of the volcano has been eroded away.

Walks from other sites provide other perspectives. Belougery Split Rock Track from Camp Wambelong is a steep 5 km circuit to BELOUGERY SPLIT ROCK TRACHYTE DOME, where outer trachyte breccias encase massive trachyte. Danu Saddle–Mount Exmouth Track is a 10 km return trip from Camp Wambelong. West from Danu Saddle the track ascends through trachyte flows, tuffs and breccias into basalt tuffs, breccias and flows which cap Mt Exmouth. Cathedral Arch Track, a 1 km return trip east of Danu Saddle, features dykes, domes and lavas of rhyolite including a 50 m thick rectangular jointed flow at Cathedral Arch. Siding Springs Mountain Road from John Renshaw Drive passes up through trachyte flows, dykes, volcanic breccias and mud flows. Just below the sharp bend west onto the capping basalts there is an evolved basalt flow of mugearite rock. The summit of Siding Springs, past the Anglo-Australian telescope, returns into trachyte and gives panoramic views of the surrounding volcanic sequence.

QUEENSLAND

This large State offers a smorgasbord of volcanic features along its eastern seaboard. Numerous central volcanoes contribute many scenic peaks and ranges; older andesitic volcanoes enter the scene in southern and central sections. Young basaltic lava fields emerge in the tropical northern section.

F. SUTHERLAND

BINNA BURRA: SOUTHEAST WALKS

(Map p.220) Walks from Binna Burra enter Lamington National Park, on the northeast flank of the TWEED SHIELD VOLCANO. Binna Burra, 100 km south of Brisbane, is reached by taking the Pacific Highway to Nerang, then through Beechmont. The walking tracks offer close views of volcanic horizons within the huge volcano.

From the National Park sign, a track to the swimming hole on Coomera River descends through beds of rhyolite tuffs. These contain carbonised tree trunks in the basal cliffs. The WHITE CAVES track leads to delicately wind-sculptured beds of soft white tuffs. The track continues to a short tunnel in brecciated glassy lava linking KWEEBANI CAVE and the ABORIGINAL COOKING CAVE. This probably marks a volcanic pipe of obsidian, now broken up and altered to grey perlite.

From the parking and picnic area at the end of Binna Burra Road, longer walks lead to several features of geological interest. A 12 km return walk to Ballanjui Falls passes Koolanbiba Lookout and Cave, providing distant views of volcanic horizons and a close view of horizontal flow-banded rhyolite. BALLANJUI FALLS marks a rhyolite intrusion where flow layers dip at an angle to the main rhyolite flows. A 13 km return walk to SURPRISE ROCK meets a rhyolite dyke showing well-developed horizontal joint columns. A 19 km circuit to the rhyolite cliffs at SHIPS STERN gives views of TURTLE ROCK, formed in underlying rhyolite tuffs, and EGG ROCK, a rhyolite neck. An 18 km return walk south to the State border lookout gives splendid views into the erosional caldera of TWEED VOLCANO and the central MT WARNING PLUG.

BRUCE HIGHWAY: SOUTHERN STRETCH

(Map p.220) The highway skirts past the Glass House Mountains, 55–70 km north of Brisbane. These engaging stumps of trachyte and rhyolite volcanoes, 25–27 million years old, were left standing after the coastal plain eroded away. Closer views are gained by turning off the freeway for 5 km to Beerburrum township.

MT BEERBURRUM VOLCANO, a bulky trachyte peak outside the town, is climbed by a track to the summit. A network of roads leads to other peaks, such as the twin rhyolite peaks of MT TUNBUBUDLA VOLCANO to the west. North from

VOLCANIC EARTH

Beerburrum, a 5 km road to Glass House Mountains township passes Mt Tibrogargan volcano, a rhyolite peak. From Glass House Mountains a road leads 2 km east to Mt Ngun Ngun volcano, where diverse rhyolite flow structures and vertical jointing are well exposed. A 7 km easterly excursion leads to a picnic area at the base of Mt Beerwah volcano, an imposing trachyte intrusion, and gives a close view of Mt Coonowrin volcano's vertical rhyolite spine. Access back to the Bruce Highway is made through Beerwah township, 4 km north of Glass House Mountains township.

A delightful side detour through the rolling basalt hills of the Maleny area is made by turning east through Landsborough, 6 km north of Beerwah. These basalts are a few million years older than the Glass House intrusions and a flow

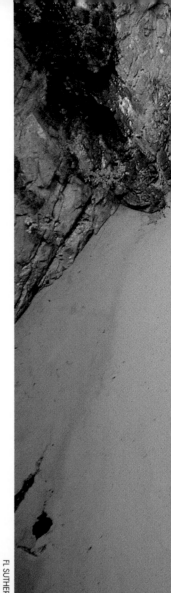

▷ *Bottom of Hypipamee crater, Atherton Tablelands, Queensland, March 1994.*

◁ *The Blow Hole, Double Head, Yeppoon coast, Queensland, July 1987. Inverted fan of cooling columns in rhyolite intrusion.*

overlying sandstones can be seen on the escarpment 9 km west of Landsborough. Vantage points south of Maleny near Mary Cairncross Park offer sweeping views of the Glass House Mountains (photograph p.119). Weathered flow tops in the basalt sequence and overlying thick red soils appear in cuttings along the road north of Maleny to Mapleton. The road then descends east through the basalts and rejoins the Bruce Highway at Nambour.

Noosa National Park lies 21 km east of the Bruce Highway from Cooroy, 123 km north of Brisbane. From the park entrance, 0.8 km from Noosa, a 4 km track leads to Dolphin Point, Granite Bay and Hells Gates. Dykes of basalt, andesite and granite cut the Triassic sandstones along the shoreline. These dykes fed eroded lavas and some are banded by multiple injections of molten material. At Granite Bay a larger intrusion of diorite rock has baked the sandstones to quartzite. At Laguna Lookout, southeast of Noosa, a quarry shows a diorite sill with large dark amphibole crystals. This invaded and baked dark shales. The Noosa intrusions represent older, andesitic Cretaceous and Triassic volcanism.

BRUCE HIGHWAY: TROPIC OF CAPRICORN TOURS

(**Map p.221**) The tropic passes through Rockhampton and the Fitzroy River, 640 km north of Brisbane. Mt Archer, 8 km northeast of Rockhampton, offers an unimpeded view. Plugs resembling those of the Glass House Mountains rise from the low plain north of Rockhampton. They belong to significantly older central volcanoes, 70–80 million years old.

The road to the Yeppoon coast turns off Bruce

Highway 5 km from Rockhampton and passes through this plug field. The largest peak, 393 m high, forms MT WHEELER RHYOLITE PLUG, 23 km northeast of Rockhampton. The plugs are best inspected in coastal exposures south of Yeppoon. DOUBLE HEAD RHYOLITE PLUG, south of Rosslyn Harbour, shows sloping columns in a large quarry beyond the ferry terminal (photograph p.204). Bluff Point National Park to the south gives fine views of the Rosslyn Bay intrusion. A similar intrusion here shows cooling columns formed at right angles to its steep contact. Another intrusion further south extends to the Causeway, 7 km north of Emu Park, where an alternative route returns to Rockhampton.

The Capricorn Highway leaves Bruce Highway south of Rockhampton and is dubbed the Gemfields Highway. MT HAY VOLCANO, 41 km west of Rockhampton, supports a gemstone tourist park. Here fossickers can dig and slice their finds of thunder eggs (spherulites with agate and quartz-filled interiors), recovered from an altered rhyolite below the mountain. The highway leads to Emerald, 263 km west, and then 42 km to Anakie, the entrance to the gemfields. Sapphire and Rubyvale lie 10 km and 16 km north of Anakie. These towns are small centres for sapphire mining and gem cutting. Gemstones are recovered from alluvial deposits, and tourists can visit surface and underground mines. The gem sapphires and zircons came from basaltic volcanoes, now left as plugs forming many small peaks through the district. The nearest to Rubyvale is POLICEMANS KNOB PLUG. The Willows Gemfield, another sapphire field reached by the Capricorn Highway, is 71 km west of Emerald and 11 km south of the turnoff.

For more extended volcanic viewing, alternative routes can be taken back to the Bruce Highway from Emerald. A southern route passes the scenic Zamia Range in the SPRINGSURE CENTRAL VOLCANO, 100 km along the Dawson Highway (photograph p.169). A northern route passes the picturesque Peak Range of the CLERMONT CENTRAL VOLCANO, 100 km along the Gregory and Peak Downs Highways.

BRUCE HIGHWAY: NORTHERN TABLELAND TOUR

(Map p.221) The basaltic tablelands of the young Atherton volcanic field lie west of the Bruce Highway between Innisfail and Cairns. Over 50 volcanoes are known, including lava shields, cinder cones, maars and a blast crater. The field is reached by turning off the highway 5 km north of Innisfail to meet the Palmerston Highway at PIN GIN HILL VOLCANO, a basalt cone with a 780,000 year old lava apron. The highway rises westward to Crawfords Lookout. From here an old basalt gorge filling is seen lying between Douglas Creek and North Johnstone Rivers, which join north of the lookout. A track leads into the Douglas Creek gorge 1.8 km west along the highway from the lookout. Here the Tehupala Falls and Wallacha Falls descend over basalt flows. Attractive waterfalls cascading over massive columnar basalt are also seen 30 km further west along the highway at Milla Milla Falls and 60 km further west at Millstream Falls beyond Ravenshoe. Cuts along the road 3 km west of Ravenshoe expose much older volcanics of the region and reveal fine specimens of flow banded rhyolites 300 million years old.

The road from Ravenshoe via Malanda to Yungaburra passes several young volcano vents. HYPIPAMEE CRATER, 2 km north of Ravenshoe, is a steep open vent with vertical walls, probably made by a single violent explosion (photographs p.182, 204). BROMFIELD SWAMP MAAR, 6–7 km south west of Malanda on Upper Barron Road, is the largest maar in the field. MT QUINCAN CONE AND CRATER, 3 km south of Yungaburra, has quarry works on its south side. Small volcanic bombs and pieces of mantle peridotite are found here, but permission for entry should be sought. The Seven Sisters cinder cones are an aligned group of cones crossing Atherton Road, 3 km west of Yungaburra. LAKE EACHAM MAAR (photograph p.100) is joined by a side road, 5 km east of Yungaburra, while LAKE BARRINE MAAR lies beside the Yungaburra–Gordonvale Road, 7 km east northeast of Yungaburra. GILLIES CRATER is a breached, swamp-filled crater, 16 kms from Gordonvale. The explosive crater beds contain pieces of basalt, mantle peridotite and slate, and schist from the underlying metamorphic basement.

The Bruce Highway is rejoined at Gordonvale. GREEN HILL VOLCANO forms a grass-covered vent to the east, 13 km southwest of Cairns. Its eruption may have deflected the Mulgrave River southward from a former outlet closer to Cairns. The highway terminates at Cairns, 1,717 km north of Brisbane.

Lyttelton volcano, Banks Peninsula, South Island, New Zealand, November 1992.

◁ Drowned erosion caldera, looking southwest.

▽ Basalt dyke cutting through explosive deposits and feeding overlying lava flow, Summit Road.

SOUTH ISLAND, NEW ZEALAND

No active volcanoes menace the South Island, but basalt lavas 2–3 million years old are seen at Timaru on the Christchurch–Dunedin highway. Large eroded central volcanoes provide fine exposures around harbours near Christchurch and Dunedin. Other older, classic volcanic exposures along the intercity highway include basalt breccias full of xenocrysts and xenoliths at Kakanui South Head, and pillow lavas and breccias at Oamaru, 109 km and 124 km from Dunedin respectively.

BANKS PENINSULA: DOUBLE VOLCANO TOUR

(Map p.222) Two large volcanoes formed here between 15 and 5 million years ago. The sea has recently penetrated them to form Lyttelton and Akaroa harbours. LYTTELTON VOLCANO, the older and more complex vent, is 15–9 million years old, but was reactivated 8–6 million years ago. AKAROA VOLCANO is a shield 9–8 million years old. The original volcanoes covered 1,200 km², stood 1,500 m above sea level and were connected to the mainland only 20,000 years ago. A network of roads from Christchurch into the volcanoes gives their essential features good coverage. Two full day tours and two half-day tours are described.

The Port Hills Tour explores LYTTELTON VOLCANO. The Christchurch road meets Bridle Path Road at the mouth of Heathcote River. A'a lavas with rubbly tops and bases lie in the south embankment. MOA-BONE POINT CAVE, past Beachville Road

turnoff, is a sea cave eroded into a'a lava. SHAG ROCK, at the Avon– Heathcote Estuary mouth, is a sea stack of columnar jointed basalt resting on volcanic ash. As Bridle Path Road ascends above Lyttelton, tuffs with black pyroxene crystals outcrop west of Mt Pleasant, 500 m before Bridle Path carpark. The carpark offers sweeping views of Lyttelton volcano. The road continues through volcanic mud flow deposits to a carpark next to CASTLE ROCK, a trachyte lava dome.

◁ Spectacular coastal erosion in phonolitic flow and explosive beds, Spit Beach, North Head, Otago Harbour, New Zealand, February 1983.

▽ Dunedin volcano, from north side looking towards Port Chalmers vent, South Island, New Zealand, March 1986.

At The Sign of the Kiwi at Dyers Pass, Summit Road continues for 400 m to a parking area. Here, four vertical dykes of trachyte and basalt radiate from the central vent and intrude one of the lava fissures. The Sign of the Bellbird carpark at Cass Peak gives superb views across Lyttelton vent. Further on, GIBRALTAR ROCK marks a trachyte dyke. CONICAL HILL, 1 km before the Summit Road–Gebbie Pass junction, is a rhyolite dome with radiating cooling columns. Turning east to Teddington and then north 3 km leads to cuttings of white rhyolite carrying smoky quartz crystals. Christchurch is reached by continuing along Gebbie Pass or by turning east past Governors Bay to Lyttelton tunnel.

Quail Island Tour takes a summer ferry service from Lyttelton Harbour and follows the shores at low tide. North of New Jetty basalts intrude and lie over volcanic mud flows. Along the northern shore columnar jointed basalts form cliffs, followed by altered basalts cut by dykes and then by conglomerate beds marking old streams between basalts. Shag Point exposes two sets of basalts, the younger ones containing fragments of country rocks. These basalts continue along the west shore to the south, where older rhyolites appear. A lava dome cut by trachyte dykes is passed on the way to Waterski Beach. From there, rhyolites continue via Old Jetty back to New Jetty and the ferry.

Mount Herbert Walkway, a 16 km trek, intersects some of the youngest basalt lavas of Banks Peninsula. From Orton Bradley Park at Charteris Bay the walk passes underlying sandstones that are 50–60 million years old. BIG ROCK, 2 km along, is an eroded andesite dome, an early phase of the volcano. Some 3 km along the track, irregularly jointed basalt underlies explosive beds. Wavy and dune-like beds in these

deposits typify base surges from lava-water explosions. MT HERBERT summit, about 7 km along the track, looks out on columnar jointed basalt capping MT BRADLEY to the west. Lava flows descend from Mt Herbert along the walkway down to Diamond Harbour, 6 km by road from Orton Bradley Park.

Onawe Peninsula Walk enters the heart of AKAROA VOLCANO. A turnoff from the Christchurch–Akaroa road between Settler's Cheese Factory and Hotel De Pecheurs leads to a carpark. The walk is best made at low tide. A tuff cone exposed under the carpark contains volcanic bombs showing spindle, breadcrust and cowpat forms. Trachyte dykes cut basalt lavas 200 m south. A vent breccia intrudes the lavas 600 m south and carries angular blocks that have been ripped off vent walls. Airfall tuffs appear below the lavas 1 km south. On the east side, a trachyte sill has intruded between the tuffs and a lower basalt flow. The main intrusive core of the volcano forms the highest end of the peninsula, and consists of massive coarse-grained syenite rock weathered into rounded kernels. A sliver of dark gabbro is found at the contact. The trig station on top has views of the harbour and coastal villages in the volcano.

DUNEDIN: HARBOURSIDE TOUR

(Map p.222) Otago Harbour is a long 20 km inlet entering the 13–10 million year old DUNEDIN VOLCANO. Dunedin city lies on the southwest flank of the volcano. Lavas, explosive beds and intrusions are exposed over distances of 30 km and there is a central vent near Port Chalmers. Good sections through the volcanic sequence are found along the highway travelling northeast from Dunedin to Port Chalmers and through to North Head.

Scott Monument and Lookout overlook the PORT CHALMERS VENT. This site is reached from a turnoff on the Dunedin–Port Chalmers highway at Sawyers Bay. The vent contains volcanic breccia full of blocks of phonolite, pieces of basalt and underlying basement schists torn off its walls. The rock was used as a building stone in Dunedin. From Port Chalmers, the road proceeds past Rocky Point, made of the oldest known basalt and trachyte lavas; Pulling Point, a slabby jointed nepheline hawaiite flow; and Taylor Point, a nepheline benmoreite intrusion with subvertical flow planes. The road then cuts through Otafelo Point, a trachyte dyke, to reach Aramoana at North Head.

Tiers of lavas and explosive beds forming North Head can be followed from Aramoana northwards along cliffs above Spit Beach. There are 25 different flows identified in these sections. The trachytes and trachyte breccias of an initial minor phase form the lower cliff at Aramoana. They are followed by flows 2 to 13 which marked the first main eruptive phase. At Spit Beach, flow 14 is exposed and marks the start of the second main eruptive phase. It is a prominent phonolite flow halfway up the section and descends northwards to beach level. It is overlain by phonolite scoria and then thinner basaltic lavas forming flows 15 to 25. Flows and intrusions of a third main eruptive phase are encountered elsewhere, along a detour from Port Chalmers north along Purakanui Road and then westward along Blueskin Road. Mt Mihiwaka, Mihiwaka South and Mt Kettle are phonolite dome intrusions. A carpark on the road about 9 km from Port Chalmers and 10 km from Dunedin leads to a walkway to the Organ Pipes at Mt Holmes and on to Mt Cargill. Mt Holmes is a columnar alkali basalt and Mt Cargill is a nepheline benmoreite dome. Examples of the varied rocks of Dunedin volcano can be viewed in the University of Otago, Geology Department.

NORTH ISLAND, NEW ZEALAND

North Island is renowned for its active volcanoes and myriad geothermal stirrings. Four active volcanoes include the Mt Tarawera and White Island volcanoes in the Rotorua–Bay of Plenty thermal region, and the imposing Ngauruhoe and Ruapehu volcanoes in the Tongariro region. The last two volcanoes require considerable climbs to reach their summits and enquiries about accommodation and park facilities should be made beforehand. Other areas also support vents that may have erupted in the last 250 to 800 years — for example, Mt Egmont, and Rangitoto Island in Auckland Harbour. The more accessible volcanoes around Auckland and Rotorua are described here.

AUCKLAND: CITY OF VOLCANOES TOUR

(Map p.223) Auckland is called 'the city of volcanoes' because its 48 vents add a natural excitement to the harbour and its environs. Some 34 vents form tuff rings, tuff cones and maars due to lava-water interactions, and 23 of the tuff rings enclose cones built by fire fountaining. Altogether 26 lava fields are capped by one or more scoria cones. In 15 lava fields and 10 minor flows, lava has spread beyond the original vent, while 7 flows remained in their vents. The largest volcano is Rangitoto Island and other prominent volcanoes include Mt Eden, One Tree Hill, Three Kings and Crater Hill. The volcanoes form one-off features, erupted over the last 60,000 years.

A few volcanoes form islands. Rangitoto volcano is linked by tracks to Motutapu Island. The track crosses the central scoria cone and circles around the southern shoreline over the lava fields. Rugged a'a lava flows show ridges curving in the direction of lava flow. The rubbly basalt is full of gas holes, and there are cavernous lava tubes southeast of the summit. Gritty dust from Rangitoto's eruption before AD 1200 showered Motutapu Island and buried the oldest known Maori settlement. Motukorea Island volcano is a smaller scoria cone-capped lava field southeast of Rangitoto. The compound scoria cone shows a well-developed central crater and part of a tuff ring.

Puketutu volcano in Manukau Harbour is linked to land by constructed causeways. Mt Mangere volcano to the northeast is only just attached to land. Mt Mangere scoria cone has a crater breached by outpouring lava and plugged by a lava dome. Large wrinkles in pahoehoe lava are seen on the foreshore.

One Tree Hill volcano forms the hub of the field, 6 km south of the city centre. It is reached from Manukau Road via ring roads and a walkway. The walkway from the south side passes between the two horseshoe-shaped craters, formed when copious lava carried away scoria. It ascends to the monument for John Campbell, a former mayor of Auckland, who is buried there. The monument overlooks the main crater on the west. Most of Three Kings Volcano, west of One Tree Hill, has been quarried, and only one of the original five scoria cones remains. However, a 10 km lava flow from the volcano flowed down into Waitemata Harbour where it extends out as Te Tokaro Reef.

Crater Hill volcano to the south, near Papatoetoe, is a complex volcano, with tuff rings, cones and lavas. Quarry operations reveal its explosive beds, of particular interest to specialists. The outer tuff ring shows lighter-coloured tuffs from steam explosions interbedded with darker basalt lapilli beds formed by Hawaiian-style fire fountaining. These are followed by basalt scoria deposited from Stromboli-style explosions. Faults indicate that the inner crater wall collapsed during eruptions, which include two phases of steam explosions and four phases of basaltic airfall eruptions. Manurewa volcano, another nearby southern volcano, is quarried for scoria at Wiri and has an interesting lava cave.

Those with limited time can stroll to the DOMAIN VOLCANO TUFF RING, beside the museum and close to the city centre, or visit the adjacent MT EDEN VOLCANO (next page). The MT EDEN CONE is terraced by Maori fortifications and imposing cooling columns in basalt are exposed in an old quarry behind Auckland Grammar School.

ROTORUA AND BAY OF PLENTY: GEOTHERMAL TOUR

(Map p.222) Rotorua is linked by domestic flights from international airports at Auckland, Wellington and Christchurch, or is reached via State Highway 5. This tourist centre near Sulphur Bay on south Lake Rotorua is known for its 'sulphurous' smell and therapeutic thermal baths. The adjacent WHAKAREWAREWA GEOTHERMAL FIELD boasts some famous features (photographs, p.40 and 212). PAREKOHORU THERMAL SPRING, a classic alkaline chloride spring, was called the Champagne Pool before it lost temperature and sparkle. POHUTU GEYSER still has some virility and the PRINCE OF WALES' FEATHERS GEYSER erupts in sympathy with Pohutu. Other geysers here include WAIKAROHINI GEYSER, MAHANGA GEYSER, KERERU GEYSER and WAIKITE GEYSER, the latter inactive since 1965. GEYSER FLAT SINTER DEPOSIT is built from these 7 vents, which are aligned along a fault. Many springs discharge along PUARENGA THERMAL STREAM, and NGAMOKAIAKOKO THERMAL MUD POOL fascinates visitors.

MOUNT TARAWERA VOLCANO is reached from Rotorua by taking State Highway 5 past Earthquake Flat and then turning east onto State Highway 38. A further 4 km to a turnoff east leads around Lake Rerewhakaaitu and 10,000 year old steam explosion caters, then descends the rim of HAROHARO CALDERA to Crater road. From here, tracks for 4-wheel-drive vehicles or walkers climb up RIDGE LAVA DOME and across KANAKANA LAVA DOME to the edge of the 1886 ERUPTION FISSURE (photograph, p.14). A northeast view to WAHANGA LAVA DOME exposes the dark basaltic deposits which in 1886 overlaid the paler rhyolite eruptives of these domes. This 1886 eruption created the Tarawera fissure and devastated some native villages and thermal

▽ *Conglomerates and pillow lavas of old uplifted submarine volcanoes exposed on the Waitakere Range coast, near Auckland, North Island, New Zealand, February 1986.*

tourist sites. The fissure craters can be skirted east past RUAWAHIA LAVA DOME, CRATER LAVA DOME and TARAWERA LAVA DOME. Descents into and ascents from the various 1886 craters can be made by any fit person who chooses the routes with care.

WAIMANGU GEOTHERMAL FIELD and TE WAIROA BURIED VILLAGE lie southwest of the 1886 Tarawera volcanic chasm near Lake Rotomahana. From Rotorua, State Highway 5 proceeds to Earthquake Flat and on to Waimangu Forest. Waimangu thermal reserve is entered from the southern tearooms carpark and the track passes several craters formed during the 1886 eruptions. SOUTHERN CRATER contains a cold lake. ECHO CRATER holds Frying Pan Lake, a large hot pool which discharges acid sulphate water at 55° C (photograph, p.58). INFERNO CRATER has a small lake linked to the discharge of Frying Pan Lake. Its level fluctuates by about 15 m and its temperature varies by nearly 50° C every six weeks or so. The track continues to LAKE ROTOMAHANA, which occupies several craters formed in the 1886 fissure eruption.

Lake Rotomahana is viewed by boat and the STEAMING CLIFFS lie near the famous Pink Terrace site that was destroyed by the eruption. Crossing the southwest flank of Mt Tarawera gives a view of the volcanic chasm. Another boat crosses Lake Tarawera to Te Wairoa, once the departure point for tours to the celebrated Pink and White Terraces on the original Lake Rotomahana. The falling mud from the 11 km high 1886 Lake Rotomahana eruption column destroyed Te Wairoa village. Excavations have exposed the sites of the hotel, flour mill, bakery, blacksmith's shop and several Maori huts.

A plane or helicopter flight from Rotorua around WHITE ISLAND VOLCANO (photograph, p.20–21) is an effortless way to salute this offshore vent. Depending on budget, flights can be extended to view the other volcanoes of the Taupo–Tongariro zone. Remember, flights for close views may be cancelled during major eruptions!

▷ Terraces deposited by thermal waters, Whakarewarewa, Rotorua, North Island, New Zealand, February 1986.

EARTH VOLCANIC

213

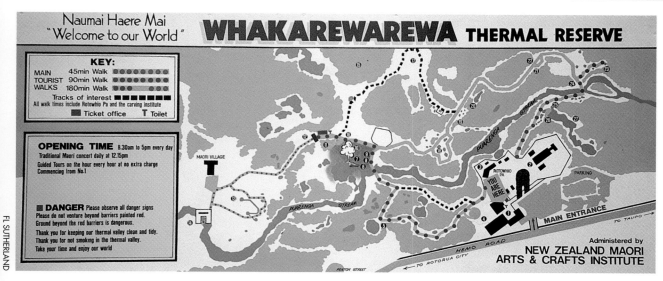

◁ Mt Eden crater, looking north, with Auckland city in background, North Island, New Zealand, February 1968.

△ Sign at Whakarewarewa Thermal Reserve, Rotorua, North Island, New Zealand, February 1986.

CHAPTER 13

APPENDIX 1: MAPS OF AUSTRALIA & NEW ZEALAND VOLCANIC AREAS

LIST OF MAPS

Eastern Australian volcanic trends	215
Volcanic features of the Australasian region	216
Western Victoria	217
Mt Gambier, South Australia	217
Devonport to Stanley, northwest coast of Tasmania	218
Hobart district, Tasmania	218
Warrumbungles area, New South Wales	219
Lord Howe Island, New South Wales	220
Glass House Mountains area, Queensland	220
Binna Burra area, Queensland	220
Rockhampton–Yeppoon area, Queensland	221
Atherton area, Queensland	221
Banks Peninsula, South Island, New Zealand	222
Dunedin/Otago Harbour, South Island, New Zealand	222
Rotorua area, North Island, New Zealand	222
Auckland area, North Island, New Zealand	223

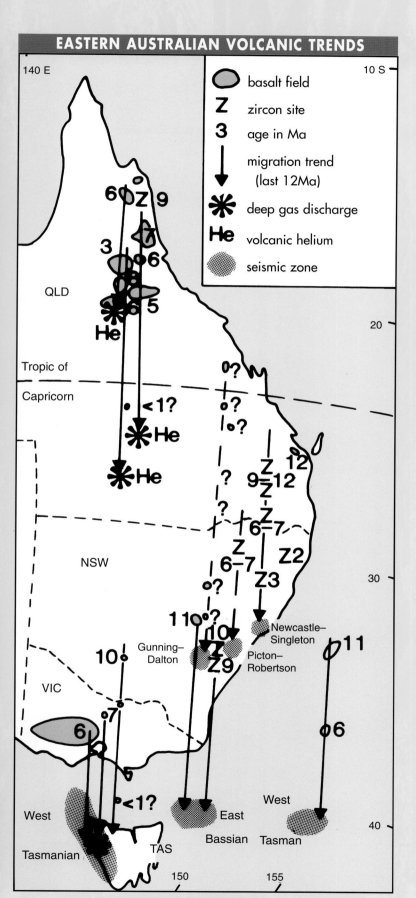

This map show posssible migration trends over the last 12 million years (Ma), present seismic zones, and deep discharges of volcanic gases over potential hotspot zones.

VOLCANIC FEATURES OF THE AUSTRALASIAN REGION

Areas covered in the maps on the following pages are indicated as small rectangles, with the relevant page numbers in this section given in red.

EARTH — VOLCANIC

WESTERN VICTORIA

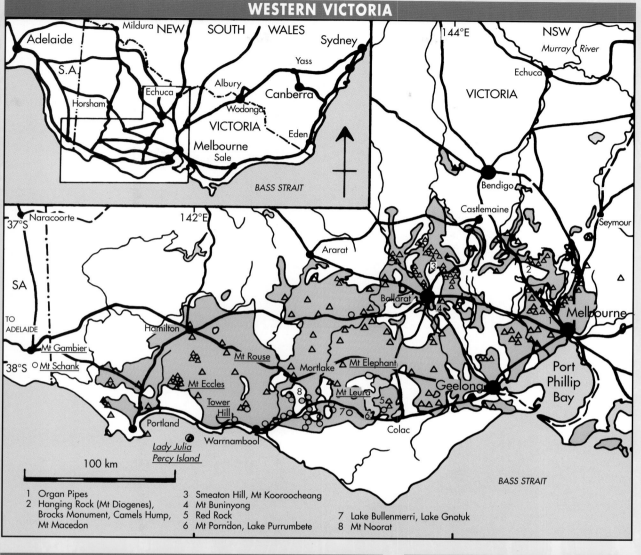

1. Organ Pipes
2. Hanging Rock (Mt Diogenes), Brocks Monument, Camels Hump, Mt Macedon
3. Smeaton Hill, Mt Kooroocheang
4. Mt Buninyong
5. Red Rock
6. Mt Porndon, Lake Purrumbete
7. Lake Bullenmerri, Lake Gnotuk
8. Mt Noorat

MT GAMBIER, SOUTH AUSTRALIA

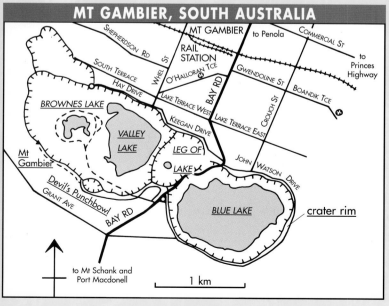

KEY

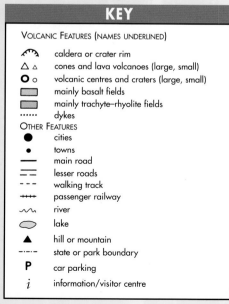

VOLCANIC FEATURES (NAMES UNDERLINED)
- caldera or crater rim
- cones and lava volcanoes (large, small)
- volcanic centres and craters (large, small)
- mainly basalt fields
- mainly trachyte–rhyolite fields
- dykes

OTHER FEATURES
- cities
- towns
- main road
- lesser roads
- walking track
- passenger railway
- river
- lake
- hill or mountain
- state or park boundary
- P car parking
- i information/visitor centre

APPENDIX 1

217

VOLCANIC EARTH

APPENDIX 1

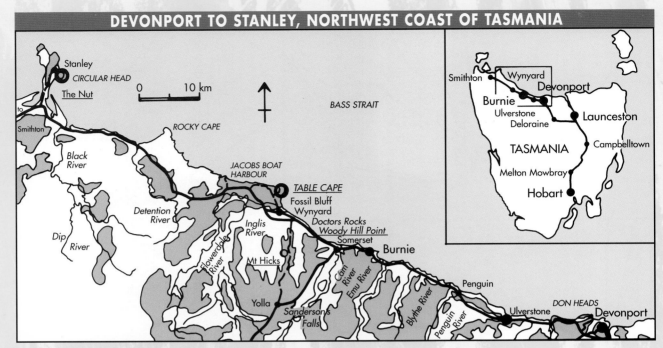

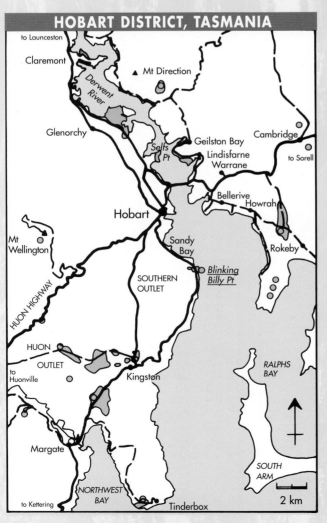

KEY

VOLCANIC FEATURES (NAMES UNDERLINED)
- caldera or crater rim
- △ △ cones and lava volcanoes (large, small)
- ○ ○ volcanic centres and craters (large, small)
- mainly basalt fields
- mainly trachyte–rhyolite fields
- ⋯⋯ dykes

OTHER FEATURES
- ● cities
- • towns
- ⎯ main road
- = lesser roads
- -- walking track
- ┼┼┼ passenger railway
- river
- lake
- ▲ hill or mountain
- — · — state or park boundary
- P car parking
- i information/visitor centre

218

WARRUMBUNGLES AREA, NEW SOUTH WALES

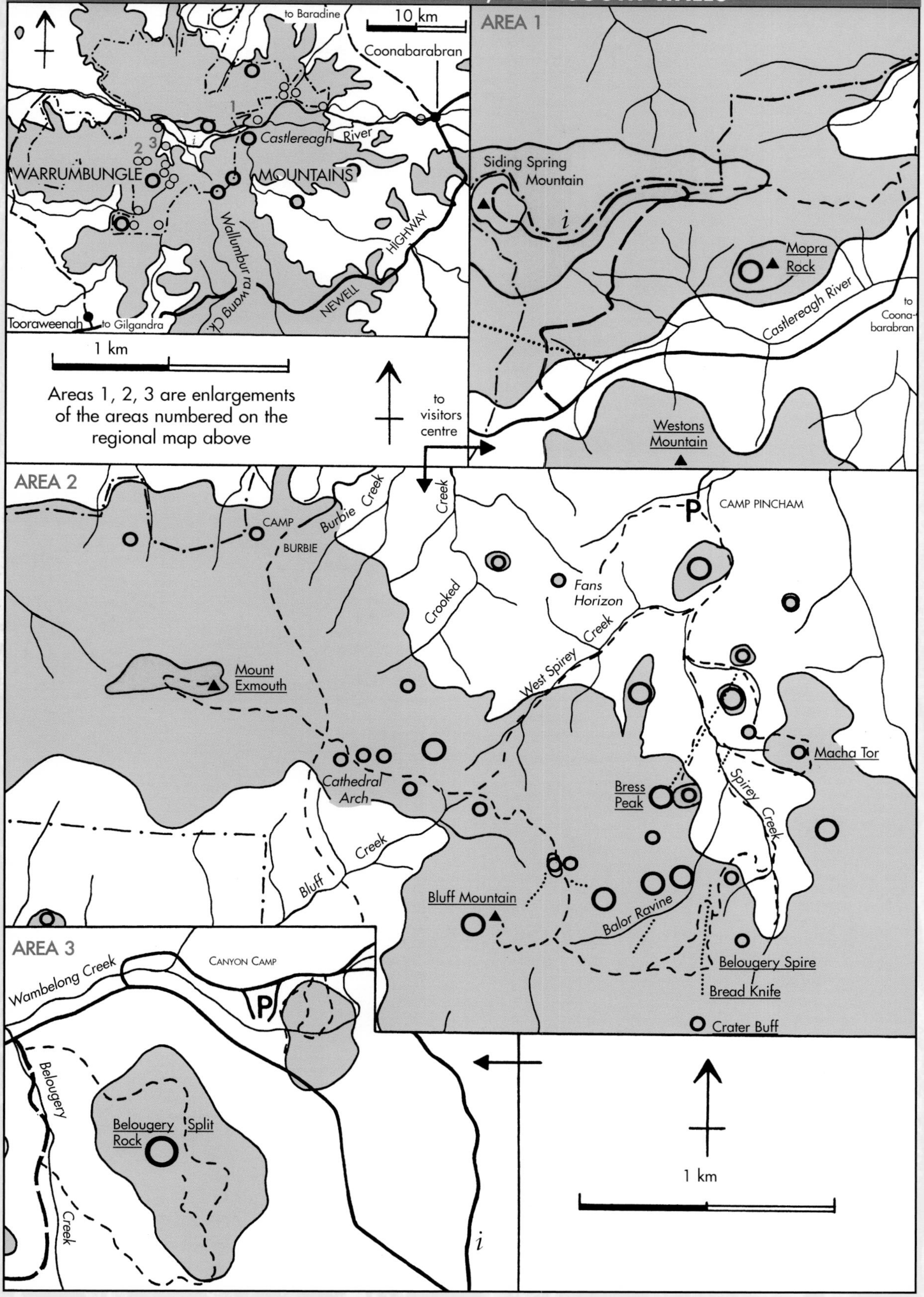

AREA 1
- Siding Spring Mountain
- Mopra Rock
- Castlereagh River
- to Coonabarabran
- Westons Mountain

Regional map labels:
- to Baradine
- Coonabarabran
- Castlereagh River
- WARRUMBUNGLE MOUNTAINS
- Wallumburrawang Ck
- NEWELL HIGHWAY
- Tooraweenah
- to Gilgandra

10 km

1 km

Areas 1, 2, 3 are enlargements of the areas numbered on the regional map above

to visitors centre

AREA 2
- CAMP BURBIE
- Burbie Creek
- Crooked Creek
- Fans Horizon
- West Spirey Creek
- CAMP PINCHAM
- Mount Exmouth
- Macha Tor
- Cathedral Arch
- Bress Peak
- Spirey Creek
- Bluff Creek
- Bluff Mountain
- Balor Ravine
- Belougery Spire
- Bread Knife
- Crater Buff

AREA 3
- Wambelong Creek
- CANYON CAMP
- Belougery Creek
- Belougery Split Rock

1 km

EARTH VOLCANIC

219

APPENDIX 1

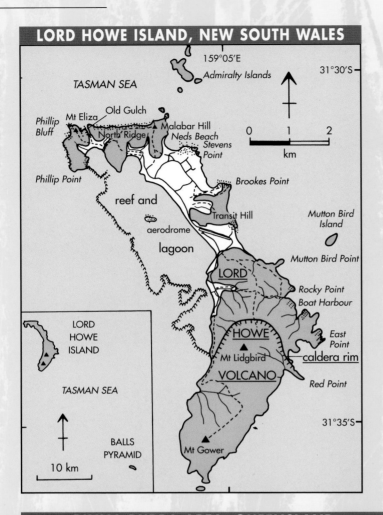
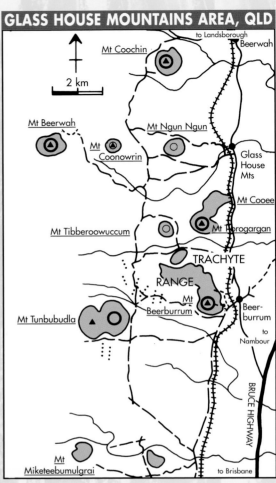

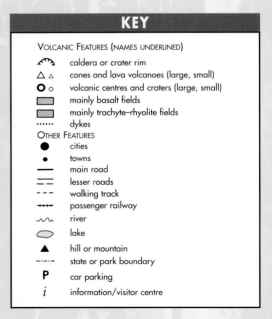

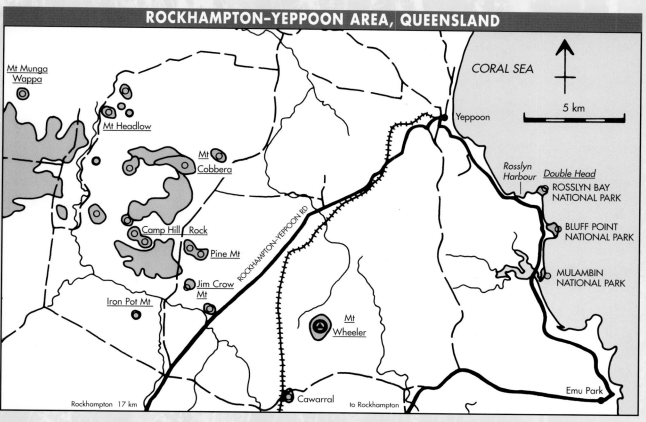
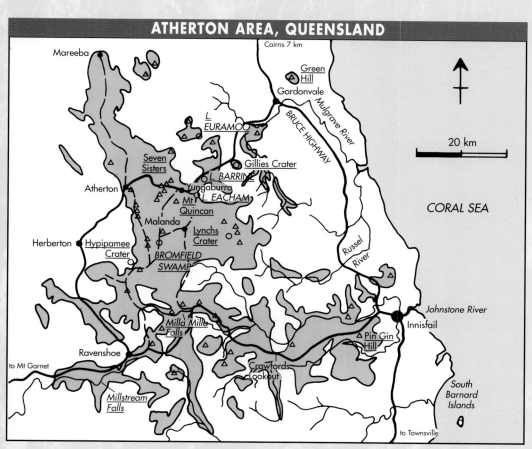

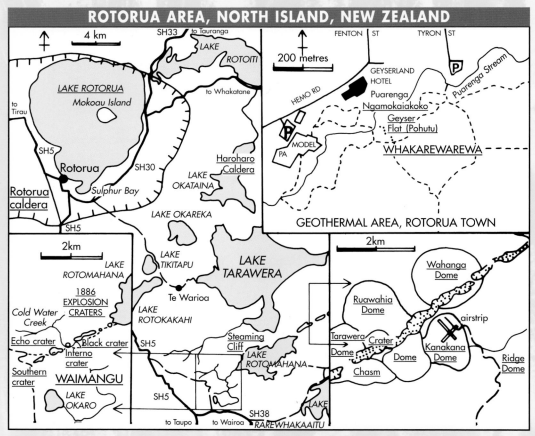

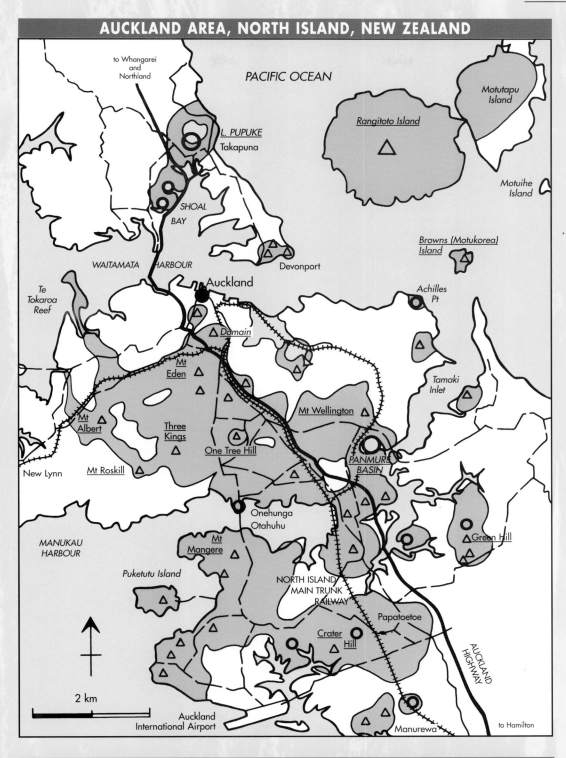

APPENDIX 2: SOME SIGNIFICANT ERUPTIONS

DATE	VOLCANO, COUNTRY	TYPE OF ERUPTION	DEATHS	MAIN EFFECTS
circa 1620 BC	Santorini Island, Greece	major explosive, caldera forming	unknown 30,000 fled	buried Akrotiri, a Minoan city on possibly founded Atlantis legen
Aug 24 AD 79	Mt Vesuvius, Italy	major explosive, caldera forming	2,000 30,000 fled	devastated and buried Pompeii, He and Stabiae, until first uncovered
circa 186	Lake Taupo, New Zealand	major explosive, caldera forming	unknown, uninhabited?	50 km high eruption column, a flow covered 20,000 square km
circa 300	Ilopango, El Salvador	major explosive, caldera forming	unknown people fled	displaced highland Mayan Indians thousands to hundreds of thous
Feb 4 1169	Mt Etna, Italy	explosive fissure, lava flows	16,000	ash falls, ash flows, lavas
circa 1586	Kelud, Indonesia	explosive, crater lake eruption	10,000	mud flows from deep lake eject
Dec 16 1631	Mt Vesuvius, Italy	major explosive, lava flows	6,000	ash flows, avalanches, mud flows, Torre del Greco and Torre Annur
circa 1638	Raung, Indonesia	explosive and hydrological	1,000 to thousands	mud flows and ash flows
Mar 11 1669	Mt Etna, Sicily	major explosive, lava flows	20,000	ash falls, ash flows, lavas partly destroyed Catania and villages
Aug 4 1672	Merapi, Indonesia	explosive, glowing clouds	3,000	ash flows
Dec 10 1711	Awu, Indonesia	explosive, crater lake eruption	3,177	mud flows
June 15 1742	Cotopaxi, Ecuador	explosive, melting ice	800	mud flows, floods
Sept 22 1760	Makian, Indonesia	explosive, lava flows	2,000	mud flows
Aug 11 1772	Papandajan, Indonesia	explosive, steam explosions	2,957	avalanche, mud flows
July 26 1783	Asama, Japan	explosive, glowing clouds	1,200	ash and block flows, mud flows
June 8 1783	Laki, Iceland	major fissure, lava flows	10,521 from famine	15 cubic km lava, 280,000 livest killed, fishing and climate affect
May 21 1792	Unzen, Japan	earthquake, dome collapse	14,524 many drowned	avalanche caused huge wave, sw ing peninsula and opposing coa
Apr 10 1815	Tambora, Indonesia	major explosive, caldera forming	66,000 many indirect	ash flows devastated Sumbawa a Caused 1816 'Year without a Su
Oct 8 1882	Galunggung, Indonesia	explosive	4,011	ash flows, villages destroyed
Aug 27 1883	Krakatoa, Indonesia	major explosive, caldera formed	96,400 many drowned	Krakatoa blasted apart, 30 m hig swept Sunda Strait, global clima
Mar 2 1856	Awu, Indonesia	explosive, crater lake eruption	2,806	mud flows
June 10 1886	Mt Tarawera, New Zealand	major fissure, lake eruption	100	ash and mud fall, Te Wairoa villa tourist geothermal features dest
July 15 1888	Bandai-San, Japan	steam blast, avalanche	461	villages buried, lakes formed be hummocky terrain of avalanche
June 7 1892	Awu, Indonesia	explosive, crater lake eruption	1,532	mud flows
May 7 1902	La Soufrière, St Vincent	explosive, glowing cloud	1,600	ash and block flow, mud flow
May 8 1902	Mt Pelée, Martinique	explosive, glowing cloud	29,000	ash and block flows, ash hurric St Pierre and harbour destroye
Oct 24 1902	Santa Maria, Guatemala	major explosive flank eruption	thousands	28 km high column, ash fall ove square km, coffee industry rui
Jan 27 1911	Taal, Philippines	explosive, crater lake eruption	1,334	outward blast destroyed village bank, ash covered Volcano Isla

EARTH VOLCANIC

DATE	VOLCANO, COUNTRY	TYPE OF ERUPTION	DEATHS	MAIN EFFECTS
June 6 1912	Novarupta, Alaska, USA	major explosive, caldera forming	none, uninhabited	ash fall, Kodiak blacked out, ash flow created Valley of Ten Thousand Smokes
May 19 1919	Kelud, Indonesia	explosive, crater lake eruption	5,110	mud flows from lake ejection, over 100 villages destroyed
circa 1931	Merapi, Indonesia	explosive	1,369	ash flow
May 29 1937	Rabaul Papua Nuigini	explosive, steam explosion	505	ash flows from Vulcan cone, steam blasts from Tavurvur fissure
Feb 20 1943	Parícutin, Mexico	explosive, lava flows	none	400 m high cone, Parícutin and San Juan Parangicutaro buried by lava, evacuations
circa 1951	Hibok-Hibok, Philippines	explosive	500	ash flows, mud flows
Jan 21 1951	Mt Lamington, Papua Nuigini	explosive, glowing clouds	2,942	ash and block flows, devastating 180 sq km
Dec 24 1953	Ruapehu, New Zealand	collapsed crater lake rim	157	mud flow, railway bridge swept away, derailing express train
Mar 30 1956	Bezymianny, Kamchatka, Russia	major explosion, flank collapse	none	mud flows, ash fall, avalanche, devastated 500 square km
Mar 17 1963	Mt Agung, Indonesia	explosive, glowing clouds	1,184	ash flows, mud flows
July 29 1968	Arenal, Costa Rica	explosive, block and bomb showers	78	area between Tabocon and Pueblo devastated 4,000 evacuated
July 6 1975	Ploskiy Tolbachik, Kamchatka, Russia	explosive fissure, caldera forming	none	13 km high cloud, cones and lavas, eruption date predicted
May 18 1980	Mt St Helens, Washington, USA	major explosion, flank collapse	60	660 square km blast, avalanche, mud flows, ash flows, ash fall
Mar 28 1982	El Chichón, Mexico	major explosive 3 explosions	2,000	ash flow destroyed El Naranjo, 25 km high cloud, global climate impact.
Apr 5 1982	Galunggung, Indonesia	major explosive	none	ash flows, ash falls, 80,000 evacuated, 20 km high column affected air flights
Jan 3 1983	Pu'u'O'o, Hawaii, USA	fissure vent and lava flows	1	lavas invaded visitors centre, cultural sites, Kalapana and Kaimu Bay, still flowing 1994
July 23 1983	Colo Indonesia	explosive evacuations	none	ash flows, 7,000 evacuated, coconut plantations destroyed
Sept 1984	Mayon, Philippines	explosive, lava flows	none	73,000 evacuated
Nov 13 1985	Nevado del Ruiz Colombia	small explosion, melt waters	22,000	mud flow down Lagunellas River, Amero township buried
Aug 21 1986	Lake Nyos, Cameroon	gas cloud from lake overturn	1,700	carbon dioxide cloud, people and 3000 cattle suffocated
June 3 1991	Mt Unzen, Japan	explosive, glowing cloud	42	ash and block flows, partly destroyed Kamikoba, 10,000 evacuated
June 15 1991	Mt Pinatubo, Philippines	major explosive, caldera forming	435	30 km high column, ash flows, mud flows, 200,000 evacuated, global climate impact
Aug 12 1991	Mt Hudson, Chile	major explosive, lava flow	none	18 km high column, ash falls, devastated livestock, global climate impact, evacuations
Dec 14 1991	Mt Etna, Italy	flank fissure, lava flow	none	flowed for 473 days until Mar 30 1993, the longest flow at Etna for 300 years
Feb 2 1993	Mayon, Philippines	explosive	70	ash flow, over 60,000 evacuated
Sept 19 1994	Rabaul Papua Nuigini	explosive, from two vents	5	ash falls, base surges, floating pumice, 50,000 evacuated, Rabaul part-destroyed
Nov 23 1994	Merapi, Indonesia	explosive, lava flows	40	steam blast, 5,000 evacuated

APPENDIX 3: TYPICAL MINERALS OF VOLCANIC ROCKS

MINERALS	COMPOSITION	TYPE*	MAIN ROCKS
NATIVE ELEMENTS			
sulphur	sulphur	Reg	mostly volcanic vents
OLIVINE GROUP			
olivine	magnesium, iron silicate	Reg	basalts, nephelinites, kimber...
monticellite	calcium, magnesium silicate	Odd	olivine melilitites
PYROXENE GROUP			
CLINOPYROXENE SUBGROUP			
diopside	calcium magnesium silicate	Reg	most volcanic rocks
augite	calcium, magnesium, iron, alumino-silicate	Reg	most volcanic rocks
aegirine	sodium, iron silicate	Par	evolved alkaline rocks
pigeonite	magnesium, iron, calcium silicate	Par	andesites, dacites, basalts
ORTHOPYROXENE SUBGROUPS			
enstatite	magnesium silicate	Odd	basalts
hypersthene	magnesium, iron silicate	Par	basalts, andesites, dacites
FELDSPAR GROUP			
PLAGIOCLASE SUBGROUP			
plagioclase (calcic)	calcium, sodium, alumino-silicate	Reg	basalts, andesites, dacites
plagioclase (sodic)	sodium, calcium, alumino-silicate	Reg	andesites, dacites, evolved r...
ALKALI FELDSPAR SUBGROUP			
sanidine	potassium sodium, alumino-silicate	Par	evolved alkaline rocks
anorthoclase	sodium, potassium, alumino-silicate	Par	evolved alkaline rocks
FELDSPATHOID MINERALS			
nepheline	sodium, potassium, alumino-silicate	Par	basanites, nephelinites, evo... rocks
leucite	potassium, alumino-silicate	Par	leucitites, lamproites, shosh...
melilite	calcium, magnesium, iron, alumino-silicate	Odd	melilitites, nephelinites
sodalite	sodium, alumino-chloro silicate	Odd	phonolites, nephelinites, me...
analcime	sodium, alumino-hydro silicate	Sec	basalts, analcimites, phono...
SPINEL GROUP			
magnetite	iron oxide	Reg	most volcanic rocks
ulvospinel	iron, magnesium oxide	Reg	most volcanic rocks
spinel	magnesium, iron, aluminium oxide	Par	basalts, nephelinites
IRON-TITANIUM OXIDE MINERALS			
ilmenite	iron, titanium oxide	Reg	most volcanic rocks
QUARTZ MINERALS			
high quartz	silicon oxide (formed above 573°C)	Par	trachytes, dacites, rhyolites
low quartz	silicon oxide (formed below 573°C)	Sec	altered volcanic rocks
chalcedony	silicon oxide	Sec	altered volcanic rocks
opal	silicon hydro oxide	Sec	altered volcanic rocks
ZIRCONIUM MINERALS			
zircon	zirconium silicate	Par	evolved alkaline rocks
AMPHIBOLE GROUP			
hornblende	calcium, magnesium, iron, alumino-hydro silicate	Par	andesites, evolved rocks, lamprophyres
kaersutite	calcium, magnesium, iron, titanium-hydro silicate	Par	basalts, evolved rocks, lamprophyres
riebeckite	sodium, iron hydroxy silicate	Par	evolved alkaline rocks

EARTH VOLCANIC

MINERALS	COMPOSITION	TYPE*	MAIN ROCKS
MICA GROUP			
phlogopite	potassium, magnesium, iron, alumino-hydro silicate	Par	andesites, evolved rocks, lamprophyres
biotite	potassium, iron, magnesium, alumino-hydro silicate	Par	basalts, evolved rocks, lamprophyres
APATITE GROUP			
fluorapatite	calcium fluoro-phosphate	Reg	most volcanic rocks
CALCITE GROUP			
calcite	calcium carbonate	Odd	carbonatites
		Sec	altered volcanic rocks
SIDERITE GROUP			
siderite	iron carbonate	Sec	altered volcanic rocks
DOLOMITE GROUP			
dolomite	calcium, magnesium carbonate	Odd	carbonatites
		Sec	altered volcanic rocks
ARAGONITE GROUP			
aragonite	calcium carbonate	Sec	altered volcanic rocks
ZEOLITE GROUP			
laumontite	calcium, alumino-hydro silicate	Sec	altered volcanic rocks
heulandite	calcium, sodium, alumino-hydro silicate	Sec	altered volcanic rocks
stilbite	calcium, sodium, alumino-hydro silicate	Sec	altered volcanic rocks
stellerite	sodium, calcium, alumino-hydro silicate	Sec	altered volcanic rocks
phillipsite	calcium, sodium, potassium, alumino-hydro silicate	Sec	altered volcanic rocks
natrolite	sodium, alumino-hydro silicate	Sec	altered volcanic rocks
thomsonite	sodium, calcium, alumino-hydro silicate	Sec	altered volcanic rocks
chabazite	calcium, sodium alumino-hydro silicate	Sec	altered volcanic rocks
gmelinite	sodium, calcium alumino-hydro silicate	Sec	altered volcanic rocks
gonnardite	sodium, calcium, alumino-hydro silicate	Sec	altered volcanic rocks
offretite	potassium, calcium, alumino-hydro silicate	Sec	altered volcanic rocks
levyne	calcium, sodium, potassium, alumino-hydro silicate	Sec	altered volcanic rocks
erionite	potassium, calcium, sodium, alumino-hydro silicate	Sec	altered volcanic rocks
ferrierite	sodium, potassium, magnesium, alumino-hydro silicate	Sec	altered volcanic rocks
CLAY MINERALS			
kaolinite	alumino-hydro silicate	Sec	altered volcanic rocks
nontronite	iron, magnesium, calcium, alumino-hydro silicate	Sec	altered volcanic rocks
vermiculite	magnesium, iron, alumino-hydro silicate	Sec	altered volcanic rocks
OTHER MINERALS			
chlorite	magnesium, aluminium, iron hydro silicate	Sec	altered volcanic rocks
apophyllite	calcium, potassium, fluoro-hydro silicate	Sec	altered volcanic rocks
tacharanite	calcium hydro silicate	Sec	altered volcanic rocks

*ABBREVIATIONS: **Reg** REGULARS, **Par** PART-TIMERS, **Odd** ODD ONES, **Sec** SECONDARY

APPENDIX 3

BIBLIOGRAPHY

CATEGORIES

A   FOR ADVANCED STUDENTS AND ACADEMIC READERS.
B   FOR BEGINNING STUDENTS AND GENERAL READERS.
C   FOR CHILDREN AND CASUAL READERS.

INTRODUCTORY

B   Bevens D (ed) 1992. *On the Rim of Kilauea: Excerpts from the Volcano House Register 1865–1955*. Hawaii Natural History Association, Hawaii, 168p.

B   BMR Palaeogeographic Group 1990. *Australia: Evolution of a Continent*. Bureau of Mineral Resources, Australia, 97p.

A   Brown G C & Mussett A E 1993. *The Inaccessible Earth*. 2nd edn, Chapman & Hall, London, 276p.

B   Bruce C 1980. *Eugen von Guérard*. Australian Gallery Directors Council, Canberra, 139p.

A   Cattermole P 1990. *Planetary volcanism: A study of volcanic activity in the solar system*. Ellis Horwood, Chichester, 443p.

B   Chaikin A 1994. While we weren't watching: Apollo's Scientific Exploration of the Moon. *The Planetary Report*, May/June, 14(3), 6–9.

B   Clark I F & Cook B J 1983. *Geological Science Perspectives of the Earth*. Australian Academy of Sciences, Canberra, 651p.

B   Descoeudres J-P 1994. Pompeii Revisited: The Life and Death of a Roman Town. Meditarch, University of Sydney, 183p, 16pl.

B   Doig F (ed) 1988. Tracks Through Time: the story of human evolution. *Australian Natural History* Supplement 2. The Australian Museum Trust, Sydney, 58p.

B   Etienne R 1992. Pompeii The Day a City Died. New Horizons, Thames & Hudson, London, 215p.

B   Heller S & Wallin D 1981. *Volcano. The most terrifying disaster of all...* Hamlyn, London, 301p.

B   Horgan J 1991. In the Beginning ... *Scientific American*, Feb, 264(2), 100–9.

B   Johanson D C & Edey M A 1990. *Lucy: The beginnings of humankind*. Penguin, London, 413p.

C   Joyce B 1992. *Volcanoes*. Bookshelf Publishing Australia, Gosford, Sydney, 32p.

A   McElhinny M W 1979. *The Earth: Its Origin, Structure and Evolution*. Academic Press, London, 597p.

C   McManus P (ed) 1988. *Abracadabra Earthquakes & Volcanoes*. Sphere Books, London, 63p.

B   Morrison R & M 1988. *The Voyage of the Great Southern Ark. The 4 Billion Year Journey of the Australian Continent*. Lansdowne Press, Sydney, 334p.

B   Prunster U 1983. *Shay Docking: The Landscape as a Metaphor*. A H & A W Reed, Sydney, 144p.

B   Renaut R W & Tiercelin J-J 1993. Lake Borgia, Kenya: soda, hotsprings and about a million flamingoes. *Geology Today*, Mar/Apr, 9(2), 56–61.

B Scott A C 1991. Geology on stamps: mountains of fire. *Geology Today*, Jan/Feb, 7(1), 28–9.
B Smith D G (ed) 1982. *The Cambridge Encyclopedia of Earth Sciences*. Cambridge University Press, Cambridge, 496p.
B Somerville D 1994. El Salvador's buried roots. *New Scientist*, 21 May, 30–3.
B Sutherland F L 1991. *Gemstones of the Southern Continents*. Reed Books, Sydney, 256p.
A Torrence R, Specht J, Fullagar R & Bird R 1992. From Pleistocene to Present: Obsidian Sources in West New Britain, Papua New Guinea. *Records of the Australian Museum Supplement*, 15, 83–98.
B Trevelyan R 1976. *The Shadow of Vesuvius: Pompeii* AD *79*. Folio Society, London. 128p.
A Veevers J J 1984. *Phanerozoic earth history of Australia*. Clarendon, Oxford, 418p.
B Vickers-Rich P & Rich T H 1993. *Wildlife of Gondwana*. Reed Books, Sydney, 276p.
B White M E 1991. *Time in our Hands*. Reed Books, Sydney, 191p.
B White M E 1994. *After the Greening the Browning of Australia*. Kangaroo Press, Kenthurst, 288p.

GENERAL VOLCANOES

B Anderson I 1987. Volcanoes in Paradise. *New Scientist*, 2 July, 115(1567), 50–4.
B Bolt B A (Introducer) 1980. Earthquakes and volcanoes. *Scientific American*, W H Freeman & Co, San Francisco, 154p.
A Bullard F M 1977. *Volcanoes of the Earth*, Rev. ed. University of Queensland Press, Brisbane, 579p.
A Cas R A F & Wright J V 1987. *Volcanic Successions Modern and Ancient*. Allen and Unwin, London, 528p.
B Coffin M F & Eldholm O 1993. Large Igneous Provinces. *Scientific American*, 269(4), 26–33.
B Daniels G G (ed) 1982. *Planet Earth Volcanoes*. Time Life Books, Amsterdam, 176p.
B Decker R & Decker B (Introducers) 1982. Volcanoes and the Earth's interior. *Scientific American*, W H Freeman & Co, San Francisco, 141p.
B Decker R W & Decker B B 1991. *Mountains of Fire. The Nature of Volcanoes*. Cambridge University Press, Cambridge, 198p.
A Francis P 1993. *Volcanoes: A Planetary Perspective*. Clarendon Press, Oxford, 443p.
B Grove N 1992. Volcanoes: Crucibles of Creation. *National Geographic*, Dec., 182(6), 5–41.
B Gudmundsson A T & Kjartansson H 1984. *Guide to the Geology of Iceland*. Bókaútgafan Örn og Örlygur hf, Reykjavik, Iceland, 88p.
A Johnson R W 1981. Cooke-Ravin Volume of Volcanological Papers. *Geological Survey of Papua New Guinea Memoir 10*, 265p.
B Krafft M 1993. *Volcanoes: Fire from the Earth*. Discoveries, Harry N Abrams, New York, 207p.
B Krafft M & Krafft K 1980. *Volcanoes: Earth's Awakening*. Hammonds, Maplewood, N J, 160p.
B Krauss B 1992. *Birth by Fire: A guide to Hawaii's volcanoes*. Island Heritage Publishing, Hawaii, 90p.
A LeMasurier W E & Thomson J W (eds) 1990. *Volcanoes of the Antarctic Plate and Southern Oceans*. Antarctic Research Series 48. American Geophysical Union, Washington DC, 487p.
A Lockley M G & Rice A (eds) 1990. Volcanism and fossil biotas. *Geological Society of America Special Paper 244*, Boulder, Colorado, 125p.
B Ollier C 1969. *Volcanoes*. Australian National University Press, Canberra, 177p.
B Oppenheimer C 1993. Mines in the sky. *Geology Today*, Mar/Apr 9(2), 66–8.
B Rittmann A & Rittmann L 1976. *Volcanoes*. Orbis Publishing, London, 128p.
A Roberts W L, Campbell T J & Rapp G R Jr. 1990. *Encyclopedia of Minerals* (2nd edn). Van Nostrand Reinhold Company, New York, 979p.
C Rose S van 1993. *Volcano Eyewitness Guides*. Dorling Kindersley, London, 64p.
B Rose S van & Mercer I 1991. *Volcanoes*. British Museum (Natural History), London, 60p.
B Simpkin T, Tilling R I, Taggart J N, Jones W J & Spall H 1989. *This Dynamic Planet, World Map of Volcanoes, Earthquakes and Plate Tectonics*. US Geological Survey & Smithsonian Institution, Boulder Co & Washington DC.
A Sutherland F L 1994. Volcanism around K/T boundary time – its rôle in an impact scenario for the K/T extinction events. *Earth-Science Reviews*, 36, 1–26.
A Wood C A & Kienle J (eds) 1990. *Volcanoes of North America: United States and Canada*. Cambridge University Press, Cambridge, 354p.

SPECIAL VOLCANOES

B Clague D A & Heliker C 1992. The Ten-Year Eruption of Kilauea Volcano. *Earthquakes & Volcanoes*, 23(6), 244–54.
A Clough B J, Wright J V & Walker G P L 1981. An unusual bed of giant pumice in Mexico, *Nature*, 289, 49–50.
A De Vivo B, Scandone R & Trigila R (eds) 1993. Mount Vesuvius Special Issue. *Journal of Volcanology and Geothermal Research*, 58, 1–381.
A Fedotov S A & Markhinin Ye K 1983. *The Great Tolbachik Fissure Eruption geological and geophysical data 1975–1976*. Cambridge University Press, Cambridge, 341p.
B Findley R 1981. St Helens. Mountain with a death wish. *National Geographic*, Jan, 159(1), 2–65.
B Fink J 1993. Down under the volcano. *Nature*, 366, 108.
A Gillot P-Y, Lefèvere J C & Nativel P-E 1994. Model for the structural evolution of the volcanoes of Réunion Island. *Earth and Planetary Science Letters*, 122, 291–302.
B Hoffer W 1982. *Volcano: The Search for Vesuvius*. Summit Books, New York, 189p.
B Johnson R W & Threlfall N A 1985. *Volcano Town The 1937–43 Rabaul Eruptions*. Robert Brown & Associates, Burand, Qld, 151p.
A Koyaguchi T & Tokuno M 1993. Origin of the giant eruption cloud of Pinatubo, June 15, 1991. *Journal of Volcanology and Geothermal Research*, 55, 85–96.
A Lipman P W & Mullineaux D R (eds) 1981. The 1980 Eruptions of Mount St Helens, Washington. *US Geological Survey Professional Paper 1250*. Washington, DC, 844p.
B Monteath C 1993. Erebus The Ice Dragon Revisited. *GEO Australia*, 15(3), 122–35.
B Moore J G, Normark W R & Holcomb R T 1994. Giant Hawaiian Underwater Landslides. *Science* 264, 46–7.
A Sheridan M F, Barberi F, Rosi M & Santacroce R 1981. A model for Plinian eruptions of Vesuvius. *Nature*, 289, 282–5.
B Simkin T & Fiske S 1983. *Krakatau 1883: the volcanic eruption and its effects*. Smithsonian Institution Press, Washington DC, 464p.
B Tazieff H 1975. *Nyiragongo The Forbidden Volcano*. Cassell, London, 287p.
A Thordarson T & Self S 1993. *The Laki (Skaftár Fires) and Grímsvötn eruptions in 1783–1785*. Bulletin of Volcanology, 55, 233–63.
A Walker G P L 1992. Puu Mahana Near South Point in Hawaii is a Primary Surtseyan Ash Ring, Not a Sandhills-type Littoral Cone. *Pacific Science*, 46(1), 1–10.
B Yanagi H, Okada H & Ohta K 1992. *Unzen Volcano, The 1990–1992 Eruption*, The Nishinippon & Kyushu University Press, Fukuoka, 137pp.

VOLCANIC HAZARDS

A   Aramaki S 1991. Hazardous volcanic eruptions in Japan. *Episodes*, 14(3), 264–8.

B   Berreby D 1991. Barry versus the Volcano. *Discover*, 12(6), 60–7.

A   Blong R J 1984. *Volcanic hazards: a sourcebook on the effects of eruptions*. Academic Press, New York, 424p.

A   Boyd F R Jun (ed) 1984. *Explosive Volcanism: Inception, Evolution, and Hazards. Studies in Geophysics*. National Academy Press, Washington DC, 176p.

B   Dvorak J J 1992. Tracking the Movement of Hawaiian Volcanoes: Global Positioning System (GPS) Measurements. *Earthquakes & Volcanoes*, 23(6), 255–267.

A   Francis P, Oppenheimer C & Stevenson D 1993. Endogenous growth of persistently active volcanoes. *Nature*, 366, 554–7.

A   Glaze L, Francis P W & Rothery D A 1989. Measuring thermal budgets of active volcanoes by satellite remote sensing. *Nature*, 338, 144–6.

A   McCall G J H, Laming D J C & Scott S C (eds) 1992. *Geohazards*. Chapman & Hall, London, 227p.

B   Oppenheimer C 1993. I spy with my infrared eye. *New Scientist*, 5 June, 138(1876), 27–31.

B   Rothery D 1991. Japan: Eruption of Unzen claims lives, *Geology Today*, 7(5), Sept/Oct, 170–171.

A   Smith I E M & Allen S R 1993. Volcanic Hazards at the Auckland Volcanic Field. *Volcanic Hazards Information Series No.5*, Ministry of Civil Defence, Wellington, 32p.

A   Takano B, Ohsawa S, Glover R B 1993. Surveillance of Ruapehu Crater Lake, New Zealand by aqueous polythionates. *Journal of Volcanological and Geothermal Research*, 60, 29–57.

B   Torrence R, Specht J & Fullagar R 1991. Pompeiis in the Pacific. *Paradise*, 87, 11–4.

B   Wright T L, Takahashi T J & Griggs J D 1992. *Hawai'i Volcano Watch. A Pictorial History, 1779–1991*. University of Hawaii Press, Hawaii, 162p.

ERUPTIONS AND CLIMATE

A   Ackerman M, Lippens C & Lechevallier M 1980. Volcanic material from Mount St Helens in the stratosphere over Europe. *Nature*, 287, 614–15.

A   Arnold F, Bührke Th & Qiu S 1990. Evidence for stratospheric ozone-depleting heterogeneous chemistry on volcanic aerosols from El Chichón. *Nature*, 348, 49–50.

A   Brasseur G & Granier C 1992. Mount Pinatubo Aerosols, Chlorofluorocarbons, and Ozone Depletion. *Science*, 257, 1239–42.

A   Bekki S, Toumi R & Pyle J A 1993. Role of sulphur photochemistry in tropical ozone changes after the eruption of Mount Pinatubo. *Nature*, 362, 331–3.

A   Bernard A, Demaiffe D, Mattielli N & Punongbayan R S 1991. Anhydrite-bearing pumices from Mount Pinatubo: further evidence for the existence of sulphur-rich silicic magmas. *Nature*, 354, 139–40.

B   Gerlach T 1991. Etna's greenhouse pump. *Nature*, 351, 352–3.

A   Gleason J F, Bhartia P K, Herman J R, McPeters R, Newman P, Stolarski R S, Flynn L, Labow G, Larko D, Seftor C, Wellemeyer C, Komhyr W D, Miller A J & Planet W 1993. Record Low Global Ozone in 1992. *Science*, 260, 523–6

B   Gribbin J 1993. Hot dust threatens ozone above tropics. *New Scientist*, 2 January, 137(1854), 14.

A   Handler P 1989. The effect of volcanic aerosols on global climate. *Journal of volcanology and Geothermal Research* 37, 233–49.

B   Harrington C R 1992. *The year without a summer? World Climate in 1816*. Canadian Museum of Nature, Ottawa, 576p.

A   Hofmann D J, Oltmans S J, Harris J M, Solomon S, Deshler T & Johnson B J 1992. Observation and possible causes of new ozone depletion in Antarctica in 1991. *Nature* 359, 283–7.

B   Johnson R W 1993. Volcanic Eruptions & Atmospheric Change. *AGSO Issues Paper* No.1, Australian Geological Survey Organisation, Canberra, 36p.

B   Kerr R A 1989. Research News: Volcanoes Can Muddle the Greenhouse. *Science*, 245, 127–8.

B   Kerr R A 1993. Out of Fire, Ice? *Science*, 260, 1725.

A   Luhr J F 1991. Volcanic shade causes cooling. *Nature*, 354, 104–5.

A   Minnis P, Harrison E F, Stowe L L, Gibson G G, Denn F M, Doelling D R & Smith W L Jr. 1993. Radiative climate forcing by the Mount Pinatubo Eruption. *Science*, 259, 1411–15.

B   Pang K D 1991. The legacies of eruption. Matching traces of ancient volcanism with chronicles of cold and famine. *The Sciences*. New York Academy of Sciences, Jan/Feb, 31(1), 30–5.

A   Symonds R B, Rose W I & Reed M H 1988. Contribution of Cl- and F-bearing gases to the atmosphere by volcanoes. *Nature*, 334, 415–18.

A   Tabazadeh A & Turco R P 1993. Stratospheric Chlorine Injection by Volcanic Eruptions: HCl Scavenging and Implications for Ozone. *Science*, 260, 1082–5.

B   Waters T 1992. Earth Beat, Global Cooling. *Earth*, Nov, 16–8.

A   Zielinski G A, Mayewski P A, Meeker L D, Whitlow S, Twickler M S, Morrison M, Meese D A, Gow A J & Alley R B 1994. Record of Volcanism Since 7000 BC from the GISP 2 Greenland Ice Core and Implication for the Volcano–Climate System. *Science*, 264, 948–51.

VOLCANIC ROCKS, MINERALS AND INCLUSIONS

A   Augustithis S S 1978. *Atlas of the Textural Patterns of Basalts and their Genetic Significance*. Elsevier, Amsterdam, 323p.

B   Birch W D (ed) 1989. Zeolites of Victoria. *Mineralogical Soc. of Victoria Special Publ.* No.2. Mineralogical Soc. of Victoria, Melbourne, 110p.

A   Fitton J G & Upton B G J 1987. Alkaline Igneous Rocks. *Geological Society Special Publ.* 30. Blackwell, Oxford, 568p.

A   Hekinian R 1982. Petrology of the Ocean Floor. *Elsevier Oceanographic Series* 33. Elsevier, Amsterdam, 393p.

A   Hughes C J 1982. Igneous Petrology. *Developments in Petrology* 7. Elsevier, Amsterdam, 551p.

B   Joplin G A 1971. *A Petrography of Australian Igneous Rocks*. Rev. edn. Angus and Robertson, Sydney, 253p.

A   LeMaitre R W (ed) 1989. *A Classification of Igneous Rocks and Glossary of Terms*. Blackwell Scientific Publications, Oxford, 193p.

A   McPhie J, Doyle M & Allen R 1993. *Volcanic Textures: a guide to the interpretation of textures in volcanic rocks*, University of Tasmania, Hobart, 198p.

A   Mitchell R H 1986. *Kimberlites, Mineralogy, Geochemistry and Petrology*. Plenum Press, New York, 442p.

A   Mitchell R H & Bergman S C 1991. *Petrology of lamproites*. Plenum Press, New York, 447p.

A   Nixon P H 1987. *Mantle Xenoliths*. John Wiley & Sons, Chichester, 844p.

A   Rock N M S 1991. *Lamprophyres*. Blackie & Son, Glasgow, 285p.

A   Sutherland F L, Raynor L R & Pogson R E, 1994. Spinel to garnet lherzolite transition in relation to high temperature palaeogeotherms, eastern Australia. *Australian Journal of Earth Sciences* 41, 205–20.

A   Wilson M 1989. *Igneous Petrogenesis: A Global Tectonic Approach*. Unwin Hyman, London, 466p.

TECTONICS AND VOLCANISM

A Batiza R, Smith T L & Niu Y 1989. Geology and Petrologic Evolution of Seamounts Near the EPR Based on Submersible and Camera Study. *Marine Geophysical Researches* 11, 169–236.

B Bonatti E 1994. The Earth's Mantle below the Oceans. *Scientific American*, Mar, 270(3), 26–33.

A Carey, S W 1976. *The Expanding Earth*. Elsevier, Amsterdam, 488p.

A Decker R W, Wright T L & Stauffer P H 1987. Volcanism in Hawaii Vols 1 & 2. *US Geological Survey Professional Paper* 1350, US Government Printing Office, Washington, 1667p.

A Deming J W & Baross J A 1993. Deep-sea smokers: Windows to a subsurface biosphere. *Geochemica et Cosmochimica Acta*, 57, 3219–30.

B Dvorak J J, Johnson C & Tilling R I 1992. Dynamics of Kilauea Volcano. *Scientific American*, Aug, 23(6), 18–25.

B Fryer P 1992. Mud Volcanoes of the Marianas. *Scientific American*, Feb, 266(1), 26–32.

A Haymon R M, and others 1993. Volcanic eruption of the mid-ocean ridge along the East Pacific Rise crest at 9°45–52'N: Direct submersible observations of seafloor phenomena associated with an eruption event in April, 1991. *Earth and Planetary Science Letters*, 119, 85–101.

A Keith M L 1993. Geodynamics and mantle flow: an alternative earth model. *Earth-Science Reviews*, 33, 153–337.

A Le Bas M J 1982. Quaternary to Recent volcanicity in Japan. *Proceedings of the Geologists Association*, 83(2), 179–94.

B Lutz R A & Haymon R M 1994. Rebirth of a deep-sea vent. *National Geographic*, Nov, 186 (5), 115–126.

A Owen H G 1983. *Atlas of Continental Displacement, 200 million years to the present*. Cambridge University Press, Cambridge, 159p.

B Paton T R 1986. *Perspectives on a Dynamic Earth*. Allen & Unwin, London, 142p.

B Walker G P L 1990. Review Article Geology and Vulcanology of the Hawaiian Islands. *Pacific Science*, 44(4), 315–47.

AUSTRALIAN AND NEW ZEALAND VOLCANOES

A Atkinson A, Griffin T J & Stephenson P J 1975. A Major Lava Tube System from Undara Volcano, North Queensland. *Bulletin Volcanologique*, 39, 1–28.

B Birch W D 1994. *Volcanoes in Victoria*. Royal Society of Victoria, Melbourne, 36p.

B Cox G J 1989. *Slumbering Giants: The volcanoes and thermal regions of the central North Island*. Collins, Auckland, 28p.

B Cox G J 1989. *Fountains of Fire: The story of Auckland's volcanoes*. Collins, Auckland, 28p.

B Cox G J 1994. *Mountains of Fire: The volcanic past of Banks Peninsula*. Canterbury University Press, Christchurch, 34p.

B Hill C 1991. The Legacy of Undara's rage. *Australian Geographic*, April–June 22, 50–7.

A Johnson, R W (ed) 1976. *Volcanism in Australasia*. Elsevier, Amsterdam, 405p.

A Johnson R W (Compil & ed) 1989. *Intraplate Volcanism in Eastern Australia and New Zealand*. Cambridge University Press, Cambridge, 408p.

A Kermode L 1992. *Geology of the Auckland urban area*. Institute of Geological & Nuclear Sciences Geological Map 2, Scale 1:50,000. Institute of Geological & Nuclear Sciences Ltd, Lower Hutt, 1 sheet + 63p.

A McDougall I, Embleton B J J & Stone D B 1981. Origin and evolution of Lord Howe Island, Southwest Pacific Ocean. *Journal of the Geological Society of Australia*, 28, 155–76.

A Nichol R 1992. The eruption history of Rangitoto: reappraisal of a small New Zealand myth. *Journal of the Royal Society of New Zealand*, 22, 159–80.

A Smith I E M (ed) 1986. Late Cenozoic volcanism in New Zealand. *Royal Society of New Zealand Bulletin* 23. Royal Society of New Zealand, Wellington, 371p.

A Sutherland F L 1993. Late thermal events based on zircon fission track ages in northeastern New South Wales and Southeastern Queensland: Links to Sydney Basin seismicity? *Australian Journal of Earth Sciences* 40, 461–70.

A Sutherland F L 1994. Tasman Sea evolution and hotspot trails. In van der Lingen G J, Swanson K M & Muir R J (eds) *Evolution of the Tasman Sea Basin*. Balkema, Rotterdam, 35–51.

AUSTRALIAN VOLCANIC GUIDES

A Brown M C, McQueen K G, Roach I C & Taylor G 1993. IAVCEI Canberra 1993 Excursion Guide Monaro Volcanic Province. *Aust. Geol. Survey Org. Record*, 1993/61, 23p.

B Duggan M B & Knutson J 1993. The Warrumbungle Volcano. A geological guide to the Warrumbungle National Park. *Aust. Geol. Survey Org.*, Canberra, 51p.

A Duggan M B, Knutson J & Ewart A 1993. IAVCEI Canberra 1993 Excursion Guide Warrumbungle, Nandewar and Tweed Volcanic complexes. *Aust. Geol. Survey Org. Record*, 1993/70, 40p.

A Ferguson J 1986. *4th International Kimberlite Conference Pre-Conference Field Excursion Guide to Southeastern Australia, 1–10 August*. Geological Society of Australia, Sydney, 109p.

A Gunn M J (Compl) 1986. *4th International Kimberlite Conference Post-Conference Field Excursion Guide to the Lamproites of the Kimberley Region Western Australia, 17–23 Aug*. Geol. Soc. of Australia, Sydney, 68p.

A Nicholls I A, Greig A G, Gray C M & Price R C 1993. IAVCEI Canberra 1993 Excursion Guide Newer Volcanics Province — Basalts, Xenoliths and Megacrysts. *Aust. Geol. Survey Org. Record*, 1993/58, 48p.

B Sheard M J 1983. *Volcanoes of the Mount Gambier Area*. South Australia Department of Mines and Energy, Mineral Information Series, 12p.

A Sheard M J 1990. A Guide to Quaternary Volcanoes in the Lower South-East of South Australia. *Mines and Energy Review South Australia*, 157, 40–50.

B Stevens N C 1984. *Queensland Field Geology Guide*. Geol. Soc. of Aust. (Queensland Div), Brisbane, 112p.

B Thompson D, Bliss P & Priest J 1987. *The Geology of Lord Howe Island*. NSW Department of Mineral Resources brochure, Lord Howe Island Board.

B Orth K & King R 1990. *The Geology of Tower Hill*. Department of Industry, Victoria, 17p.

B Willmott, W F 1986. *Rocks and Landscapes of the Gold Coast Hinterland. Geology and excursions in the Albert and Beaudesert Shires*. Geol. Soc. of Aust. (Qld Div), Brisbane, 38p.

NEW ZEALAND VOLCANIC GUIDES

A Cole J W, Houghton B F, Lloyd E F, Nairn I A, Simpson B M, Wood C P & Wilson C J N 1986. Intra-Congress Tours B2 & B3. *International Volcanological Congress Auckland-Hamilton-Rotorua New Zealand 1–9 February 1986. Handbook*. International Assoc. of Volcanology and Chemistry of the Earth's Interior & Royal Society of New Zealand, Auckland, 84–135.

B New Zealand Geological Survey 1987. *Scenery and Geology from the Summit Road Christchurch*. Brochure NZ Geol. Survey, DSIR, Lower Hutt.

B New Zealand Geological Survey 1988. *Volcanic landscape: A guide to the Auckland Volcanic Field*. Brochure NZ Geol. Survey, DSIR, Lower Hutt.

B Weaver S, Sewell R & Dorsey C 1990. *Extinct Volcanoes: A Guide to the Geology of Banks Peninsula*. Revised Guidebook 7. Geol. Soc. of New Zealand, Lower Hutt, 48p.

GLOSSARY

A'a A variety of lava made of jagged stony clinkers. A Hawaiian term.

Accretionary lapilli Pellets of ash made by concentric addition of ashy material around condensing water droplets or solid particles in eruption clouds.

Aerosol Substance suspended in air as a colloid because of its small particle size. Volcanic aerosols that reach the stratosphere are mostly sulphate particles.

Agate Banded or moss-like varicoloured variety of the mineral chalcedony, a microcrystalline form of quartz. Often cut and polished as a decorative stone.

Almandine A common mineral species in the garnet group; has an iron and aluminium silicate chemical composition.

Amphibole A member of a widespread group of extremely complex hydrous silicate minerals that crystallise in many igneous and metamorphic rocks.

Amygdale A round or almond-shaped lava cavity filled with minerals of later formation, commonly carbonate, silica or zeolite.

Analcime A mineral species in the zeolite group; has a sodium hydrous aluminium silicate composition.

Andesite Fine-grained volcanic rock formed in and named after the Andes Mountains. It typically contains plagioclase feldspar minerals combined with minerals of the clinopyroxene, orthopyroxene and amphibole groups.

Angiosperm Flowering plant distinguished from other seed plants because it produces seeds fully enclosed by fruits.

Anhydrite A mineral species composed of calcium sulphate. Commonly formed by evaporation processes; also formed in volcanic eruptions.

Ankaramite A type of basaltic rock of alkaline affinities; contains large, abundant crystals of clinopyroxene and olivine minerals.

Anorthoclase A mineral species in the feldspar group; has a sodium, potassium aluminium silicate chemical composition.

Apatite A member of a mineral group, the species of which consists of calcium phosphate combined with fluorine, chlorine or hydroxyl elements.

Aquagene volcanism Volcanic eruptions in which rocks form when lava is extruded into a watery environment.

Aragonite A mineral species composed of calcium carbonate. It differs from calcite in belonging to the orthorhombic crystal system.

Arc See **Island arc**.

Argon An inert gas found in the atmosphere but also formed by the radioactive breakdown of elements such as potassium.

Asthenosphere A zone in the earth's upper mantle region that underlies the rigid lithosphere. The peridotite rocks here deform by plastic flow in response to stresses imposed on them and they act as a more fluid medium.

Avalanche A rapid, often destructive snow or rock flow that can take place on volcanic slopes.

AVHRR Advanced Very High Resolution Radiometer, an instrument on board polar-orbiting satellites for measuring incoming radiation in five different wave length bands; useful for detecting volcanic ash clouds in the atmosphere.

Basalt A dark volcanic rock usually composed of minerals of the plagioclase feldspar and clinopyroxene groups along with iron oxide, olivine and orthopyroxene minerals. Basalts form Earth's most common volcanic rock.

Batholith A large complex intrusion of coarse igneous rocks in the crust, usually made up of several separate intrusions, (called plutons) that form below volcanic regions and eventually become exposed by erosion.

Bauxite Rock rich in hydrated aluminium oxide minerals, formed by the deep weathering and leaching of parent rocks.

Benioff zone A zone of deep earthquakes triggered when a slab of lithosphere descends below a region of colliding crust and mantle rock. Named after the geophysicist Hugo Benioff.

Benmoreite A variety of basalt rock highly enriched by aluminium and alkali metals and dominated by minerals such as sodium-rich plagioclase and alkali feldspars. Named after Ben More in Scotland.

Bennettitalean Belonging to an extinct order of gymnosperms, plants that carried seeds exposed in cone scales and lived during the Triassic to Cretaceous Periods.

Bomb See **Volcanic bomb**.

Boninite An andesite volcanic rock characterised by high magnesium levels. It forms in island arcs, such as the Bonin Islands.

Brachiopod A phylum of invertebrate marine animals, characterised by bivalved shells that resemble oil lamps. They have lived on earth since Cambrian times.

Breached crater A crater with an incomplete rim; the gap is often made by a later eruption or sometimes by erosion.

Breccia Rock made up of large angular fragments deposited close to the source area. It can form from sedimentary deposition, igneous activity or rock movements. Volcanic breccias contain angular pieces of erupted lava and country rocks formed within or close to the vent.

Butte A small isolated flat-topped hill formed by the extensive erosion of horizontal rock layers. It sometimes has a hard cap provided by a volcanic flow.

Cainozoic A geological era from 65 million years ago to the present day.

Calcalkaline A special term used for certain rocks usually associated with volcanic activity in unstable regions. Their particular chemistry contains 56–61 wt % of silicon oxide, when the calcium oxide content matches the alkali oxide (sodium oxide and potassium oxide) content. These rocks grade into calcic suites when silicon oxide exceeds 61 wt %, and into alkalic suites when silicon oxide falls below 51 wt %.

Calcite A mineral species composed of calcium carbonate. Unlike aragonite, it belongs to the trigonal crystal system.

Caldera A roughly circular depression formed by the collapse of a volcano into its underlying chamber during a major eruption.

Cambrian A geological period about 570 million to 510 million years ago.

Cap A general term for a volcanic remnant that forms a topping on older rocks.

Carbon dating See **Radio-carbon dating**.

Carbonatite Relatively rare igneous rock, rich in carbonate minerals such as calcite and dolomite. Carbonatites usually form intrusive masses or dykes and accompany alkaline igneous rocks. They often include minerals rich in rare earth elements and in barium, niobium, thorium and phosphorus elements.

Carboniferous A geological period about 355 million to 300 million years ago in the Palaeozoic era, when rich coal beds were laid down in the northern hemisphere.

Cauldron A volcanic crater formed when a volcano subsides into a large underlying chamber (cauldron) along a ring fracture, and sometimes intruded by a ring dyke. Its formation resembles a caldera.

Chabazite A common mineral species in the zeolite group; a calcium, sodium hydrous aluminium silicate.

Clinopyroxene A member of the pyroxene group of minerals that belongs to both the monoclinic and triclinic crystal systems.

Coimadai fauna Australian fossil fauna containing marsupial remains found below volcanic horizons of Pliocene Age in the Coimadai area, about 8 km northwest of Bacchus Marsh, Victoria.

Colonnade A set of vertical or slightly inclined columns formed by inward jointing in lava flows as they cool down from the molten state. Colonnades form in upper and lower parts of flows, often separated by an entablature of less regular cooling joints.

Columbite A mineral species in which the rare metal elements niobium and tantalum are associated with oxygen that is combined with iron and manganese.

Complex A diverse but distinctive and interrelated body of rocks, which may be largely volcanic in a volcanic complex, or largely intrusive in a plutonic complex.

Cone See **Volcanic cone**.

Coulée flow A very thick, relatively short, blocky lava flow, usually made of viscous rhyolite or dacite lava. Its shape is something between a lava flow and a lava dome.

Cowlesite A rare mineral species in the zeolite group; has a calcium hydrous aluminium silicate chemical composition.

Corundum A common aluminium oxide mineral well-known for its gem varieties — sapphire and ruby.

Crater Circular depression formed by explosive volcanic eruption (volcanic crater), or by an impact on Earth by a body from space (meteoritic crater).

Crater row A term for several aligned craters caused by eruptions along a fissure.

Cretaceous A geological period about 145 to 65 million years ago at the end of the Mesozoic era. It is noted for chalk deposits in Europe and it ended with the extinction of many life forms, including dinosaurs.

Crust The outermost solid layer of the Earth, extending from a thickness of 5 km under oceans (basaltic crust) to a thickness of about 60 km under continents (granitic crust).

Cyathealean Belonging to the Cyathealeo order of ferns which are gymnosperm plants.

Cycad A gymnosperm plant bearing woody cones. The group includes extinct and living species.

Dacite A volcanic rock, usually light-coloured when crystallised, containing plagioclase, alkali feldspar, quartz, dark mica and amphibole minerals. It has 63–70 wt % silicon oxide in its chemical composition.

Diatreme A volcanic vent formed by explosive action and filled with angular fragments injected by the action of gas and other fluids. The vent tapers downwards to a feeder zone, which usually tapped deep-seated, volatile, rich alkaline melts. The explosive activity is usually caused by the rising melt interacting with groundwater.

Diorite A coarse-grained igneous rock made of plagioclase feldspars, pyroxenes, amphiboles and sometimes mica and quartz. It is the intrusive equivalent of andesite.

Diprotodont A marsupial belonging to a group of browsers, most of them quite large, that became extinct in the Pleistocene epoch.

Dodecahedral A crystal form composed of 12 rhomb-shaped faces that reflect the symmetry of the cubic crystal system.

Dolerite A medium grained igneous rock containing plagioclase feldspar, pyroxene and iron oxide minerals. It is the intrusive equivalent of basalt.

Dome A mound of viscous lava, usually rhyolite trachyte or dacite, growing over a volcanic vent.

Doming Upwarping in the form of a dome; it can take place over a large area.

Downwelling A downward movement of colder fluid materials; these include low-viscosity materials like the atmosphere and oceanic water masses as well as highly viscous materials such as slowly flowing mantle peridotite rocks.

Drainage The passage of water over and through the land surface from its source to the sea, eventually forming a system.

Dyke A tabular intrusion that cuts discordantly and usually steeply across the surrounding rocks (see also **Ring dyke**).

Echinoid A free-living class of animal belonging to the phylum Echinodermata. The animal is contained in a casing of calcareous plates, that are often heart-shaped or cushion-shaped.

Eclogite A coarse metamorphic rock made of garnet and pyroxene of specific chemical compositions and formed by the conversion of basaltic rocks under high temperatures and pressures.

El Niño An equatorial climatic anomaly fed and reinforced by the upwelling of warm eastern Pacific equatorial water. This produces torrential rainfall on the Pacific coast of South America, from where the term came. This produces a converse effect — drought in Australia. See also **La Niña**.

Entablature The central, jointed part of a lava flow that shows a more chaotic development of cooling joints than the upper and lower columnar parts or colonnades do, and produces various fan-shaped, rosette-shaped or irregular structures.

Erionite A rare mineral species in the zeolite group; has a potassium, calcium, sodium hydrous aluminium silicate chemical composition.

Eruption The volcanic release of lava and gas from the Earth's interior onto the surface and into the atmosphere. There are many kinds of eruption.

Escarpment A steep slope or cliff below the edge of a flat or gently sloping area. The abbreviated term is 'scarp'.

Feeder The general term for a conduit that brings magma up to erupt at the surface. It may take the form of a plug, dyke or sill.

Feldspar The general name for a common group of silicate minerals characterised by an aluminium silicate framework that is balanced by metals such as calcium, sodium and potassium.

Feldspathoid The general name for a group of silicate minerals characterised by an aluminium silicate framework that is balanced by metals such as sodium and potassium. They have less silicon for each chemical unit than feldspars do.

Ferrierite A rare mineral species of the zeolite group; has a sodium, potassium, magnesium hydrous aluminium silicate chemical composition.

Fission track dating A technique that dates minerals and glasses by counting the minute tracks left by the spontaneous discharge of charged particles during the radioactive breakdown of uranium 238 in the material.

Fissure volcano A linear fracture on the surface through which lava, pyroclastics and gas erupt.

Flank crater A crater left by an eruption on the side of a larger volcano, away from the main central vent.

Flow A run of lava that erupts from a vent and flows down the volcanic slopes. It sometimes travels along local drainages for a long way before solidifying.

Folding The buckling and contortion of geological layers by earth movements; it occurs when they are compressed or when they slump during stretching. When these processes involve a large area, it is called a fold belt.

Fumarole A small vent, in an active volcanic area, that discharges steam or hot gases, usually at temperatures of 100–1,000° C.

Gabbro A coarse-grained igneous rock mainly composed of plagioclase feldspar and pyroxene minerals. It is the intrusive equivalent of basalt and coarser than dolerite.

Gastropod A class of invertebrate animals belonging to the phylum Mollusca, which includes snails and slugs. Many species have a coiled shell that is usually preserved in fossil forms.

Geodetic levelling Measurements that correlate levels of land with the shape of the Earth's surface.

Geothermal Concerns heat flow through the Earth. Areas of relatively high heat flow and heated groundwaters are called geothermal fields.

Geyser A surface vent that periodically spouts a fountain of boiling water. The vent is often built up by mineral deposits precipitated from this hot water.

Ginkgophyte A gymnosperm plant related to the living maidenhair tree, *Ginkgo biloba*.

Gmelinite A rare mineral species in the zeolite group; it is a sodium, calcium hydrous aluminium silicate in the hexagonal crystal system.

Gneiss Coarse metamorphic rock containing bands of dark minerals such as mica, amphibole and pyroxene separated from bands of light minerals such as feldspar and quartz.

Gonnardite A rare mineral species in the zeolite group; it is a sodium, calcium hydrous aluminium silicate in the orthorhombic crystal system.

Granite A common, coarse igneous rock; alkali feldspar, quartz and mica are its main mineral constituents and it is found in large intrusions.

Granodiorite A common, coarse igneous rock; plagioclase feldspars, quartz and mica are its main mineral constituents and it usually occurs in large intrusions.

Granulite A coarse, granular metamorphic rock containing bands of minerals such as feldspar and pyroxene, and sometimes quartz and garnet.

Greenhouse effect An increase in air temperature in the lower atmosphere brought about as a result of the absorption and re-radiation of heat by clouds and gases. Water vapour, carbon dioxide, methane and chlorofluorocarbons are the main greenhouse gases.

Gymnosperms Vascular plants that carry seeds (spores) exposed in cone scales, e.g. Conifers.

Hackly A sharp or jagged fracture, typical of brittle metals.

Halogen A member of a group of elements that includes fluorine, chlorine, bromine and iodine. These elements react with metals to form mineral halides.

Hawaiian style Volcanic eruptions characterised by spectacular fire fountains and fluid flows of lava, typical of those of Hawaii.

Hawaiite A variety of basalt in which the plagioclase feldspar, and hence the rock, is slightly enriched by sodium. Named after Hawaii, where it is a common type.

Heulandite A mineral species in the zeolite group; composed of sodium and calcium hydrous aluminium silicate.

Hominid A member of the primate family which includes humans and their immediate ancestors.

Hornito A small cone or chimney built of lava splatter, found on the surface of lava flows.

Hotspot An area of high volcanic activity over a deep stationary magma source in the asthenosphere. Movement of the Earth's plate over the hotspot produces a chain of hotspot volcanoes.

Hyalotuff Explosive volcanic rock made of fine-grained shards of glassy lava. Typically formed by the explosive quenching of hot lava by water.

Hydrothermal Concerning the activity of very hot water derived from the interactions of molten material below the surface.

Icelandite A variety of basalt of tholeiitic affinities enriched in sodium. It was first identified in Iceland.

Igneous One of the three main groups of rocks; the 'fire rocks' that form from molten magma (underground) or lava (aboveground).

Ignimbrite Poorly sorted volcanic rock formed from the deposition of a pumice-bearing pyroclastic flow. Fragments in the lower layers may be flattened and welded together by high temperatures.

Ilmenite An iron titanium oxide mineral, widespread in many igneous rocks.

Intrusive A body of igneous rock that has invaded the Earth's upper layers as molten material before solidifying there.

Island arc A series of volcanoes that lies over a subducting oceanic slab and commonly forms an arc-like chain in plan. Island arc volcanoes typically erupt lavas that are chemically related to the andesitic suite. They are prone to large earthquakes triggered by the descending slab.

Isotope dating A means of determining the age of materials by using the measured proportions of isotopes of elements; one isotope is produced by the radio-active decay of a parent element and accumulates with time changing the isotopic ratios.

Jökulhlaup An Icelandic term for a sudden violent discharge of meltwaters from a glacier or ice-cap, sometimes due to underlying volcanic activity.

Jurassic A geological period between about 210 and 145 million years ago, in the Mesozoic era. Named after beds in the Jura mountains in Europe.

Kimberlite A fragmentary igneous rock, poor in silicon, rich in alkali elements such as sodium and potassium, and commonly altered by fluid interaction. It carries fragments and crystals from deep levels and sometimes includes diamonds.

La Niña The counterpart to the El Niño climatic effect, when cooler upwellings in the equatorial east Pacific waters produce periods of dry stable weather off the South American coast.

Laccolith An intrusion of igneous rock that domes up the overlying rocks but retains a flat floor over the underlying rocks.

Lahar The catastrophic mud flow on the flanks of a volcano that causes most volcanic fatalities. It often brings down large blocks of rock with it.

Lamproite A special variety of igneous rock related to the lamprophyres and characterised by very high potassium and relatively low silicon contents. Some lamproites contain mantle materials, which may include diamond.

Lamprophyre A distinctive group of igneous rocks usually found in small intrusions. They include a variety of porphyritic rocks containing conspicuous crystals of mica, amphibole and pyroxene minerals, while feldspar minerals are mostly confined to the groundmass.

Lapilli Fragments of volcanic rock, usually 2–64mm, deposited by explosive eruptions.

Lateral slip faulting The fracturing and displacement of rocks by a sideways movement.

Laterite Rock rich in hydrated iron and aluminium oxide minerals formed by the weathering and leaching of parent rocks.

Laumontite A mineral species in the zeolite group; composed of calcium hydrated aluminium silicate.

Lava Molten rock erupted from a volcano, often as a lava flow. The solid rock usually contains cavities left by escaping gas.

Lava lake A molten pool of lava held in a sizeable depression, often the crater of the volcano erupting the lava.

Lavicle A small protrusion on the roof and walls of lava tubes, formed by dripping lava.

Lead An alluvial deposit formed in a drainage channel. It carries heavy minerals washed in from other rocks sometimes in sufficient concentrations to mine. A lead buried by later volcanic activity is termed a deep lead.

Leucite A mineral species in the feldspathoidal group; composed of potassium aluminium silicate.

Leucitite A volcanic rock containing leucite among its minerals; relatively poor in silicon, but high in alkali metals, particularly potassium.

Levyne A rare mineral species in the zeolite group; composed of a calcium, sodium, potassium hydrated aluminium silicate.

Lithosphere The upper rigid part of the Earth, made of crustal rocks and the uppermost part of the mantle. It is divided into a dozen or so areas by geological activity along the mid-ocean ridge or collision zones that form the Earth's plates.

Lobate Made up of lobes of material, typically formed in flowing pahoehoe lava.

Low velocity zone A zone in the upper mantle, best observed beneath oceanic floors. It retards the progress of earthquake waves and suggests that molten material is present at levels ranging 40–160 km deep.

Maar A volcanic crater, often occupied by a lake and surrounded by a low ring of ejected material. Maars are usually formed by the explosive interaction of erupting lava with groundwater.

Magma Hot molten rock, usually of silicate but sometimes of carbonate or sulphide composition. It rises when rocks melt in the Earth's interior. On eruption, magma degasses to form lava.

Magnesio-wüstite A magnesium-rich variety of the mineral species wüstite, an iron oxide.

Mantle The zone that lies between the Earth's crust and core and extends down to about 2,900 km. Its upper part is mostly peridotite rock but its lower part may transform into rocks containing other minerals.

Massif A massive topographic and structural feature, usually formed of rocks more rigid than those of surrounding areas.

Melilite A mineral species composed of calcium magnesium aluminium silicate.

Melilitite A type of basalt rock containing the mineral melilite. It is lower in silicon, but higher in calcium than is normal basalt.

Melt The liquid part of silicate magma, formed by melting silicate source rocks. It may contain suspended crystals.

Mesa A flat-topped hill usually formed by a hard capping rock that overlies softer layers. The hard cap can be formed by a lava flow.

Mesolite A mineral species in the zeolite group; a sodium, calcium hydrated aluminium silicate.

Mesozoic A geological era about 250 million to 65 million years ago.

Metapyroxenite A metamorphic rock composed mainly of pyroxene minerals; formed by the recrystallisation of igneous pyroxenites.

Mica A member of a group of hydrous silicate mineral species, characterised by layered silicate sheets combined with alkali and other metals. This gives mica its flaky but flexible structure.

Mid-ocean ridge A long, linear elevated ridge formed by volcanic action along ocean floors; the active zone that makes the floors spread apart.

Miocene A geological epoch extending from 25 to 5 million years ago, during the Tertiary period. The term means 'less new'.

Möberg An Icelandic term for a ridge of chilled pillow lavas and breccias formed by subglacial eruptions.

Mode The percentage by volume of each separate mineral in a particular rock.

Monogenetic volcano A volcano built up in a single episode, although this may include both explosive and flow activity.

Monzonite A coarse igneous rock, intermediate in character between syenite and diorite. Named after Monzoni in Italy.

MORB Mid-Ocean Ridge Basalt.

Mud flow See **Lahar**.

Mugearite A variety of basalt that is rich in alkali elements and contains plagioclase feldspar enriched with sodium. Named after Mugeary in Scotland.

Natrolite A common mineral species in the zeolite group; a sodium hydrated aluminium silicate.

Neck A vertical volcanic pipe-like intrusion forming part of an eroded volcanic vent.

Nepheline A mineral species in the feldspathoid mineral group; a sodium aluminium silicate.

Nephelinite A volcanic rock related to basalt but containing the mineral nepheline instead of plagioclase feldspar, and lower in silica content.

Niobium A rare, heavy metal element with an atomic number of 41 and atomic weight of 93.

Nuée ardente A fast turbulent gas cloud erupted from a volcano, sometimes incandescent ('glowing cloud'). It forms a deposit of ash and blocks and is often created by the collapse of a growing lava dome.

Obduction The over-riding by oceanic lithosphere onto a continental margin along a collision zone between different plates. The oceanic plate is forced up rather than downwards, as in subduction.

Obsidian A dark volcanic glass usually composed of rhyolite or dacite. Its surface is often curved and fractured.

Oceanite A variety of basalt containing conspicuous and copious large crystals of the mineral olivine.

Octahedral An eight-sided crystal shape; each face is an equilateral triangle, which gives double-ended joined pyramids. This is a typical 'diamond' crystal form.

Offretite A rare mineral species in the zeolite group; a potassium, calcium hydrated aluminium silicate.

Ogive A curved arcuate structure formed repeatedly and pointing down the slope of viscous lava flows or glaciers. It results from a faster flow in the middle of a generally slow moving flow.

Olivine A common mineral species; a magnesium iron silicate.

Orthopyroxene A member of the pyroxene group of silicate minerals, which belong to the orthorhombic crystal system.

Ozone A form of oxygen composed of 3 atoms; it becomes concentrated in the atmosphere at 15–30 km altitude and forms the ozone layer.

Pahoehoe A Hawaiian term for a smooth, ropy or lobe-like variety of lava.

Palaeozoic A geological era about 570 million to 250 million years ago.

Palagonite Pale yellow to yellow-brown hydrated volcanic glass, formed by the alteration of basalt glass by interaction with water.
Peridot The gem variety of the mineral olivine, usually a transparent green.
Permian A geological period about 300 million to 250 million years ago in the Palaeozoic era during which sedimentary beds were deposited in the Perm area of Russia.
Petrologist A scientist who studies rocks (petrology) in relation to their minerals and textures (petrography) or origin (petrogenesis). Derived from a Greek term for rock.
Phacolite A variety of the zeolite mineral chabazite with a complex twinning characteristic that gives its crystals an almond shape.
Phenocryst A large crystal, often with a well-formed crystal shape that lies in a finer-grained body of igneous rock. Rocks containing phenocrysts are termed porphyritic in texture.
Phillipsite A common mineral species in the zeolite group; a potassium, sodium, calcium hydrated aluminium silicate.
Phonolite A fine-grained volcanic rock related to the basalts but containing abundant alkali feldspar and feldspathoidal minerals. The name is derived from a Greek term for 'sound', because it often gives a ringing note when struck with a hammer.
Pillow An elongated cylindrical tube or bulb of lava, commonly formed by the chilling that occurs when lava flows into or under water. It usually shows a rounded shape when broken in cross-section, with glassy outer crusts. Stacked pods of pillows are called pillow lava.
Pipe A vertical channel through which lava flowed or exploded on its way up to the surface. It is usually circular or ovoid.
Plagioclase A member of a mineral series in the feldspar group; it extends from sodium to calcium-rich aluminium silicates in chemical composition.
Plate An area of the earth outlined by strong earthquake or volcanic activity along its active margins. These areas act as rigid segments of the lithosphere, while deformations and seafloor spreading take place along their borders.
Pleistocene A geological epoch about 2 million to 10,000 years ago, after which the Holocene epoch began.
Plinian Major eruptions in which the pyroclastic column reaches approx 55 km high and sheds over an area of 500–5,000 km^2. Named from Plinius' description of the AD 79 Vesuvius eruption.
Plug A roughly circular, resistant peak of volcanic rock that indicates a cylindrical body of cooled magma and a former feeder of a volcano, now exposed by erosion.
Plume A column of ascending hot rock below the lithosphere that causes overlying melting, and a subsequent rise of magma into the crust and onto the surface. A plume acts as a hotspot and gives rise to hotspot volcanoes.
Polycarp A cone-like fruiting body found in cycad plants.
Porphyritic A texture in igneous rocks produced by large well-crystallised minerals (phenocrysts) held in a finer-grained groundmass. An intrusive rock containing over 25% of phenocrysts is often termed a 'porphyry'.
Primary magma See **Primitive**.
Primitive In igneous terms, an unevolved kind of magma directly formed from its source rock. In biological evolutionary terms, it is an early form of an evolving organism.
Proterozoic The more recent division of Precambrian time which forms an eon extending from about 2,500 million years ago to the start of the Cambrian Period in the Palaeozoic era about 570 million years ago.
Pulse In igneous geology, a sudden increase in the intrusion or eruption of magma within regularly recurring intervals of magmatic intrusion or eruption.
Pumice A light-coloured glassy volcanic rock, usually a rhyolite, dacite, trachyte or phonolite. It is riddled with gas holes, that make it light enough to float on water.
Pyroclastic A volcanic rock made of fragments produced by explosive action. The term means 'fire broken'. A wide range of pyroclastic rocks are produced by many different kinds of explosive activity.
Pyrope A common mineral species in the garnet group; a magnesium, iron, aluminium silicate.
Pyroxene A widespread group of silicate minerals, with a chain-like silicate structure, bound by various metal elements.
Quaternary A geological period about 2 million years ago to the present day.
Radio-carbon dating A method used for dating organic materials. From the isotope ratios of carbon 12 and carbon 14 it estimates the residual carbon 14 left after the death of the organism. This isotope is not renewed after death, so its decay indicates the time elapsed, back to about 70,000 years.
Ring dyke A ring-like intrusion that cuts steeply through the surrounding rocks.
SAGE Stratospheric Aerosol and Gas Experiment—-a satellite instrument for measuring the amount of solar radiation cut out by aerosols in the Earth's atmosphere.
Scoria Loose, rubbly fragments of basalt that erupt from a vent and build up a scoria cone during Strombolian activity.
SeaMARC A sonar technique used in oceanic surveys, where a super side scan images the sea floor in a swathe about 10 km wide, on either side of the towing ship.
Seamount A submarine mountain rising at least 1,000 m from the sea floor.
Seismic Concerning earthquake activity. The more seismic an area, the more likely an earthquake will be felt there.
Shale A common sedimentary rock, essentially compacted mud (mudstone); fine-grained, flaky minerals give it a natural parting along the beds.
Shield volcano A broad, low-profile volcano, usually built up from fluid lavas so that it resembles a round, fighting shield.
Shoshonitic Belonging to a suite of volcanic rocks between trachyte and andesite in composition and characteristically enriched with the element potassium. Named after the Shoshone River, Wyoming, USA.
Silica A short term for silicon dioxide, the constituent of quartz, the Earth's most common mineral. It is usually quoted as the measure of silicon in rocks, and in volcanic rocks this varies from less than 10% (carbonatites) to over 80% (rhyolites).
Siliceous A rock abundant in silica, and particularly in free quartz rather than just silicate minerals.
Silk A term used for gemstones that contain microscopic needles or plates of titanium or iron oxide minerals. This causes reflections from within the gemstone, giving it a silky sheen.
Sill An intrusive igneous rock that extends as a tabular body and can form sheets. The molten rock commonly invades horizontal sedimentary layers and insinuates itself between beds.
Slab A part of the Earth's oceanic lithosphere that is displaced by collisions at plate boundaries. Subducting slabs descend and obducting slabs override.
Somma A circular or crescent-shaped ridge left after a volcano's catastrophic eruption and collapse. A new cone growing inside it is termed a somma stage of eruption. Named after Mt Somma, flanking Mt Vesuvius in Italy.
Spherulite A sphere-like aggregate of radiating fibrous crystals found in silicon-rich volcanic rocks, formed when glassy rocks devitrify. Quartz and feldspars commonly crystallise in a spherulite.
Spheruloid A sphere-like body that also occurs in altered silicon-rich volcanic rocks, but lacks the radial structure of spherulites.
Spinel A term that applies to a group of minerals or to one of its members. The group consists of a series of oxides of metals such as iron, aluminium, magnesium, chromium and zinc. Spinel as a member mineral is largely magnesium, aluminium oxide.
Spreading The generation of new sea floor at a mid-ocean ridge. Older sea floor is shouldered sideways so that continuously enlarging sea floor spreads away on each side of the ridge.
Stellerite A rare mineral species in the zeolite group; a calcium hydrous aluminium silicate similar to stilbite.
Stilbite A common mineral species in the zeolite group; a sodium calcium hydrous aluminium silicate.

Stock An intrusion of igneous rock, roughly circular when eroded, and usually less than 20 km² in surface area.

Stratosphere A higher layer of the Earth's atmosphere, where temperatures average about minus 60° C between 10 to 50 km above the Earth's surface. It overlies the troposphere and only materials from larger volcanic eruptions reach the stratosphere.

Stratovolcano A volcano built up of alternating layers (strata) of lavas and pyroclastic deposits. Sometimes termed a 'composite volcano'.

Strombolian deposits Volcanic material ejected by activity similar to that of Stromboli volcano off Italy's coast. Basalt lava continually exploding upwards in small bursts, punctuated by lava flows, builds up a steep cone.

Subduction A process in which slabs of oceanic floor are forced below a zone of colliding plates and become consumed through melting and metamorphic conversion.

Summit crater A crater excavated by explosive activity at the top of a volcanic mountain.

Surge A turbulent flow of erupted gas and pyroclastic materials travelling over the ground, often at speed. These diluted pyroclastic materials have various origins; cold, wet surges called basal surges develop through water-lava steam explosions; hot, dry ground surges develop from the head of pyroclastic flows. Ash cloud surges develop from gas and ash spilling over pyroclastic flows.

Surtseyian A type of volcanic eruption where water flooding the vent causes high-energy steam explosions. Eruption columns reach up to 20 km, as they did when Surtsey Island rose out of the sea off Iceland.

Syenite A coarse igneous rock composed mainly of minerals of the alkali feldspar, pyroxene, amphibole and mica groups. It is the intrusive equivalent of trachyte lava.

Syncline A trough-shaped fold, in layered rocks, with either symmetrical or asymmetrical limbs on either side of it. Synclines often close off at the ends into a basin-like structure.

Synclinorium A series of synclines separated by anticlinal arches; the whole series is folded down into a major regional synclinal structure.

Tacharanite A rare mineral species; a calcium, aluminium hydrated silicate. The name comes from a Gaelic word for the mineral's tendency to change when exposed to air.

Talus An accumulation of coarse rock fragments that have fallen down from steep-sided mountains or cliffs.

TAS The chemical classification of volcanic rocks according to their Total Alkali (TA) content and Silica (S) content.

Tectonic plate One of Earth's lithospheric plates, defined by active seismic and volcanic borders formed along mid-ocean ridges or collision zones.

Tertiary A geological period about 65 million to 2 million years ago in the Cainozoic era, before the Quaternary period.

Thermo-luminescent dating A method of dating previously heated materials by measuring the natural thermo-luminescence of volcanic lavas, buried sands or pottery. The energy they emit depends on, among other factors, the time that has elapsed since their heating.

Tholeiitic A term for the mineral and chemical affinity of the more silica-rich basalt rocks. Some quartz and/or orthopyroxene appear among the constituent minerals; the term comes from Tholey in Germany.

Thorite A rare mineral species; a thorium silicate.

Thunder egg An ovoid cavity in rhyolitic volcanic rocks usually filled by silica minerals, such as chalcedony (agate), opal and quartz. The term comes from Oregon, USA.

TOMS Total Ozone Mapping Spectrometer, an instrument used by the US Nimbus 7-satellite to map the amount of ozone in Earth's atmosphere.

Trachyte A volcanic rock rich in alkali feldspar minerals, which often show a flow alignment of the crystals. Its chemical composition is intermediate between basalt and rhyolite.

Transform fault A special type of fracturing in ocean floors, where movements take place at right angles to the mid-ocean ridges and indicate the direction of spreading.

Trench A deep oceanic depression, up to 11 km deep and 50–100 km wide, which lies off an island arc or active continental margin. It usually marks the region of a descending oceanic slab.

Triassic A geological period extending from about 250 million to 210 million years ago in the Mesozoic era. Named after a threefold division of rocks in Germany.

Troposphere The atmospheric layer between the Earth's surface and the stratosphere; its upper boundary is marked by the tropopause. The troposphere shows a general decrease of temperature with altitude, and varies in height with season and latitude.

Tsunami A sea wave produced by a catastrophic event such as a submarine earthquake, landslide or volcanic eruption. It can be very destructive when it strikes land. The term comes from Japan.

Tuff Compacted volcanic ash beds; their fragments are mostly less than 2 mm across. Lapilli tuff contains more than 10% of slightly larger fragments.

Tumuli Small mounds or domes on the crust of a lava flow, pushed up by underlying fluid lava. Unlike lava blisters they are solid structures.

Tuya A flat-topped or steep-sided volcano, built up by eruption into meltwaters produced by eruption under glacial ice.

Twinning A crystal that grows in a mineral symmetrically in opposite directions. Twinning may be simple or multiple.

Upwelling The term can mean the rising of a large amount of molten material to the Earth's surface, or quiet, low-energy eruptions of lava and volcanic gases. When applied to oceanic currents, it refers to cold heavy water rising from deeper levels to the surface.

Uranopyrochlore A rare mineral species; an hydroxy oxide of uranium, calcium, cerium, niobium and tantalum.

Vent In volcanic terms, an opening of the Earth's surface through which lava and gases are discharged from underground sources.

Vesicle A lava cavity produced by gas bubbles; when lava erupts bubbles are released which leave cavities when it solidifies.

Vesicular The texture of volcanic rocks containing numerous gas cavities.

Viscous A sluggish flow in a material that resists movement. Silicate lavas, rich in silicon, tend to be viscous and flow only short distances before solidifying.

Volcanic bomb A lump of lava, more than 32 mm wide, thrown out of a volcano. It may show aerodynamic forms or crusty lava that has cracked during flight and it may have a spindle, ropy, ribbon, cannonball or breadcrust shape.

Volcanic cone A conical mound of volcanic ejecta formed around an eruption vent. It usually has slopes up to 30° steep and a crater at the summit.

Volcano A natural vent or fissure that erupts molten rock and gases, either explosively or effusively. It can exist in an active, dormant or extinct state.

Volcanology The geological science that studies volcanism, its causes and effects. Practitioners are called volcanologists.

Vulcanian A type of explosive eruption typified by Vulcano volcano off Italy's coast. Violent gas discharges from viscous magma in the vent vigorously eject a large load of older solid volcanic material from the volcano walls.

Within-plate Volcanic or earthquake activity in the fairly inactive interior parts of a tectonic plate. This volcanism, unrelated to mid-ocean ridges or subduction, is due to the uplift of mantle hotspots.

Xenocryst A crystal that did not crystallise from its igneous rock host when molten, but was brought up as an accidental 'stranger'.

Xenolith Like a xenocryst in origin, but brought up as an accidental rock fragment.

Zeolite A widespread group of hydrous silicate minerals with a silicon-aluminium framework structure held together by metals. Gaps in the structure allow the easy exchange of metals, water and gases from solutions. The term comes from the ready loss of water on heating.

Zircon A widespread mineral species; a zirconium silicate. It usually contains some uranium, which is used in age dating.

INDEX

Page numbers in **bold** refer to illustrations, diagrams and maps. Maps inside front and rear covers have been referred to as **front map** and **rear map**. Place names of lakes and mountains beginning with Mt or Lake have been listed under the next word in the name, such as:
    Egmont, Mt
    Barrine, Lake
Volcano and volcanic site names are listed in their own right and under the territory (state in the case of Australia) or ocean in which they are located.

a'a lava **5**, **85**, 123, 124, 207, 210, 232
Aboriginal Cooking Cave, Binna Burra, Qld 203
Aboriginal myths 6, 99–100
accretionary lapilli 106, 232
Admiralty Islands, NSW 201, **map 220**
    see also Papua Niugini
Advanced Very High Resolution Radiometer (AVHRR) 51, 232
aerosol 232
aftermaths, eruption **64**
agate **48**, 127, 138, 232
Agate Creek, Qld **126**, 127
agate-filled spheruloids **126**
agglomerates 177, 198
Agung, Mt, Indonesia **front map**, 225
air falls 12
Airly Mountain, NSW 134
Akaroa, Sth Is, NZ 18, 207, 209, **map 222**
Akutan, Alaska, USA **front map**, 55
Alaska see USA
Alba Petera, Mars 7
Alcove, The, Mt Eccles, Vic 197
Aleutian Islands, Alaska, USA 54
Aleutian trench 92
Aleutian-Asian margin 186
Aliad, Kurile Islands, Russia **front map**, 55
alkali basalts 72, 92, 95, 150, **151**, 169, 170, 186
alkali feldspars 147, 148, **149**, **154**, 155, 226
    sanidine 226
alkaline lavas 79, 84, 85, 86, 87, 89, 92, 199
alkaline rocks 118, 148
alkalis 72
Allingham Creek, Qld 143
alluvial deposits 129, 157
almandine 232
almandine garnets 161
Alpha Regio highlands, Venus 7
Alpha ridge, Canadian Basin 96
Alpine Fault, Sth Is, NZ 37
Alum Mountains, NSW 122
aluminium 70, 72, 79, 84
aluminium hydroxides 43
aluminium silicate 147
aluminium-rich minerals 161
alumino-silicates 48, 148
Alvie, Mt, Vic 108
Alvin (submersible craft) 28
Amero, Colombia 64
amethyst 48
ammonium 61
amphibole 79, 147, 148, **149**, 163, 205, 226, 232
    hornblende 226
    kaersutite 226
    riebeckite 226
amygdale 127, 232
Anak Krakatoa, Indonesia **front map**, 21, **224**
Anakena, Easter Island (Chile) **44**
Anakie, Mt, Vic 161
Anakie, Qld 117, 160, 206
analcime 152, 153, 226, 232
Andean volcanoes 83–5
andesite 72, 79, 82, 84, 85, 86, **87**, 88, 115, 121, 123, 124, 154, 156, **178**, **179**, 205, 209, 232

andesite volcanoes 29, 181, 203
andesitic ignimbrite 177
Andrew, Mt, Qld **149**
angiosperm 232
Angus, George F. 104
anhydrite 50, 232
Aniakchak, Alaska, USA **front map**, 55
animals see fauna
ankaramite 150, 232
ankaratite 151
anorthoclase 226, 232
anorthoclase feldspars **154**, 161
Antarctic plate **71**, 176
Antarctica
    Balleny Islands **front map**, 172, **map 173**, 174, 189
    Crater Hill **174**
    Erebus, Mt **front map**, 17, 22, 61, **map 173**, 174, **174**, 189
    future volcanism 189
    Hut Point Peninsula, Ross Island **174**
    James Ross Island 23
    Melbourne, Mt **front map**, 23, **map 173**, 174, 189
    ozone hole 54, **map 54**
    Pleiades, The **front map**, 189
    Ross Sea 61
    Ross Sea volcanoes 25
    subglacial volcanoes 23
    Takahe, Mt **front map**, 23
    Terror, Mt **175**
    Transantarctic Mountains **95**, 115
    Vega Island 23
anticlines 123
Antipodes Islands (NZ) 181, 189
Antrim plateau basalts 157
Anvils Peak, Qld 128, 129
apatite 38, 148, **149**, 163, 226, 232
    fluorapatite 227
apophyllite 153, 227
aquagene pillow breccias 153
aquagene volcanism 136, 232
Arabanga, Qld 179
aragonite 147, 227, 232
Aramoana, Sth Is, NZ 209, **map 222**
Aratiatia Rapids, Nth Is, NZ 132
arc see island arc
Arch, The, Qld 126
Archaean greenstones, WA **156**, 157
archaeobacteria 41
Arenal, Costa Rica **front map**, 225
Argentina **53**, 84
argon 232
argon 39. 36–7
argon 40. 34, 36–7
Argyle diamond mine, WA 47, **47**
Argyle pipe, WA 47, 157, 158, **159**, 163, 165
Armidale-Walcha region, NSW 186
artefacts, volcanic rock 3, 45
artistic depictions of volcanoes 101, 104–5, **104–5**
Asama, Japan **front map**, 55, 224
Ascension Island (UK) 94
ash cloud hurricanes 10
ash falls 67
Aschcroft, William 3
Asian plates 70
Askja, Iceland **front map**, 63
asteroids 7
asthenosphere 69, 70, 72, 73, 76, 79,

82–3, 88, 168, 232
Atherton, Qld 184, 187, **map 216**, **map 221**
Atherton Tablelands, Qld 136, 139, **182**
Atherton volcanic field, Qld 188, 206, **map 221**
Atlantic volcanoes 26
atmospheric pressure 162
Auckland, Nth Is, NZ 32, 33, 36, 62, 64, 181, 189, **map 216**
    volcanic tour 210–11, **map 223**
Augustine, Alaska, USA **front map**, **ii–iv**, 55
Australia
    Bass Basin **121**
    Bass Strait 122, 198–9
    Dowar Island **108**
    Heard Island **front map**, 23, 94, 96, 176, **176**, 189
    Macquarie Island **70**, **74**
    Maer Island **108**, 110
    Murray Islands **108**, 110
    see also places listed under names of states eg Tasmania, Victoria
Australian margin volcanoes 86–8
Australian plate **28**, 70, **70**, 79, 82, 83, 86, 94, 162–3, 168, 172, 186
Australian synthesis 155–7
Australian-Antarctic Discordance 76
Australian-Antarctic downwelling 184, 188
Australian-Antarctic Mid-Ocean Ridge 83
Australian-Antarctic spreading ridge 76, 172, **map 173**
Australian-Indian-Antarctic rift junction 176
auto-breccias 124
Avachinsky, Russia **front map**, 55
avalanche slides 116
avalanches 232
    and explosions 10–12, 21
    glowing 86
AVHRR see Advanced Very High Resolution Radiometer
Awu, Indonesia **front map**, 55, 224
axial rift volcanoes 71
Azores (Portugal) 92, 94
    Caldeira **front map**, 16
    Capelinhos **front map**, 22
    Sao Miguel Island 49

Babuyan, Philippines **front map**, 55
Bacchus Marsh, Vic 142
back arc basins **79**, 82
Badger Corner, Flinders Island, Tas 138
Baffin Island, Canada 96
Bahia de los Angeles, Mexico **89**
Bald Knobs, NSW **128**
Bali see Indonesia
Ball, Mt, Qld 117
ball and socket columns 127
Ballanjui Falls, Binna Burra, Qld 203, **map 220**
Ballarat, Vic 132, 139, 196, **map 217**
Balleny Islands, Antarctica **front map**, 172, **map 173**, 174, 189
Balls Pyramid, (NSW) Tasman Sea 201, **201**, **map 220**
Bam, Papua Niugini **front map**, 79
Bandai-San, Japan **front map**, 21, 224
Banks, Mt, NSW 191, **map 192**
Banks Peninsula, Sth Is, NZ 181, **207**, 207–9, **map 216**, **map 222**

Banksia 139
Baralaba, Qld **154**, 174
Bardabunga, Iceland **front map**, 55
Barkers Cave, Qld 126, **126**
Barrine, Lake, Qld 99, 107, 206, **map 221**
Barrington, NSW 136, 174, **175**
Barrington Tops, NSW **149**, 161
basalt 72, **125**, **128–9**, **148–9**, 154, 177, 200, 203, 204, 205, 206, **207**, 208, 209, 211, 232
basalt dykes **70**, 110
basalt fields **map 168**, 169, 184
    Nth Is, NZ 181
basalt gorge fillings 206
basalt lapilli 197
basalt lava flows **39**
basalt plugs 117
basalt rock 70, **144**, 146, **146**
basalt soils 42–3
basalt spatter 102, 103
basalt-filled gorges 134
basaltic andesites 79, 84, 121, 124, 154, 177
basaltic clan of volcanoes 29
basaltic icelandite 151
basaltic lapilli deposits **14**
basaltic lavas 175, 179, **179**
basaltic volcanoes 181
basalts 74, 79, 84, 85, 86, 111, 113, 117, 140, 142, 150–1, 153, 156
    extraterrestrial 7
basanite 92, 150, 199
base surges **123**, 106, 107, 108, 109, 209
Bass Basin, Australia 121
Bass Strait, Australia 122
    volcanic tour 198–9, **map 218**
batholith 85, 232
Bathurst-Oberon area, NSW 134
Batur, Indonesia **front map**, 19, **82**
bauxite 43, 233
Beauty Point, Tas 128
Beerburrum, Mt, Qld 118, 203, **map 220**
Beerwah, Mt, Qld 118, 204, **map 220**
Belanglo, NSW 119
Bell Bay, Tas 128
Bellarine Peninsula, Vic 191, **map 192**
Belmore, Mt, NSW 115
Belougery Spire, NSW 115, 202, **202**, **map 219**
Belougery Split Rock, NSW 115, 203, **map 219**
Ben Lomond, NSW 147, **147**, 152
Benioff zone 78, 233
benmoreite 150, 233
bennettitalean 233
Bennettitalean cycad 138
Berwick, Vic 131, 139
Bevendale flow, NSW 133
Bezymianny, Russia **front map**, 12, 225
Bezymianny activity 10–12
Big Ben, Heard Island (Australia) **front map**, 23, 176, **176**
Big Rock, Sth Is, NZ 209
Billy Mitchell, Solomon Islands **front map**, 55
Biloela, Qld 129
Bindook, NSW 123
Bingara, NSW **135**, 165
Binna Burra, Qld 203, **map 216**, **map 220**
Bioko, (Equatorial Guinea) 25
biscuit rock 118, **155**
Bismarck Sea Basin 82
Biu, Nigeria 25
Black Peak, Alaska, USA **front map**, 55
Black Pyramid, King Island (Tas) 110
black smokers 25
Blackall, Qld 187
Blackwood, Mt 116, 117
Blaggard Creek, Qld 143
Blanch Bay, Papua Niugini 80
blast craters 206
blind dome **11**

Blinking Billy Point, Tas 113, 199, **map 218**
block lava 124
blocks 197
Blot, C. 65
Blow, The, Tas **122**
Blow Hole, Double Head, Qld **204**
Blowhard, Mt, Vic 106
Blowhole, Kiama, NSW **120**, 121
Blue Lake, SA **front map**, **16**, 17, 109, 198, **map 217**
Blue Mountains, NSW **map 192**, 187, 188, 190–1
Blue Tier, Tas 38
Bluff, The, SA 113
Bluff Downs fauna 143
Bluff Mountain, NSW 115
Bluff Point National Park, Qld 206, **map 221**
Boat Harbour, Lord Howe Island (NSW) 201, **map 220**
Boat Harbour, Tas 160
Boat Mountain, Qld 140
Bobby Dazzler Mine, Rubyvale, Qld **160**
Bolivia 84
Bombo, NSW 128, **180**
bombs *see* volcanic bombs
Bondi, NSW 119
Bondi dyke and pipe, NSW 118
boninite 82, 233
Bo'ok, Mortlake, Vic 100
boomerangs *see* volcanic boomerangs
Bootlace Cave, Mt Gambier, SA 198
boron 61
Borrodaile Plains, Tas 135
Bouvet Island (Norway) 94
Bow Hill, Tas 163, 164, **164**
Bowral, NSW 175
Box Hill, The, NSW 136
brachiopod 233
Bradley, Mt, Sth Is, NZ 209, **map 222**
Bradys Lookout, Tas **129**
Brazil, amethyst mining 48
breached crater **11**, 12, 233
Breadknife, The, NSW **98**, 115, **200**, 202, **map 219**
breccia pipes **114**, 116
breccia zones 127
breccias 111, 112, 119, **119**, 122, 177, 199, 201, 203, 209, 233
Breises tin mine, Tas **129**
Bress Peak, NSW 202–3, **map 219**
Brick Clay Creek, NSW 164
Bridget Peak, NSW 202
Bridgewater, Cape, Vic 113
Brigooda, Qld 102, 161, 170–1, **map 171**, 187, **188**
Brigooda-Oberon boomerang 170–1, **map 171**, 185
Brisbane, Qld 187
Brittons Swamp, Tas 142
Broadlands-Ohaaki geothermal field, Nth Is, NZ 49, 88
Brocks Monument, Vic 196, **map 217**
Broken Hill, NSW **46**, 47
bromine 51
Bromo, Mt, Indonesia **front map**, 19
Bromfield Swamp, Qld 107, 206, **map 221**
Browne's Lake, SA 17, 109, 198, **map 217**
Bruce, Mt, Qld **149**
Bruce Highway, Qld 203–6
Bucca range, Qld 177
Buckland, Qld 136
Buckland shield, Qld **136**, 188, **190**
Bugaldi, NSW 142
Bullenmerri, Lake, Vic 108, 164, 197, **map217**
Bullenmerri Crater, Vic 37, 108, **map 217**
bulls-eyes (gravity) **map 169**
Bunbury basalt, WA 176
Bunbury-Kerguelen plume 176
Bundaberg, Qld 102, 103, 187

Bundaberg-Gin Gin region, Qld 113
Buninyong, Mt, Vic 196, **map 217**
Bunya Ranges, Qld 161
buried volcanic fields 119–23
Burnie, Tas 127, 128, 198, **map 218**
Burning Mountain, NSW 100, **101**
Burr, Mt, SA 113, 198
Burramys 142
Bushy Park, Tas **130**
buttes 129, 233
Byduk Caves, Vic 126
Byrock, NSW 169

Cainozoic era 233
Cairns, Qld 180
calcalkaline volcanoes 177–9
calcalkaline lavas 79, 84, 87, 233
calcite 147, 152, 227, 233
calcium carbonates 32, 147
Caldeira, Azores (Portugal) **front map**, 16
caldera craters 201
caldera floor **11**
caldera lakes 17–18
caldera rim **11**, **13**
caldera volcanoes 18, 111–12
calderas 15–16, 88, 177, 179, 180, 233
    drowned 207
Cam River, Tas 135
Cambrian period 123, 233
Camels Hump, Vic 196, **map 217**
Cameron West, Mt, Tas 128
Cameroon, Mt, Cameroon **front map**, 25
Cameroon Volcanic Line 25, 88
Camp Hill Rock, Qld **map 221**
Campaspe River, Vic 127
Campbell Plateau, Sth Is, NZ 181
Campbells Folly, Qld 115
Camperdown, Vic 197
Campi Flegrei, Italy **front map**, 67
Canada
    Baffin Island 96
    Ellesmere Island 96
    Meager Mountain 49
Canary Islands (Spain) 94
Canobolas, Mt, NSW 111, **111**, 136, 170
cap 233
Cape Barren Island, Tas **184**
Cape Verde Islands (Republic of Cape Verde) 94
Capelinhos, Azores (Portugal) **front map**, 22
Capella, Qld 139
capping lava flows 110, 111
Capricornia Basin 174
caps 203
carbon 12. 32
carbon 13. 32
carbon 14. 32
carbon dating *see* radio-carbon dating
carbon dioxide 51, 61
carbonaceous bands 139
carbonates 122, 127, 147, **147**, 148, 157
carbonatite lavas 148
carbonatites 176, 233
Carboniferous period 122, 180, 233
carbonised tree trunks 203
Carey, S.W. 97
Cargellico, NSW 169
Cargill, Mt, Sth Is, NZ 210, **map 222**
Carnarvon Basin, WA 176
Carnarvon Gorge, Qld 136, **136**
Cascades volcano observatory, Washington, USA 60
Castle Rock, Sth Is, NZ 20, **map 222**
Castlereagh River, NSW 133–4
Casuarina fossil **130**, 138
Cathedral Arch, Warrumbungles, NSW 203, **map 219**
Cathedral Rocks, Nth Is, NZ **58**
Cato Fracture, Tasman Sea 174
Cato Trough 172
cauldron 233
central vents 209

central volcanoes 111, 167, 168, **168**, 203, 205
Cerberus plains, Mars 7
Ceren, El Salvador 2–3
CFCs *see* chlorofluorcarbons
chabazite 152, 153, 227, 233
chains, moving **91**, 91–2, 94
chalcedony 226
Chalmers, Port, Sth Is, NZ 209, **map 222**
Charleville, Qld 187
Chatham Island, NZ 181
Chatham Rise, Sth Is, NZ 181
Chile 84
    Hudson, Mt **front map**, 50, 51, 52, **52–3**, 54, 55, 225
    Lascar **front map**, 61
    Licancábur **front map**, 20
    Nevado Ojos del Salado **front map**, 23
    Osorno **front map**, **85**
    Socompa 21
    Villarrica **front map**, **84**, **85**
Chinchilla, Qld 187
Chinese chronicles, ancient 3
chisel marks 124
chlorine 51, 61
chlorites 147, 227
chlorofluorocarbons (CFCs) 51, 54
chlorohydrocarbons 51
Christchurch, Sth Is, NZ 189, 207, 208, **map 222**
Christmas Island (Australia) 176, 189
chromium 161
Chudleigh, Qld 161, **162**, 187
Cibotium tasmanense 138
cinder cones **13**, 102, 206
Circular Head, Tas 199, **199**, **map 218**
Claremont, Tas 199, **map 218**
Clarke Hills, Qld 184, 187
classification of rocks 146, 150
clay swelling 42–3
clays 122, 127, 139, 146, 147, 157, 227
    kaolinite 227
    nontronite 227
cleft lava 124
Cleopatra crater, Venus 7
Clermont, Qld 206
climate 49–57
clinopyroxenes 148, **149**, **151**, 162, **164**, 226, 233
    aegirine 226
    augite 226
    diopside 226
    pigeonite 226
coal 118, 138
coal seam combustion 100
Coanjula, NT 158
Cobbera, Mt, Qld **map 221**
Cocked Hat Hill, Tas 135, 138
Coimadai fauna 142, 233
Colac, Vic 196–7, **map 217**
Colac, Lake, Vic 108, 196
Colima, Mexico **front map**, 55
collapsed caldera **14**
collapsing column **13**
collision, plate 71
Colo, Indonesia **front map**, 225 synonym of Una Una 55
Colombia 84
    Amero 64
    Galeras **front map**, 60, 61
    Nevado del Ruiz **front map**, 64, 225
    Ruiz 55
colonnade 233
Columbia River basalt 94
columbite 233
columnar jointed basalts 206, 207, 208, 209, 210
columns 127–8, **128**, **180**, 198, 201, 203, 206, 207, 208, 211
Comboyne, NSW 136, 172
complex 233
complex volcanoes **13**, 18
composite cones **11**, **14**, **84**

composite volcanoes 18
compression 186
Condobolin, NSW 169
cone sheets **10**, 115, 116
cones, volcanic 18, 197, 198, 210, 237
conglomerates **211**
Conical Hill, Sth Is, NZ 208, **map 222**
conical volcanoes 20
Conjuboy, Qld 184, 187
continental crust 70, **73**, **79**
continental heat flow 164–5
continental margin upwelling theory 97
continental-ocean subduction volcanoes **79**, 83–6
Coochin, Mt, Qld 118, **map 220**
Cooee, Mt **map 220**
Coode Island, Vic 193
Cook (submarine volcano), Solomon Islands **front map**, 22
Cooktown, Qld 184, 187
Cooktown field, Qld 188
Coolum, Mt, Qld 118
Cooma, NSW 175
Coonabarabran, NSW 152, 200, **map 216**, **map 219**
Coonowrin, Mt, Qld 118, 204, **map 220**
Cooran, Mt, Qld 118
Cooroora, Mt, Qld 118
Cooroy, Mt, Qld 118
Copeton, NSW 158, **158**, **159**
copper mining 46–7
copper-rich sulphide ore deposits 123
Coragulac, Lake, Vic 108
Coral Sea 78, 160, 164, 172, 174
Corangamite, Lake, Vic 108, 196, 197
Coricudgy, Mt, NSW 136
Corio Bay, Vic **map 192**, 193
Cornelius, R.R. 65
Coromandel Peninsula, Nth Is, NZ 189
corundum 157, 233
Cosgrove, Vic 169
Costa Rica
  Arenal **front map**, 225
Cotopaxi, Ecuador **front map**, 23, 224
coulée flow 124, 233
Coulston Lakes, Qld 102
cover rocks 162
cowlesite 153, 233
Cox, Geoffrey 101
Crater Dome, Nth Is, NZ **front map**, 212, **map 222**
Crater Hill, Antarctica **174**
Crater Hill, Nth Is, NZ 32, 210, **map 223**
Crater Lake, Nth Is, NZ 62
crater lakes **11**, 17–18, 197, 198
crater rims **84**, **102**, 188, 198, **map 217**
crater row **10**, 12, 15, 233
craters 12–18, 187, 198, 210, 212, **212**, 233
Craters of the Moon, Idaho, USA 15
Cretaceous period 56–7, 95, 96, 122, 134, 139, 140, 177, 179, 205, 233
cross-cutting dykes 115, 210
crust 233
  continental and oceanic 70
crystallisation 146, 152–3
Cumberland Islands, Qld 177
Cyathealean 233
cycad 233
Cygnet, Port, Tas 118, **155**

dacites 72, 79, 84, 85, 88, 123, 124, 154, 156, 177, 233
Danan, Indonesia **front map**, 23
DANTE (robot) 61
dating
  fossils 131
  volcanic rocks 148
  volcanoes 31–8
David, T.W.E. 100, 139
Dawson, James 100
Daylesford, Vic 139
debris flow **11**
Deccan, India 48, 55, 56–7

Deccan plume basalts 96
Dee River, Tas 136
deep gas discharges **map 215**
deep lead **133**
deformation, geological 119–23
Delegate, NSW 165
deltaic stage 109–10
D'Entrecasteaux Islands, Papua Niugini 86, 87
denuded fields 118–19
Derby, Tas **129**
Derwent River system, Tas **132**, 135–6, 138, 199, **map 218**
Devil's Post Pile, California, USA 127
Devil's Punchbowl, SA 109, 198, **map 217**
Devonian period 123
Devonport, Tas 198, **map 216**, **map 218**
diamonds 47, 129, 134, **135**, 157–9, **158**, **159**, 163, 165, 176
diatoms and diatomite beds 142
diatremes **13**, 118, 233
dinosaur extinctions 56–7, **56**
Diogenes, Mt, Vic see Hanging Rock
diorite 85, 155, 180, 205, 233
diprotodont 233
disruptions 131–43
dissected structures 18
dissected volcano **11**
distribution of volcanoes **front map**, 25–9
Docking, Shay 104
  'Mt Elephant with Lava Flow' **10**
  paintings 104
  'Volcano, Town and Western Plains' **105**
documentation 3, 6
dodecahedral crystals 157, 158, 233
Dog Rock, Vic 143
dolerite 115, 118, **118**, 128, 154, 176, 199, **199**, 233
dolerite dyke **159**
dolerite intrusions 175
dolerite sills 174, **176**
dolomite 227
Dolphin Point, Qld 205
Domain, Auckland, Nth Is, NZ 211, **map 223**
domes 115, 118, 202, 203, 210, 233
doming 233
Don Heads, Tas 198, **map 218**
Donna Bay, Qld 141
dormant volcanoes 31
Double Head, Qld 206, **map 221**
Dowar Island, Australia **108**
downwelling 233
drainage 233
drainage disruptions 132–6
Dromedary, Mt, NSW **36**, 118
drowned calderas 17–18
DTD Seamount, East Pacific 27
Dubbo, NSW 170, **map 171**, 175
Duck Point Bay, Tas **134**
Dumbell, Mt, Qld 117
Dundee rhyolite ignimbrite 180
Dunedin, Sth Is, NZ 18, 181, 189, 207, **208**, 209, **map 216**, **map 222**
Dunedin Harbour, Sth Is, NZ 209–10, **map 222**
dyke swarms 115
dykes **10–11**, 74, 115, 116, 117, 118, **118**, 158, 176, **179**, 200, **200**, 201, 202, 203, 205, **207**, 208, 209, 233

Eacham, Lake, Qld 99, **100**, 107, 206, **map 221**
earthquake epicentres **rear map**
earthquakes 21, 62, 190
  Australian 186–7
  submarine 78
East Pacific Rise ridge 28, 37, 50, **74–5**, 75
Easter Island (Chile)
  Anakena **44**
  Rano Raraku **44**, 45

Ebor, NSW 136
Ebor Falls, NSW **39**, 128
Ebor-Dorrigo, NSW 38, **39**
Eccles, Mt, Vic **front map**, 32, 103, 125, 126, **194**, 197, **map 217**
echinoid 233
Echo Crater, Nth Is, NZ **front map**, 212, **map 222**
eclogite 158, 163, 165, 233
economic uses of volcanic products 42–9
Ecuador 84
  Cotopaxi **front map**, 23, 224
Eden, Mt, Nth Is, NZ 210, 211, **212**, **map 223**
Egg Rock, Binna Burra, Qld 203
Eglon, Mt, Kenya 48
Egmont, Mt, Nth Is, NZ **front map**, 8, 88, 189, 210
Eiffel volcanic field, Germany 12, **88**, 89
El Chichón, Mexico **front map**, 50, 51, 52, 54, 55, 63, 225
El Misti, Peru **front map**, 20
El Niño 233–4 see also La Niña
El Niño-La Niña oscillation 50, 52
El Salvador 2–3
  Ceren 2–3
  Ilopango **front map**, 55, 224
Elaeocarpus 139
Eldgjá, Iceland **front map**, 18, 55
electrons 32
Elephant, Mt, Vic 8, **10**, 19, 102–3, 197, **map 217**
Eliza, Mt, Lord Howe Island (NSW) **94**, 201, **map 220**
Ellendale pipes 158
Ellesmere Island, Canada 96
emerged volcano **13**
emergent volcanoes **12**, 109–11
emerging volcano **12–13**
Emmaville, NSW 139
Emperor seamounts, North Pacific 92
Emperor-Hawaiian hotspot chain 172
Emydura 142
entablature 127, 234
environments, volcanic 41–57
Erebus, Mt, Antarctica **front map**, 17, **22**, 61, **map 173**, 174, **174**, 189
erionite 153, 227, 234
eroded vents 113–15
eroded volcanoes 196
Eromanga Basin 177
erosion 113–19
erosional caldera **15**, 16
erosional landforms 128–9
Erskine Park, NSW 119
eruption clouds **14**, 50–1, **50**
eruption column **11**
eruption plume **10**
eruptions 234
  crater lake **87**
  significant 224–5
escarpment 234
Esk rift 180
Ethiopia 2
Ethiopian rift 89
Ettingshausen, C. Baron von 139
Eucalyptus 139
Eungella, Qld **149**
Euramoo, Lake, Qld 99, 107, **map 221**
Eurasian plate 70
Euryzygoma 143
evolution of life 41
evolved lava 146
Exmouth, Mt, NSW 203, **map 219**
expanding earth theory 97
Expedition Range, Qld 134, 169
explosion craters 197, 198
explosions and avalanches 10–12, 21
explosive alkaline pipes 175
explosive beds 209
explosive craters 199, 206
explosive deposits 197, 198, 200, 201, **207**

explosive pipes 118, 119
extinct volcanoes 31
extinction events 56–7
extinctions 50
  animal and plant 2
Falcon Island, Tonga **front map**, 22
FAMOUS ocean project **74**
fault **11**
fauna, fossil 140–3
feeder 234
feldspar 47, 72, 79, 113, 121, **144**, 151, **151**, 157, 226, 234
  cut **162**
feldspathoids 148, 150, **151**, 151, 226, 234
  sodalite 226
felsic rocks 151, 154
Ferdinandea see Graham Island
Fernandina, Galápagos Islands (Ecuador) **front map**, 55
ferrierite 152, 153, 227, 234
Fiji Basin 82
fish fossil **140**
fission track dating 38, 234
fission tracks **36**, 160
fissure lines 198
fissure volcanoes **13**, 18, 234
fissures 208, 211
Fitzroy trough, WA 189
Flagstaff Hill, NSW 43, **44**
flank crater 12, 234
flank eruptions 4
Flat Topped Bluff, Tas 110
Flinders Island see Tasmania
Flindersia 139
flood basalts 95–6
flora, fossil 138–9, **139**
flow bands 177
flow feet 123, **124**
flow foot breccias 110, 124, **124**, **134**, 135, 136
flow fragmentation 177
flow lines and gas 127
flow sequences 128
flow tops 127
flows 116, 187, 188, 234
fluid inclusions 160
fluorine 51, 61
Focal Peak, Qld 112, 115, 136, **166**, 172
fold belt **79**
folding 234
formulas, prediction 65
Forth River, Tas 134
fossil beds, Wynyard, Tas 198
Fossil Bluff, Tas **141**, 141–2, **142**, 198, **map 218**
fossils 137–43, **130**, **137**, **139**, 140–3
  dating 131
Four Mile Creek, Qld 134
Fox, Mt, Qld 102, 134, 188
fractionated lava 146
France
  Massif Central 111
Frankford Highway, Tas **176**
Franklin Seamount, Solomon Sea 28
Fraser Island-Barrington boomerang 172
Frying Pan Lake, Nth Is, NZ **58**, 212
Fuego, Guatemala **front map**, 55
Fuji, Mt, Japan **front map**, 8, 20, 82
fumarole 234
future Australian eruptions 183–93

gabbro 70, 74, 85, 115, 118, 154, 180, 209, 234
Gads Hill, Tas 134, 135
Galápagos Islands (Ecuador) 26
  Fernandina **front map**, 55
Galeras, Colombia **front map**, 60, 61
Gall Range, Qld 177
Galunggung, Indonesia **front map**, 50, 63, **65**, 224–5
Gambier, Mt, SA **front map**, **16**, 17, 32, 100, 109, 113, **184**, 186, 197–8, **map 216**, **map 217**

Garawilla volcanic field, NSW 122, 152
Garnet Gully, Qld 161, **188**
garnet peridotite 72, 73
garnet peridotite mantle 163
garnet pyroxenite rocks 163
garnets 47, 157, 158, 161, 162, 163, **164**
gas and flow lines 127
Gascoyne seamount, Tasman Sea **map 169**, 185
gastropod 234
Gawler Range, SA 157
Gayndah, Qld 187
Geelong, Vic 143, **map 192**, 193, **map 217**
Geilston Bay, Tas 141
gemstones and zeolites 47–8
geodetic levelling 61, 234
geothermal 234
geothermal gradient 162
geothermal power 48–9
Germany
  Eiffel volcanic field 12, **88**, 89
  Herchenberg quarry **88**
  Laacher See **front map**, 88, **88**, 90
  Rhenish shield 89
Geyser Flat Sinter deposit, Nth Is, NZ 211, **map 222**
geysers **11**, **40**, 48, 211, 234
Geysers, The, California, USA 49
Giants Causeway, N. Ireland (UK) 127
Gibraltar, Mt, NSW 118
Gibraltar Rock, Sth Is, NZ 208, **map 222**
Gillies Crater, Qld 206, **map 221**
Gillies, Mt, Qld 115
Giluwe, Mt, Papua Niugini 86
Gin Gin, Qld 187
ginkgophyte 234
Gippsland Basin, Bass Strait 78, 139, 172
Gisborne, Mt, Vic 106
Gladstone, Qld 177
glass, volcanic 146, **146**
Glass House Mountains, Qld 118, **119**, **202**, 203–5, **map 216**, **map 220**
glassy fragmental deposits 74
glassy lava 203
glassy splatter 198
glassy tuffs 199
Glen Innes, NSW 160, 161
Glenora, Tas 138
Global Positioning System (GPS) 62
  image of crustal plates **71**
  image of Pacific **28**
  image of earthquake epicentres **rear map**
Gloucester, NSW 165
Gloucester Buckets, NSW 122, **122**
glowing avalanche 86
glowing cloud **13**
gmelinite **152**, 153, 227, 234
gneiss 70, 234
Gnotuk, Lake, Vic 108, 164, 197, **map 217**
Gnotuk Crater, Vic 108, **map 217**
gold 129, 134
gold mining 46
Gondwana 138, 156, 175–6
gonnardite 153, 227, 234
Good Hope, NSW **149**, **155**
Gorda ridge Northeast Pacific 37
Gothic Cave, Vic 126
Gough Island (UK) 94
Gower, Mt, Lord Howe Island (NSW) 112, **112**, 201, **map 220**
GPS *see* Global Positioning System
Graham Island (submerged), Italy **front map**, 22
Grand Canyon, Arizona, USA 129
granite 70, 85, 123, 154, **159**, 176, 179, **179**, 180, 205, 234
Granite Bay, Qld 205
granite-metamorphic rock 119
granitic core **179**
granodiorite 85, 180, 234
Grant, Cape, Vic 113
granulite 70, 119, 234
graphite 165

gravel beds 134, **135**
  sapphire-bearing 160
gravity
  and magnetism 63
  Tasman Sea **map 169**
Great Lake, Tas 128, **134**, 136
Great Rift Zone, Idaho, USA 15
Greece
  Santorini **front map**, 3, 55, 224
Green Hill, Qld 206
greenhouse effect 51, 234
Greenland plume 96
Grim, Cape, Tas **109**, 110, **110**
Grimsvötn, Iceland **front map**, 23
Grose River, NSW 191
ground movements 61–2
groundmass minerals 145
Guatemala
  Fuego **front map**, 55
  Santa Maria **front map**, 224
Guérard, Eugen von 104
  paintings 104
  'Tower Hill' **104**
Guildford courses, Tas 135
Gula Mons, Venus 7
Gulgong, NSW 142
Gunnedah, NSW **map 175**
Gunnedah Basin, NSW 122
Gwydir River, NSW 136
gymnosperms 234
Gympie, Qld 179

hackly 234
Hagen, Mt, Papua Niugini 86
Hakone, Japan **front map**, 82
Haleakala, Hawaii, USA **front map**, 15
Halemaumau, Hawaii, USA 17, **60**, 126
halogen 234
halogen compounds 51, 54
Hamilton, Mt, Vic 132
Hamilton fauna, Vic 142
Hampden Shire Quarry, Vic 101, **102**
Hanging Rock (Mt Diogenes), Vic **30**, 37, 196, **map 217**
Harman Valley flow, Vic 32, 125
Haroharo Caldera, Nth Is, NZ **front map**, 211, **map 222**
Hawaii *see* USA
Hawaiian deposits 108, 110, 113
Hawaiian eruption **13**
Hawaiian hotspot 186
hawaiian style 74
Hawaiian Volcano Observatory **60**
Hawaiian-style fire fountains 210
hawaiite 150, 234
Hawkesbury River, NSW **map 191**, **192**
Hay, Mt, Qld 127, 177, 206
Headlow, Mt, Qld 116, **map 221**
Heard Island, Australia 94, 96
  Big Ben **front map**, 23, 176, **176**, 189
heat and fluids 61
heat flows 186, 187
Heemskirk seamount, Tasman Sea 172
Hekla, Iceland **front map**, 3, 55
helium 3. 61
helium 4. 61
Hell's Gate, Nth Is, NZ **49**
Hells Gates, Qld 205
Hellyer River, Tas 123, 138
Herbert, Mt, Sth Is, NZ 209, **map 222**
Herchenberg quarry, Germany **88**
Herculaneum, Italy 3, 66
Hercules, Tas 123
heulandite 152, 227, 234
Hibok-Hibok, Philippines **front map**, 60, 225
Hickey Falls, NSW 200
Hicks, Mt, Tas **132**, **map 218**
Hikurangi trench, Nth Is, NZ 181
Hilgenberg, O.C. 97
Hillsborough, Cape, Qld 116–17, **116–17**, 141, 168
Hivesville, Qld **42**
Hobart, Tas **114**, 115, 199, **map 216**, **map 218**

Hobsons Bay, Vic **map 192**, 193
Hogarth Range, NSW 161, **162**
Hokkaido, Japan 82
Hokusai, Katsushika 101
Hollowback, Mt, Vic 106
Holmes, Mt, Sth Is, NZ 210, **map 222**
hominid 234
Honshu, Japan 82
Hope, Mt, NSW 123
horizons (soil) **42**, 43
hornitos 125, 234
Hornsby, NSW **114**, 119, **119**
hotspot volcano 73
hotspots 76, **91**, 91–5, **96**, 160–1, 168, **map 168**, 169, 172, 184, 185, 186, 234
  continental 94–5
Hotwater Creek, Nth Is, NZ 58
Hoy, Mt, Qld 117
Hualalai, Hawaii, USA **front map**, 18
Hudson, Mt, Chile **front map**, 50, 51, 52, **52–3**, 54, 55
Hughenden, Qld 187
Hummock, Qld **102**, 103, 113
Hunter River, NSW 136
Hunter-Bowen folding 180
Hunua Ranges, Nth Is, NZ 62
Hut Point Peninsula, Ross Island, Antarctica 174
hyalotuff deposits 109, 110
hyalotuffs 135, 136, 234
hydrochloric acid 51, 61
hydrofluoric acid 51
hydrogen 61
hydrogen disulphide 54, 61
hydrothermal 234
hydrothermal activity 181
hydrous silicates 148
Hypipamee Crater, Qld **front map**, 8, **182**, **204**, 206, **map 221**

Ibusuki, Japan **front map**, 55
ice-covered vents 109
Iceland 18, 103
  Askja **front map**, 63
  Bardabunga **front map**, 55
  Eldgjá **front map**, 18, 55
  Grimsvötn **front map**, 23
  Hekla **front map**, 3, 55
  hotspot 76, 94, 96
  Katla **front map**, 23, 62
  Krafla **front map**, 23, 55, **64**, 68
  Laki **front map**, 18, 54, 55, 224
  Surtsey **front map**, 22, 109
Icelandic deposits 110
Icelandic eruption **13**
icelandite 76, 106, 150, 151, 234
igneous rocks 70, 234
ignimbrites 12, 67, 111, 132, 177, 179, 234
Illawarra region, NSW 119–21
ilmenite 157, 226, 234
Ilopango, El Salvador **front map**, 55, 224
impacts, meteorite 56, 96
India 176
  Deccan 48, 55, 56–7
Indian Ocean volcanoes 26
Indian plate 176
Indian-Australian plate 70, **71**
Indonesia
  Agung, Mt **front map**, 224
  Anak Krakatoa **front map**, **21**, 23
  Awu **front map**, 55, 224
  Batur **front map**, 19, **82**
  Bromo, Mt, **front map**, 19
  Colo, 225 **front map**, *synonym of* Una Una 55
  Danan **front map**, 23
  Galunggung **front map**, 50, 63, **65**, 224–5
  Kelud **front map**, 224–5
  Krakatoa **front map**, 3, **21**, 23, 54, 55, 224
  Makian **front map**, 224
  Merapi **front map**, 61, 65, 224–5

Papandajan **front map**, 224
Perboewatan **front map**, 23
Rakata **front map**, **21**, 23
Raung **front map**, 224
Tambora **front map**, 20, 54, 55, 224
Tengger **front map**, 19
Toba 54
Toba, Lake 21, 180
Toba Caldera 20–1
Una Una 55 *synonym of* Colo **front map**, 225
Indonesian arc 82
Inferno Crater, Nth Is, NZ **front map**, 212, **map 222**
Ingham, Qld 113
intrusions 85, 200, 203, 204, 205, 206, 209, 210
  belljar 85
intrusive rocks 154, 155, 234
Inverell, NSW **135**, 136, **157**, 160, 169, 170, **map 171**
Io (moon of Jupiter) 7
Ipswich, Qld 174
Ipswich-Brisbane area, Qld 139, 140
iridium anomaly **56**, 57
iron 70
iron hydroxides 43
iron oxides 146, 147, **149**, **151**
iron sulphide 138
Ironpot Mountain, Qld 116, **map 221**
ironstone matrix 139
iron-titanium oxides 148, 226
Irvine, Mt, NSW 191, **map 192**
island and coastal volcanoes 22–3
island arc 234
island arc volcanoes 73, **79**, 79–82
isotope dating 31–7, **36**, 234,
isotopes 31–2, 63
Italy
  Campi Flegrei **front map**, 67
  Etna, Mt **front map**, 3, 8, 18, 21, 51, 55, 62, 63, 224–5
  Graham Island (submerged) **front map**, 22
  Herculaneum 3, 66
  Lardarello 49
  Lipari Island 45
  Naples 67
  Pompeii 3, 12, **14**, 66, **66**
  Somma, Mt **front map**, **14**, 16, 66, 67
  Stabiae 3
  Stromboli **front map**, 55, 102
  Valle De Bove 21
  Vesuvius, Mt **front map**, 3, 12, **14**, 16, 55, 66–7, **67**, 101, 111, 224
  Vulcano **front map**, 55, 102
Izu-Bonin trench 82
Izu-Oshima, Japan **front map**, 6, 61, **83**

James Ross Island, Antarctica 23
Jan Mayen Island (Norway) 94
Janszoon seamount, Tasman Sea 172
Japan
  Asama **front map**, 55, 224
  Bandai-San **front map**, 21, 224
  Fuji, Mt **front map**, 8, 20, 82
  Hakone **front map**, 82
  Hokkaido 82
  Honshu 82
  Ibusuki **front map**, 55
  Izu-Oshima **front map**, 6, 61, **83**
  Kyushu 82
  Mihara-Yama **83**
  Miyakeijima **front map**, 55
  Myozin-Syo **front map**, 22
  Okinawa Trough 47
  Ontake **front map**, 21
  Oshima 55, **83**
  Ryukyu Island 82
  Sakurajima **front map**, 50
  Towada **front map**, 55
  Unzen, Mt **front map**, 6, **8**, 21, 60, 61, 63, 224–5
Japan Sea plate 82
Japanese arc 82

Java *see* Indonesia
Java trough 82
Jensen, H.I. 16
Jim Crow Mountain, Qld 116, **116**
Johnstone, R.M. 138
Johnstone River, Qld 134
joints 127–8
jökulhlaups 23, 234
Juan de Fuca ridge, East Pacific 28, 37
Juanobong, Qld 187
Jukes, Mt, Qld 116, 117, **117**
Jurassic period **114**, 115, 119, 122, 138, 139, 152, **176**, 200, 234

Kakanui South Head, Sth Is, NZ 207
Kambalda, WA **156**, 157
Kamchatka, Russia 49, 54
Kanakana Dome, Nth Is, NZ 211, **map 222**
Kandos, NSW 175
Kangaroo Island, SA 174
Kaputar, NSW 171, **171**
Karkar, Papua Niugini **front map**, 60, 79
Karymsky, Russia **front map**, 55
Katla, Iceland **front map**, 23, 62
Katmai, Mt, Alaska, USA **front map**, 55
Kayrunnera, NSW 165
Keilambete, Lake, Vic 108
Keith, M.L. 97
Kelud, Indonesia **front map**, 224–5
Kent Islands, Qld 179
Kenya 2
 Elgon, Mt 48
 Kenya, Mt **93**
 Kitum Cave 48
 Oldoinyo Lengai **front map**, 91
 Turkana, Lake 111
Kenya, Mt, Kenya **93**
Kenyan rift 89
Kereru Geyser, Nth Is, NZ 211
Kerguelen, (France)
 Ross, Mt **front map**, 176
Kerguelen Island (France) 94, 176
Kerguelen plateau, Southern Ocean 95, 176
Kerguelen Plateau-Bunbury-Rajmahal plume basalts 96
Kettle, Mt, Sth Is, NZ 210, **map 222**
Kiama, NSW 119, **120**, 121, 128
Kiandra, NSW 138
Kick 'em Jenny (submerged), Caribbean **front map**, 64
Kikai, Ryukyu Islands, Japan **front map**, 55
Kilauea, Hawaii, USA **front map**, 4–5, 18, **19**, **46**, 60, **60**, 125
Kilimanjaro, Mt, Tanzania **front map**, 8
Killarney Gap, NSW 171, **171**
Kimberley, WA 189
kimberlite 156, 158, 176, 234
King Island (Tas)
 Black Pyramid 110
Kings Plain, NSW 160
Kingston, SA 186
Kinny, P. **36**
Kinrara, Qld 36
Kitum cave, Kenya 48
Kizimin, Russia **front map**, 55
Klyuchevskoy, Russia **front map**, 51, 55
Kobakaba, Papua Niugini 62
Kohala, Hawaii, USA 18
komatiites 157
Kombiu, Mt, Papua Niugini **80**
Kooroocheang, Mt, Vic 125, **125**, **map 217**
Krafft, Katia 6, 9
Krafft, Maurice 6, 9
Krafla, Iceland **front map**, **23**, 55, **64**, 68
Krakatoa, Indonesia **front map**, 3, **21**, 23, 54, 55, 224
Kronotsky, Russia **front map**, 20
Kuai, Hawaii, USA 92
Kulnura pipe, NSW 147
Kupaianaha, Hawaii, USA **front map**, 4, **5**, **124**
Kurile Islands *see* Russia

Kurrajong, NSW 190–1, **map 192**
Kuwae, Vanuatu **front map**, 55
Kweebani Cave, Binna Burra, Qld 203, **map 220**
Kyushu, Japan 82
La Niña 234–5 *see also* El Niño
La Soufrière, St Vincent **front map**, 224
Laacher See, Germany **front map**, 88, **88**, 90
labradorite **162**
Lachlan fold belt, NSW, Vic, Tas 123, 156, 175, **179**
Lachlan River, NSW 133
Lady Barron, Flinders Island, Tas **184**
Lady Julia Percy Island, Vic 22, 110, 184, **map 217**
Laetoli, Tanzania 2
lahar 200, 235 *see also* mudflows
Lake *see under names of individual lakes*
Lake Terrace, SA 197
Lakeland Downs, Qld 160
Laki, Iceland **front map**, 18, 54, 55, 224
Lamington, Mt, Papua Niugini **front map**, 86, 225
Lamington National Park, Qld 203, **map 220**
Lamont seamounts 27
lamproite 156, 158, 235
lamprophyre 155, 158, 235
Landsat Thematic Mapper per (TM) 61
Landsborough, Mt, Qld 113, **149**
Langila, New Britain, Papua Niugini **front map**, 79
lapilli 235
lapilli tuffs **88**, 106, 119
Lardarello, Italy 49
Lascar, Chile **front map**, 61
lateral slip faulting 71, 235
lateral twinning 134
laterite **42**, 43, 235
latite 155, **180**
Lau Basin, Tonga 47, 82
laumontite 152, 227, 235
Launceston, Tas 134–5, **184**
lava 72, 146, 209, 235
 submarine 75
lava aprons 197, 206
lava caps 191
lava caves 126
lava domes 10, **11**, 18, 207, 208, 211, 212
lava driblets 126
lava fields 132, 133, 203
lava floods 1–2
lava flows **11**, 18, 123–8, 131, **132**, **132**, 133, 197, 198, 199, 203, 204, **207**, 209, 210
lava fountain **viii**, **4**, **13**
lava lakes **11**, 16–17, 61, 235
lava levees 125
lava plains 129, 132, **132**
Lava Plains, Qld 160
lava shields 206
lava tubes 125, 126, 210
lava types 5, 79, 82, 92, 123–4
lava volcanoes **13**, 18, 103, 106–9
lavicles 126, 235
Lawalul, Vic 106
lead 235, *see also* deep lead
lead 206.35, 37 also Pb206: Pb204.35
lead 207.35 also Pb207: Pb204.35
lead 208.35 also Pb208: Pb204.35
lead-zinc sulphide ore deposits 123
Leap, The, Qld 117
LeBrun, Mt, Qld 102
Leg of Mutton Lake, SA 17, 109, 198, **map 217**
Lesser Antilles 64
leucite 155, 226, 235
leucitite **44**, 89, 150, 151, 168, 169, 170, 235
Leura, Mt, Vic 100, **106**, 108, 117, **163**, 197, **map 217**

levyne 153, 227, 235
Liawenee, Tas 136
Licancábur, Chile **front map**, 20
Lidgbird, Mt, Lord Howe Island (NSW) 112, **112**, 201, **map 220**
Lightning Ridge, NSW 140
Lihir Island, Papua Niugini 46
Line Island seamounts, central Pacific 95
Lipari Island, Italy 45
liquid lava 16–17
lithium 61
lithosphere 69, 72, **73**, 79, 88, 168, 235
Little Murray River, NSW 134
lobate 235
Loihi (submerged), Hawaii, USA **front map**, 18, 60, 92
Loki, Io 7
Lolobau, New Britain, Papua Niugini **front map**, 79
Lombok *see* Indonesia
Long Island, Papua Niugini **front map**, 55, 79
Lord Howe Island (NSW) **94**, 112, **112**, 115, 144, 169, **map 169**, 201, **201**, **map 216**, **map 220**
 Admiralty Islands 201, **map 220**
 Boat Harbour 201, **map 220**
 Eliza, Mt **94**, 201, **map 220**
 Gower, Mt 112, **112**, 201, **map 220**
 Lidgbird, Mt 112, **112**, 201, **map 220**
 Malabar Hill 201, **map 220**
 North Beach **94**
 North Ridge 201, **map 220**
 Old Gulch 201, **map 220**
 Old Settlement Beach **94**
 Red Point 201, **map 220**
 Searles Point 201, **map 220**
 Stevens Point 201, **map 220**
 tour 201, **map 220**
 Transit Hill 201, **map 220**
Lord Howe Rise 78
Lords Table Mountain, Qld 128, 129
Lorinna, Tas 134
lower colonnade 127
lower crust **79**, 163, **165**
Luise Caldera, New Ireland, Papua Niugini **front map**, 46
Lune River, Tas 115, **137**, 138
LVZ *see* Low Velocity Zone 71
Lyell, Mt, Tas **122**, 123
Lynchs Crater, Qld 107, **map 221**
Lyttelton, Sth Is, NZ 18, 207, **207**, **map 222**

maars 12, **13**, 18, 106–9, **106**, 186, 196, 197, 198, 206, 210, 235
Maat Mons, Venus **7**
Macarthur, Vic 197
Maccullochella macquariensis **140**, 142
Macdonald Seamount, East Pacific **front map**, 28
Macedon, Vic 168, **168**
Macedon, Mt, **map 217**
Macha Tor, NSW 202, **map 219**
MacIntyre River, NSW 136
Macquarie Island, Australia **70**, 74
Macquarie Plains, Tas 138
Macquarie River, NSW 134
Macropus 142
 M. giganteus 143
Maer Island, Australia **108**, 110
Maggs Mountain, Tas 134
magma 61, 72, 78, 146, 235
magma chambers **11**, **13**, 79
magnesio-wüstite 158, 235
magnesium 70
magnesium-rich lavas 157
magnetic stripes **76**, **77**, 78
magnetism **76**, 76–8
 and gravity 63
magnetite **149**, 226
Mahanga Geyser, Nth Is, NZ 211
Main Range, Qld **166**
Main Range-Comboyne boomerang 172

Makian, Indonesia **front map**, 224
Malabar Hill, Lord Howe Island (NSW) 201, **map 220**
Malanda, Qld 136
Maleny, Qld 204, 205
Mamaku plateau, Nth Is, NZ 132
Managua, Nicaragua 2
Manam, Papua Niugini **front map**, **i**, 79
Mangere, Mt, Nth Is, NZ 210, **map 223**
Manila, Philippines **33**
mantle 70, 235
mantle peridotite 197, 198, 199, 206
Manurewa, Nth Is, NZ 210, **map 223**
marcasite 138
Margate, Tas 199, **map 218**
Marianas Islands
 Pagan **front map**, 55
Marianas trench, west Pacific 96
Maroa, Nth Is, NZ 88
Marrawah, Tas 142
Marrawah-Redpah district, Tas 110
Mars 7
 Alba Petera 7
 Cerberus plains 7
 Olympus Mons 7
 Tharsis Tholus 7
marsupial fossils 140–3
Martinique (France)
 Pelée, Mt **front map**, 10, 55, 59, 111, 115, 224
 St Pierre 10, 59, 115, 224
Maryborough, Qld 179
Maryborough Basin, Qld 177
Masaya, Nicaragua **front map**, 55
massif 235
Massif Central, France 111
massive lava 74
Matahina plateau, Nth Is, NZ 132
Matavanu, Western Samoa **front map**, 16
Maui, Hawaii, USA **15**, 92
Mauna Kea, Hawaii, USA **front map**, 18
Mauna Loa, Hawaii, USA **front map**, 18, **19**, **19**
Maybole, NSW 136
Mayon, Philippines **front map**, 20, 60, 225
Mayor Island, Nth Is, NZ 32, 189
Mazama, Oregon, USA **front map**, 55
McBride, Mt, Qld 36
McBride field, Qld **34**, 103, 126, **126**, 187, 188
Meager Mountain, Canada 49
Meiji seamount, North Pacific 92
Melbourne, Mt, Antarctica **front map**, **map 173**, 23, 174, 189
Melbourne, Vic, **map 217**
 future eruption scenario 191–3, **map 192**
melilite, 226, 235
melilitites 89, 92, 150, 151, 235
melts 146, 235
 basaltic 174–5
Merapi, Indonesia **front map**, 61, 65, 224–5
Mercury 7
Merriwah, NSW 153
Mersey River, Tas 134
Mersey-Forth Rivers, Tas 138
mesas 129, 235
mesolite **48**, 153, 235
Mesozoic era 235
metamorphic rocks 70
metapyroxenite 235
meteorites 7
methane 51, 61
methylchloride 51
Mexico
 Bahia de los Angeles 89
 Colima **front map**, 55
 El Chichón **front map**, 50, 51, 52, 54, 55, 63
 Nevado de Toluca **front map**, 86
 Parícutin **front map**, 225

Mexico continued
  Pathé 49
  Volcan de las Tres Virgines **front map**, **84–5**
  Yucatan 57
Mib Seamount, East Pacific 27
mica 79, 147, 148, **149**, 155, 157, 163, 226, 235
  phlogopite 227
  biotite 227
mica-lamprophyre rock **118**
microgranite 154
microsyenite 154
mid ocean ridge 71, **73**, **79**, 235
Mid Ocean Ridge Basalt see MORB
mid ocean ridge volcanoes 27, 29, 72–8, **73**
mid-Atlantic ridge 75
migratory volcanic chains 168–72
migratory volcanoes 180, 181
Mihara-Yama, Japan **83**
Mihiwaka, Mt, Sth Is, NZ 210, **map 222**
Mihiwaka South, Sth Is, NZ 210
Miketeebumulgrai, Mt, Qld 118, **map 220**
Milla Milla Falls, Qld 128, 206, **map 221**
Millmerrin, Qld 170, **map 171**
Millstream Falls, Qld 206, **map 221**
Minchinbury, NSW 119
mineral inclusions 160
minerals, volcanic 145, 146–9
minerals of volcanic rocks, typical 226–7
mines and ores 45–8
mining gemstones 206
Miocene epoch 110, 235
Mir (submersible craft) 28
Miyakeijima, Japan **front map**, 55
Moa-Bone Point Cave, Sth Is, NZ 207
möberg 23, 235
mode 146, 235
modified magma 146
Moho 70, 163
Moina, Tas 135
Mok Seamount, East Pacific 27
Molokai, Hawaii, USA 92
Monaro, NSW 42, 43
Monaro field, NSW 136
Mondilibi, Vic 106
monitoring volcanoes 59–67
monogenetic volcanoes 18, 235
Monotombo, Nicaragua **front map**, 55
monotreme fossils 140
Monte Christo Mine, NSW **135**
Monto, Qld 129, 169
monzonite 85, 155, 235
moon 7
moonstones 161
Mopra Rock, NSW 200, **map 219**
MORB (Mid Ocean Ridge Basalt) 72, 75, 76, 78, 82, 235
Mornington Island, Qld 180
Mornington Peninsula, Vic 193
Mortlake, Vic 197, **map 217**
Motukorea Island, Nth Is, NZ 210, **map 223**
Motutapu Island, Nth Is, NZ 210, **map 223**
Mount see under name of individual mountain
Mt Eccles National Park, Vic **194**, 197, **map 217**
'Mt Elephant with Lava Flow' (Docking) **10**
Mt Emu Creek, Vic 132
mud volcanoes 96
mudflows 64, 66, 67, 200, 203, 207
mugearite 150, 203, 235
Muirhead, Mt, SA 100, 113
Mungawoppa, Mt, Qld 116, **map 221**
Mungore, Qld 179
Murgon, Qld 140
Murray Islands, Australia **108**, 110
Murrumbidgee River, NSW **149**, **155**
museums 6

Musician seamounts, central Pacific 95
Myozin-Syo, Japan **front map**, 22
Nandewar, NSW 136, 170, 171, 171, **map 171**
Napier, Mt, Vic 103, 126
Naples, Italy 67
natrolite 152, **152**, 153, 227, 235
Natural Bridge, Mt Eccles Vic 197
Naturalist Plateau basalts 176
Nazca plate 83
Nebo, Qld **149**
necks 142, 200, 202, 203, 235
  and outliers 113
Nelson, Cape, Vic 113
neodymium 35, 63
neodymium 143. 35 also Nd143:Nd144.35
Nepean River, NSW **map 192**, 191
nepheline **149**, 150, 226, 235
nepheline benmoreite 199, 209, 210
nepheline hawaiite 150, 209
nepheline trachyte 160
nephelinite 89, 92, 150, 151, 169, 235
Nera volcanics, Qld 180
neutrons 32
Nevado de Toluca, Mexico **front map**, **86**
Nevado del Ruiz, Colombia **front map**, 64, 225
Nevado Ojos del Salado, Chile **front map**, 23
New Britain see Papua Niugini
New Britain arc 79, 82
New Britain trench 79
New England, NSW **156**, 160, 180
New England Tablelands, NSW 136
New England-Hodgkinson Basin fold belts 156
New Hebrides arc 82
New Ireland see Papua Niugini
New Mexico see USA
New Seamount, East Pacific 27
New South Wales, Australia
  Airly Mountain 134
  Alum Mountains 122
  Armidale-Walcha region 186
  Bald Knobs **128**
  Banks, Mt 191, **map 192**
  Barrington 136, 174, **175**
  Barrington Tops **149**, 161
  Bathurst-Oberon area 134
  Belanglo 119
  Belmore, Mt 115
  Belougery Spire 115, 202, **202**, **map 219**
  Belougery Split Rock 115, 203, **map 219**
  Ben Lomond 147, **147**, 152
  Bevendale flow 133
  Bindook 123
  Bingara **135**, 165
  Blowhole, Kiama **120**, 121
  Blue Mountains 187, 188, 190–1, **map 192**
  Bluff Mountain 115
  Bombo 128, **180**
  Bondi 119
  Bondi dyke and pipe 118
  Bowral 175
  Box Hill, The 136
  Breadknife, The **98**, 115, **200**, 202, **map 219**
  Bress Peak 202–3, **map 219**
  Bridget Peak 202
  Broken Hill **46**, 47
  Bugaldi 142
  Burning Mountain 100, **101**
  Byrock 169
  Canobolas, Mt 111, **111**, 136, 170
  Cargellico 169
  Castlereagh River 133–4
  Cathedral Arch 203, **map 219**
  Comboyne 136, 172
  Condobolin 169

New South Wales continued
  Cooma 175
  Coonabarabran 152, 200, **map 216**, **map 219**
  Copeton 158, **158**, **159**
  Coricudgy, Mt 136
  Delegate 165
  Dromedary, Mt **36**, 118
  Dubbo 170, **map 171**, 175
  Ebor 136
  Ebor Falls 128, **39**
  Ebor-Dorrigo 38, **39**
  Emmaville 139
  Erskine Park 119
  Exmouth, Mt 203, **map 219**
  Flagstaff Hill 43, **44**
  future volcanism 186–7
  Garawilla volcanic field 122, 152
  Gibraltar, Mt 118
  Glen Innes 160, 161
  Gloucester 165
  Gloucester Buckets 122, **122**
  Good Hope **149**, **155**
  Grose River 191
  Gulgong 142
  Gunnedah 175
  Gunnedah Basin 122
  Gwydir River 136
  Hawkesbury River 191, **map 192**
  Hickey Falls 200
  Hogarth Range 161, **162**
  Hope, Mt 123
  Hornsby **114**, 119, **119**
  Hunter River 136
  Inverell **135**, 136, **157**, 160, 169, 170, **map 171**
  Irvine, Mt 191, **map 192**
  Kandos 175
  Kaputar 171, **171**
  Kayrunnera 165
  Kiama 119, **120**, 121, 128
  Kiandra 138
  Killarney Gap 171, **171**
  Kings Plain 160
  Kulnura pipe 147
  Kurrajong 190–1, **map 192**
  Lachlan fold belt 123, 156, 175, **179**
  Lachlan River 133
  Lightning Ridge 140
  Little Murray River 134
  Macha Tor 202, **map 219**
  MacIntyre River 136
  Macquarie River 134
  Maybole 136
  Merriwah 153
  Minchinbury 119
  Monaro 42, 43
  Monaro field 136
  Monte Christo Mine **135**
  Mopra Rock 200, **map 219**
  Murrumbidgee River **149**, **155**
  Nandewar 136, 170, 171, **171**, **map 171**
  Nepean River 191, **map 192**
  New England **156**, 160, 180
  New England Tablelands 136
  Newcastle 186, **187**, 188
  Oaky Creek diamond mine 158, **159**
  Oberon 160, 170–1, **map 171**
  Old Man Canobolas 111
  Orange 111, **111**
  Paddy's Plain flow 134
  Pomeroy flow 133
  Prospect 118
  Richmond 190, 191, **map 192**
  Royal Range, Mt 136
  Rylstone 175
  sapphire mining 47–8
  Siding Springs 203, **map 219**
  Southern Highlands 133, 136
  St Michaels Cave 119
  Sydney Basin sedementary beds 55
  Timor 134, 200
  Tomah, Mt 191, **map 192**
  Tooraweenah 200, **map 219**

New South Wales continued
  Tootie, Mt 191, **map 192**
  Towac, Mt 111
  Tumbarumba 161
  Unanderra 152
  Vegetable Creek 139
  volcanic rocks 43
  volcanic soils 42
  volcanic tours 200–3
  Walcha-Nundle district 134
  Warning, Mt 16, **114**, 115, **166**, 203
  Warrumbungles **98**, 115, 133, 136, 142, 170, 171, **map 171**, 200, 202–3, **map 216**, **map 219**
  Werris Creek 48
  Wheeo Creek 133
  Wheeo flow 133
  White Cliffs 165
  Wilson, Mt 191, **map 192**
  Windsor 191, **map 192**
  Wingen 100, **101**
  Wollondilly River 133
  Wollongong 187, 188
  Yerranderie 123
  Yetholme **179**
  zeolite mining 48
  zeolites 152–3
New Zealand
  Antipodes Islands 181, 189
  Chatham Island 181
  future volcanism 189
  volcano distribution 86, 87–8
New Zealand Geological Survey 49
New Zealand (North Island)
  Aratiatia Rapids 132
  artistic depictions of volcanoes 101
  Auckland 32, 33, 36, 62, 64, 181, 189, 210–11, **map 216**, **map 223**
  Broadlands-Ohaaki geothermal field 49, 88
  Cathedral Rocks **58**
  Coromandel Peninsula 189
  Crater Dome **front map**, 212, **map 222**
  Crater Hill 32, 210, **map 223**
  Crater Lake 62
  Domain, Auckland 211, **map 223**
  Echo Crater **front map**, 212, **map 222**
  Eden, Mt 210, 211, **212**, **map 223**
  Egmont, Mt **front map**, 8, 88, 189, 210
  Frying Pan Lake **58**, 212
  geothermal tourism 48–9
  Geyser Flat Sinter deposit 211, **map 222**
  Haroharo Caldera **front map**, 211, **map 222**
  Hikurangi trench 181
  Hotwater Creek **58**
  Hunua Ranges 62
  Inferno Crater **front map**, 212, **map 222**
  Kanakana Dome 211, **map 222**
  Kereru Geyser 211
  Mahanga Geyser 211
  Mamaku plateau 132
  Mangere, Mt 210, **map 223**
  Manurewa 210, **map 223**
  Maroa 88
  Matahina plateau 132
  Mayor Island 32, 189
  Motukorea Island 210, **map 223**
  Motutapu Island 210, **map 223**
  Ngamokaiakoko thermal mud pool 211, **map 222**
  Ngatura 36
  Ngauruhoe, Mt **front map**, 19, **86**, 87, **189**, 210
  Northland 180, 189
  Okataina **front map**, 88
  Okete 36, 181
  One Tree Hill 210, **map 223**
  Panmure Hill 32 also Panmure Basin **map 223**

New Zealand *continued*
  Parekohoru thermal spring 211
  Pink and White Terraces 212
  Pohutu Geyser 49, 211, **map 222**
  Prince of Wales' Feathers Geyser 211
  Puarenga thermal stream 211, **map 222**
  Puketutu 210
  Rangitoto Island **front map**, 32, 210, **map 223**
  Rerewhakaaitu, Lake 211
  Ridge Dome 211, **map 222**
  Rotoma geothermal field 88
  Rotomahana, Lake **front map**, 212, **map 222**
  Rotorua **front map**, 48–9, **49**, 88, 211–12, **map 216**, **map 222**
  Ruapehu, Mt **front map**, 17, 19, 22, 32, 62, 87, **87**, 189, 210, 225
  Ruawahia Dome **front map**, 212, **map 222**
  Southern Crater 212, **map 222**
  Steaming Cliffs 212, **map 222**
  Taranaki, Mt **front map**, 87, 88, 181
  Tarawera, Mt **front map**, 12, **14**, 15, 19, 60, 61, 88, 210, 211–12, 224
  Tarawera Dome 212, **map 222**
  Taupo 38, 87, 88, 189, 212
  Taupo, Lake **front map**, 59–60, 189, 224
  Te Tokaru Reef 210
  Te Wairoa buried village 212, **map 222**
  Three Kings 210, **map 223**
  Tongariro, Mt **front map**, 87
  Tongariro region 210, 212
  volcanic tours 210–13, **maps 222**
  Wahanga Dome 211, **map 222**
  Waiheke Island 62
  Waikarohini Geyser 211
  Waikato River 132
  Waikiti Geyser 211
  Waimangu **front map**, **58**, 61, 88, 212, **map 222**
  Wairakei 49, 88, 153
  Waitakere Range 62, **211**
  Wellington, Mt **front map**, 33, **map 223**
  Whakarewarewa **front map**, **40**, 49, 211, **212**, **map 222**
  Whangaehu River 17
  White Island **front map**, 19, **20**, 46, 63, 87, 88, 189, 210, 212
  zeolites 153
New Zealand (South Island)
  Akaroa 18, 207, 209, **map 222**
  Alpine Fault 37
  Aramoana 209, **map 222**
  Banks Peninsula 181, **207**, 207–9, **map 216**, **map 222**
  Big Rock 209, **map 222**
  Bradley, Mt 209, **map 222**
  Campbell Plateau 181
  Cargill, Mt 210, **map 222**
  Castle Rock 207, **map 222**
  Chalmers, Port 209, **map 222**
  Chatham Rise 181
  Christchurch 189, 207, 208, **map 222**
  Conical Hill 208, **map 222**
  Dunedin 18, 181, 189, 207, **208**, 209, **map 216**, **map 222**
  Dunedin Harbour 209–10, **map 222**
  Gibraltar Rock 208, **map 222**
  Herbert, Mt 209, **map 222**
  Holmes, Mt 210, **map 222**
  Kakanui South Head 207
  Kettle, Mt 210, **map 222**
  Lyttelton 18, 207, **207**, **map 222**
  Mihiwaka, Mt 210, **map 222**
  Mihiwaka South 210
  Moa-Bone Point Cave 207
  North Head, Otago Harbour **208**, 209, **map 222**
  Oamaru 207
  Otafelo Point 209
  Otago Harbour **208**, 209, **map 216**, **map 222**

New Zealand Sth Is *continued*
  Port Chalmers vent **208–9**, **map 222**
  Pulling Point 209
  Quail Island 208, **map 222**
  Rocky Point 209
  Shag Rock 207, **map 222**
  Solander Island 88
  Southern Alps 37
  Spit Beach, Otago Harbour **208**, 209, **map 222**
  Taylor Point 209
  Timaru 207
  volcanic tour 207–10, **map 222**
Newbery, J. Cosmos 126
newberyite 126
Newcastle, NSW 188
  earthquake 186, **187**
Ngamokaiakoko thermal mud pool, Nth Is, NZ 211, **map 222**
Ngatura, Nth Is, NZ 36
Ngauruhoe, Mt, Nth Is, NZ **front map**, 19, **86**, 87, 189, 210
Ngun Ngun, Mt, Qld 118, 204, **map 220**
Nicaragua 2
  Managua 2
  Masaya **front map**, 55
  Monotombo **front map**, 55
Ninety-East Ridge, Indian Ocean 176 *see also* 95
Ninole, Hawaii, USA 18
niobium 235
nitric acid aerosols 54
nitrous oxides 51
Nive River, Tas 128
Nive-Derwent River confluence, Tas 136
Noorat, Mt, Vic **102**, 103, 197, **map 217**
normal fault 11
normal magnetism 76, **76–7**
norms 146
North American plate 94
North Arm, Qld **178**, 179
north Atlantic ridge 75
north Atlantic rift **74**
North Beach, Lord Howe Island (NSW) **94**
North Head, Otago Harbour, Sth Is, NZ **208**, 209, **map 222**
North Ridge, Lord Howe Island (NSW) 201, **map 220**
Northern Ireland
  Giants Causeway 127
Northern Territory, Australia
  Coanjula 158
Northland, Nth Is, NZ 180, 189
Norway
  Bouvet Island 94
Nothofagus 138, 139
Novarupta, Alaska, USA **front map**, 225
nuée ardente 10, 235
Nulla, Qld 187, 188
Nulla field, Qld 103
Nut, The, Tas **112**, 113, 199, **199**, **map 218**
Nyamuragira, Zaire **front map**, 90, **90**
Nyiragongo, Zaire **front map**, 16, 90

Oahu, Hawaii, USA 92
Oakwood, Qld 187
Oaky Creek diamond mine, NSW **158**, **159**
Oamaru, Sth Is, NZ 207
obduction 71, 235
  sea floor 87
Oberon, NSW **map 171**, 160, 170–1
Obi Obi gorge, Maleny, Qld **178**
observatories and networks 60
obsidian 3, 45, 111, 146, 203, 235
ocean crust 74
ocean island volcano **79**
ocean-continent subductions 78, **79**, 83–6
oceanic island volcano **73**
oceanic oozes 74
oceanic subductions 78, 79, **79**, 82
oceanic-micro continent subduction 86–8

oceanite 154, 235
octahedral 235
octahedral crystals 157, 158
offretite 153, 227, 235
ogive 124, 235
Oguracaulis
  O. banksii 138
  fossil **137**
Ok Tedi, Papua Niugini 46–7
Okataina, Nth Is, NZ **front map**, 88
Okete, Nth Is, NZ 36, 181
Okinawa Trough, Japan 47
Okmok, Alaska, USA **front map**, 55
Old Faithful geyser, Yellowstone, Wyoming, USA 48
Old Gulch, Lord Howe Island (NSW) 201, **map 220**
Old Man Canobolas, NSW 111
Old Settlement Beach, Lord Howe Island, NSW **94**
Oldoinyo Lengai, Kenya **front map**, 91
olivine 79, 113, 122, 146, **149**, 150, 151, **151**, 155, 157, 161, **164**, 197, 226, 235
  cut **162**
  monticellite 226
olivine basalt 198
olivine crystals **144**, **146**
olivine dolerite 199
olivine lamproite 158
olivine nephelinite **149**, 198
olivine tholeiites 151
Olympus Mons, Mars 7
One Tree Hill, Nth Is, NZ 210, **map 223**
Ontake, Japan **front map**, 21
Ontong Java Plateau, Pacific ocean 95
opal 226
Orange, NSW 111, **111**
ores and mines 45–8
Organ Pipes National Park, Vic 127, 196, **map 217**
Orroroo, SA 176
Orroroo kimberlite dykes, SA 176
orthopyroxenes 148, 150, **151**, 162, **164**, 226, 235
  enstatite 226
  hypersthene 226
Oshima, Japan 55, **83**
Osmundacaulis 138
  O. jonesii 138
  O. nerii 138
Osorno, Chile **front map**, **85**
Osservatorio Vesuviano 67
Otafelo Point, Sth Is, NZ 209
Otago Harbour, Sth Is, NZ **208**, 209, **map 216**, **map 222**
Ouse River, Tas 128, 136
outliers and necks 113
Owen, R. 97
Owen Stanley ranges, Papua Niugini 87
Owenia 139
oxides 157
oxygen 51, 63
ozone 51, 54, 235
ozone layer 51, 54, **map 54**
ozone volcanoes 51, 54

Pacific plate **28**, 70, **71**, 82, 92, 94, 186
Pacific Rise 27
Pacific Volcanic Rim 26
Paddy's Plain flow, NSW 134
Pagalu, Gulf of Guinea 25
Pagan, Marianas Is **front map**, 55
Pago, Papua Niugini **front map**, 19, 79
pahoehoe lava **5**, 123, **124**, 210, 235
Palaeozoic era 235
palagonite 109, 236
pallid zone soils **42**, 43
Pallimnarchus 143
Paloona, Tas 135
Palorchestes 142
Panguna, Papua Niugini 47
Panmure Hill, Nth Is, NZ 32 also Panmure Basin **map 223**
Papandajan, Indonesia **front map**, 224

Papua Niugini
  Bam **front map**, 79
  Blanch Bay 80
  D'Entrecasteaux Islands 86, 87
  Giluwe, Mt 86
  Hagen, Mt 86
  Karkar **front map**, 60, 79
  Kobakaba 62
  Kombiu, Mt **80**
  Lamington, Mt **front map**, 86, 225
  Langila, New Britain **front map**, 79
  Lihir Island 46
  Lolobau, New Britain **front map**, 79
  Long Island **front map**, 55, 79
  Luise Caldera, New Ireland **front map**, 46
  Manam **front map**, **i**, 79
  New Britain 3, **map 216**
  Ok Tedi 46–7
  Owen Stanley ranges 87
  Pago **front map**, 19, 79
  Panguna 47
  Rabalankaia 80
  Rabaul, New Britain **front map**, 18, 19, 62, 79, 80–1, **80–1**, **map 216**, 225
  Ritter **front map**, 79
  Talasea region 3
  Tavurvur, New Britain **front map**, 80, **80**, 225
  Tuluman, Admiralty Islands **front map**, 22
  Uluwan, New Britain **front map**, 19, 55, 79
  Victory, Mt **front map**, 86
  volcano distribution 86–7
  Vulcan, New Britain **front map**, 80, **80**, 225
  Waiowa **front map**, 86
  Witori Caldera, New Britain **front map**, 19
  Woodlark Basin 28, 47
Parekohoru thermal spring, Nth Is, NZ 211
Parícutin, Mexico **front map**, 225
passenger minerals and rocks 157–65
Pathé, Mexico 49
Patoa, Hawaii, USA 49
Pauzhetka, Russia **front map**, 55
Pavlov, Alaska, USA **front map**, 55
Peleean eruption **13**
Pelée, Mt, Martinique (France) **front map**, 10, 55, 59, 111, 115, 224
Perboewatan, Indonesia **front map**, 23
Peregian, Mt, Qld 118
peridot 47, 161, **162**, 197, 236
peridotite 72, 74, 78, 119, 158, 161, **163**, **164**, 165
perlite 203
Permian period 56–7, 121, 152, **179**, 180, **180**, 236
Peru 84
  El Misti **front map**, 20
Peru-Chile trench 83
Petersen Cave, Qld 126
Petrifaction Bay, Flinders Island, Tas 138
petrified forests, Tas 138
petrologist 150, 236
phacolite 153, 236
phenocrysts **144**, 145, **146**, 154, 155, 161, 236
Philippine plate 82
Philippines
  Babuyan **front map**, 55
  Hibok-Hibok **front map**, 60
  Manila **33**
  Mayon **front map**, 20, 60, 225
  Pinatubo, Mt **front map**, 31, 50, **50**, 51, 52, 54, 55, 60, 63, 225
  Taal **front map**, 17, 60, 224
Phillip Island, Vic 153
phillipsite 152, 153, 227, 236
phonolite 66, 89, 111, 150, 151, 175, 209, 210, 236
phosphates 148

photography monitoring 61
phreatomagmatic activity 106
picrites 150
picritic alkali basalts 176
pillow 236
pillow lava **24**, 74, **74**, 75, 124, 199, **211**
pillow lava breccias **109**, 110, 124
Pin Gill Hill, Qld 206, **map 221**
Pinatubo, Mt, Philippines **front map**, 31, 50, **50**, 51, 52, 54, 55, 60, 63, 225
Pine Mountain, Qld 116, **map 221**
Pink and White Terraces, Nth Is, NZ 212
Pinnacle Rock, Qld 117, **154**
Pinwill Cave, Qld 126
pipes **11**, 119, **119**, **158**, 165, 236
 volcanic 203
Pisgah, Mt, Vic 106
Pit, The, Mt Eccles, Vic 197
pitchstones 146
plagioclase 151, 226, 236
plagioclase feldspars 33, 147, **148**, 148, 150, **151**, 161
plants *see* flora
plate 236
plate boundaries **71**
plate movement 70–2, **73**, **79**, **91**
plate tectonics 29, 69–97, 168
Pleasant, Mt, Qld 117
Pleiades, The, Antarctica **front map**, 189
Pleistocene epoch 236
Plenty flow, Vic 132
Plinian activity 176
Plinian eruptions 12, **13**, **14**, 236
Plinius the Elder 3
Plinius the Younger 3
Ploskiy Tolbachik, Russia **front map**, 225
plugs 115, 116, 117, 118, **119**, **166**, 188, 199, 206, 236
plumes 96, 172, 174, 184, 236
 hot rock 92
pluton **10**
Pohutu Geyser, Nth Is, NZ 49, 211, **map 222**
Policemans Knob, Qld 206
polonium 210. 37
polycarp 236
Pomeroy flow, NSW 133
Pompeii, Italy 3, 12, **14**, 66, **66–7**
Porndon, Mt, Vic 106, 111–12, 125, 197, **map 217**
porphyritic 236
porphyritic lavas 155
porphyritic rocks 154, **155**
Port Augusta, SA 176
Port Chalmers, Sth Is, NZ **208**, **map 222**
Port Fairy, Vic 193, 197
Port Phillip Bay, Vic **map 192**, 191, 193
Portarlington, Vic **map 192**, 191, 193
Portland, Cape, Tas 118, **118**
Portland, Vic 113, **map 217**
Portugal *see* Azores 92, 94
postage stamps 3, 6
potassium 72, 79
potassium 39. 34, 36
potassium 40. 34
potassium oxide 72
potassium-argon dating 34–6, 37
potassium-rich basalts 121
potassium-rich lavas 84, 85, 86, 87, 88, 155
Precambrian era 160, 165
predicting eruptions 63–5
prehistoric man 2–3
pressure ridges 125
primary magma 146
primary minerals 148
primitive lava 146
primitive magma 236
Prince of Wales' Feathers Geyser, Nth Is, NZ 211
Proserpine, Qld 177
Prospect, NSW 118
Protemnodon 142

proterozoic 236
Proterozoic quartzites **159**
Proterozoic volcanics 157
proto-actinium 231. 37
protons 32
Pseudomys 143
pseudo-subduction line 177–9
Puarenga thermal stream, Nth Is, NZ 211, **map 222**
Puketutu, Nth Is, NZ 210
Pulling Point, Sth Is, NZ 209
pulse 236
pumice 66, 146, 236
 mining 45
 rafts 28–9
Purrumbete, Lake, Vic 107, 108, 197, **map 217**
Pu'u 'O'o, Hawaii, USA **front map**, **vii**, **viii**, 4, **4**, 225
pyrite 41
pyroclastic 236
pyroclastic cones 102
pyroclastic deposits 12, 110, 111, 113, 116, 131, **179**, 197
pyroclastics 18
pyrope 236
pyrope-almandine 161
pyroxene 79, 147, **149**, **151**, 157, 163, 207, 226, 236
pyroxenites 119, 161, 163

Quail Island, Sth Is, NZ 208, **map 222**
quarries and quarrying 43–5, 119
quartz 33, 47, 72, 138, 148, **149**, 150, 152, **154**, 157, 208, 226
 chalcedony 226
 opal 226
quartz tholeiites 151, **151**
quartz trachyte 151
quartzite 160, 205
Quaternary period 236
Que River, Tas 123
Queen Victoria Building, Sydney 43
Queensland, Australia
 Aboriginal Cooking Cave, Binna Burra 203
 Agate Creek **126**, 127
 Allingham Creek 143
 Anakie 117, 160, 206
 Andrew, Mt **149**
 Anvils Peak 128, 129
 Arabanga 179
 Arch, The 126
 Atherton 184, 187, **map 216**, **map 221**
 Atherton Tablelands 136, 139, **182**
 Atherton volcanic field 188, 206, **map 221**
 Ball, Mt 117
 Ballanjui Falls, Binna Burra 203, **map 220**
 Baralaba **154**, 174
 Barkers Cave 126, **126**
 Barrine, Lake 99, 107, 206, **map 221**
 basalt fields 101
 Beerburrum, Mt 118, 203, **map 220**
 Beerwah, Mt 118, 204, **map 220**
 Biloela 129
 Binna Burra 203, **map 216**, **map 220**
 Blackall 187
 Blackwood, Mt 116, 117
 Blaggard Creek 143
 Blow Hole, Double Head **204**
 Bluff Point National Park 206, **map 221**
 Boat Mountain 140
 Bobby Dazzler Mine, Rubyvale **160**
 Brigooda **map 171**, 102, 161, 170–1, 187, **188**
 Brisbane 187
 Bromfield Swamp 107, 206, **map 221**
 Bruce, Mt **149**
 Bruce Highway 203–6
 Buckland 136, **136**
 Buckland shield **190**

Queensland *continued*
 Buckland volcanic field 188
 Bundaberg **102**, 103, 187
 Bundaberg-Gin Gin region 113
 Bunya Ranges 161
 Cairns 180
 Camp Hill Rock **map 221**
 Campbells Folly 115
 Capella 139
 Carnarvon Gorge 136, **136**
 Charleville 187
 Chinchilla 187
 Chudleigh 161, **162**, 187
 Clarke Hills 184, 187
 Clermont 206
 Cobbera, Mt **map 221**
 Conjuboy 184, 187
 Coochin, Mt 118, **map 220**
 Cooktown 184, 187
 Cooktown field 188
 Coolum, Mt 118
 Coonowrin, Mt 118, 204, **map 220**
 Cooran, Mt 118
 Cooroora, Mt 118
 Cooroy, Mt 118
 Coulston Lakes 102
 Dolphin Point 205
 Donna Bay 141
 Double Head 206, **map 221**
 Dumbell, Mt 117
 Eacham, Lake 99, **100**, 107, 206, **map 221**
 Egg Rock, Binna Burra 203
 Eungella **149**
 Euramoo, Lake 99, 107, **map 221**
 Expedition Range 134, 169
 Focal Peak 112, 115, 136, **166**, 172
 Four Mile Creek 134
 Fox, Mt 102, 134, 188
 future volcanism 187–8
 Gall Range 177
 Garnet Gully 161, **188**
 Gayndah 187
 Gillies Crater 206, **map 221**
 Gillies, Mt 115
 Gin Gin 187
 Gladstone 177
 Glass House Mountains 118, **119**, **202**, 203–5, **map 216**, **map 220**
 Granite Bay 205
 Green Hill 206, **map 221**
 Gympie 179
 Hay, Mt 127, 177, 206
 Headlow, Mt 116, **map 221**
 Hells Gates 205
 Hillsborough, Cape 116–17, **116–17**, 141, 168
 Hivesville **42**
 Hoy, Mt 117
 Hughenden 187
 Hummock, The **102**, 103, 113
 Hypipamee Crater **front map**, 8, **182**, **204**, 206, **map 221**
 Ingham 113
 Ipswich 174
 Ipswich-Brisbane area 139, 140
 Ironpot Mountain 116, **map 221**
 Jim Crow Mountain 116, **116**, **map 221**
 Johnstone River 134, **map 221**
 Juanobong 187
 Jukes, Mt 116, 117, **117**
 Kent Islands 179
 Kinrara 36
 Kweebani Cave, Binna Burra 203, **map 220**
 Lakeland Downs 160
 Lamington National Park 203, **map 220**
 Landsborough, Mt 113, **149**
 Lava Plains 160
 Leap, The 117
 LeBrun, Mt 102
 Lords Table Mountain 128, 129
 Lynchs Crater 107, **map 221**

Queensland *continued*
 Main Range **166**
 Malanda 136
 Maleny 204, 205
 Maryborough 179
 Maryborough Basin 177
 McBride, Mt 36
 McBride field **34**, 103, 126, **126**, 187, 188
 Miketeebumulgrai, Mt 118, **map 220**
 Milla Milla Falls 128, 206, **map 221**
 Millmerrin 170, **map 171**
 Millstream Falls 206, **map 221**
 Monto 129, 169
 Mornington Island 180
 Mungawoppa, Mt 116, **map 221**
 Mungore 179
 Murgon 140
 Nebo **149**
 Ngun Ngun, Mt 118, 204, **map 220**
 North Arm **178**, 179
 Nulla 103, 187, 188
 Oakwood 187
 Obi Obi gorge, Maleny **178**
 Peregian, Mt 118
 Petersen Cave 126
 Pin Gill Hill 206, **map 221**
 Pine Mountain 116, **map 221**
 Pinnacle Rock 117, **154**
 Pinwill Cave 126
 Pleasant, Mt 117
 Policemans Knob 206
 Proserpine 177
 Quincan, Mt 102, 206, **map 221**
 Ramsay, Mt **154**
 Ravenshoe 206, **map 221**
 Ridler, Mt 118
 Rockhampton 37, **154**, 168, 174, 177, 205, **map 216**, **map 221**
 Rosslyn Bay 116, 206, **map 221**
 Rubyvale 160
 Rugged, Mt 188, **190**
 sapphire mining 47–8
 Scholfield, Mt 117
 Sebastopol, Mt **154**
 Seven Sisters 102, 206, **map 221**
 Shaw Islands 179
 Ships Stern, Binna Burra 203, **map 220**
 Silver Plains 188
 Springsure 161, **168**, 206
 St Martin, Mt 113, 164, 188
 St Peter's Dome **168**
 Staircase Range 134
 Stevens Island 110
 Stone River 113
 Stony Range 113
 Stuart, Mt **179**, 180
 Sturgeon field 187
 Surat Basin 122
 Surprise, Mt 34, 36, 126
 Surprise Rock, Binna Burra 203
 Tamborine, Mt 127
 Tararan 113, **149**
 Taylors Cave 126, **126**
 Thursday Islands 180
 Tibberowuccum, Mt 118, **map 220**
 Tibrogargan, Mt 118, **202**, 204, **map 220**
 Tinbeerwah, Mt 118
 Tingammara 140
 Toomba 32, 103
 Toomba Flow 32
 Townsville **179**, 180
 Tunbubudla, Mt 118, 203, **map 220**
 Turtle Rock, Binna Burra 203
 Tweed 16, 136, **166**, 167, 172, 203
 Tweed Shield 115, **166**, 167
 Undara **34**, 36, 103, 126, **126**
 volcanic soils 42
 volcanic tours 203–6
 Walleroo 187
 Wedge Island **116**, 141
 Wheeler, Mt 116, 206, **map 221**
 Wild River 136

Queensland *continued*
   Wildhorse Mountain 118
   Windera **148**
   Wyangapini, Mt 161
   Yeppoon 205–6, **map 221**
   Yungaburra 206, **map 221**
   York, Cape 180, 188
Queensland seamount, Tasman Sea **map 169**, 170
Quincan, Mt, Qld 102, 206, **map 221**
Quipollornis koniberi 142

Rabalankaia, Papua Niugini **80**
Rabaul, New Britain, Papua Niugini **front map**, 18, 19, 62, **62**, 79, 80–1, **80**, **map 216**, 225
radial drainages 136
radial dykes **10**, 115
radioactive breakdown in isotopes 32, 34, 35
radio-carbon dating 32, 236
Rajmahal basalts 176
Rakata, Indonesia **front map**, **21**, 23
Ramsay, Mt, Qld **154**
Rangitaiki Valley, Nth Is, NZ 132
Rangitoto Island, Nth Is, NZ **front map**, 32, 210, **map 223**
Rano Raraku, Easter Island (Chile) **44**, 45
Raung, Indonesia **front map**, 224
Ravenshoe, Qld 206, **map 221**
Razorback, Mt, SA 198
Read, Mt, Tas 123
Red Point, Lord Howe Island (NSW) 201, **map 220**
Red Rock, Vic 108, 196, 197, **map 217**
Redoubt, Alaska, USA **front map**, 50, 63
Redpa, Tas 142, 153
Redruth, Vic 139
Rerewhakaaitu, Lake, Nth Is, NZ 211
Réunion Island (France) **front map**, 21–2, 94, 96
reverse fault **101**
reverse magnetism 76, **76–7**
Reynolds Island, Tas 136
Rhenish shield, Germany 89
rhyolite 72, 84, 85, 86, 87, 88, 89, 94, 95, 111, 116, 117, 118, 122, 123, 124, 146, **149**, 150, 151, 154, **154**, 156, 171, 175, 177, **178**, 179, 203, 204, 206, 208, 209, 211
rhyolite ash **14**
rhyolite tuffs 140
rhyolite volcanoes 181
Richmond, NSW 190, 191, **map 192**
Ridge Dome, Nth Is, NZ 211, **map 222**
Ridler, Mt, Qld 118
rift valleys **10**, 26
   volcanoes **73**, 79, 88–91
   wall **89**
rift zones **map 173**, 174
ring dykes **10**, 115, **166**, 236
Ringarooma River, Tas 129, **129**, 138
Rio Grande rift, New Mexico, USA 89
Ritter, Papua Niugini **front map**, 79
rivers
   buried and dammed 134–6
   displaced 132–4
Robbins Island, Tas 110
Robe, SA 186
Rockhampton, Qld 37, **154**, 168, 174, 177, 205, **map 216**, **map 221**
rocks, volcanic 43–4, 145–6, 150–7
   weathering 42–3
Rocky Point, Sth Is, NZ 209
Rokeby, Tas 199, **map 218**
ropy lava **124**
Rose Rivulet, Tas 134
Rosebery, Tas 123
Rosevears, Tas 128
Ross, Mt, Kerguelen (France) **front map**, 176
Ross Island *see* Antarctica
Ross Sea, Antarctica 61
Ross sea volcanoes, Antarctica 25
Rosslyn Bay, Qld 116, 206, **map 221**

Rotoma geothermal field, Nth Is, NZ 88
Rotomahana, Lake, Nth Is, NZ **front map**, 212, **map 222**
Rotorua, Nth Is, NZ **front map**, 48–9, **49**, 88, **map 216**
   volcanic tour 211–12, **map 222**
Rouse, Mt, Vic **133**, **map 217**
Rowsley fault, Vic **map 192**, 193
Royal Range, Mt, NSW 136
Ruapehu, Mt, Nth Is, NZ **front map**, 17, 19, 22, 32, 62, 87, **87**, 189, 210, 225
Ruawahia Dome, Nth Is, NZ **front map**, 212, **map 222**
rubidium 87. 35, 37 *also* Rb87: Rb86.35
rubidium-strontium dating 35, 37
ruby 47, 161
ruby crystals **161**
Rubyvale, Qld 160
Rugged, Mt, Qld 188, **190**
Ruiz, Colombia 55 *abbrv.* Nevado del Ruiz
Rumbles Seamounts, SW Pacific **front map**, 28, 189
Russia
   Aliad, Kurile Islands **front map**, 55
   Avachinsky **front map**, 55
   Bezymianny **front map**, 12, 225
   Kamchatka 49, 54
   Karymsky **front map**, 55
   Kizimin **front map**, 55
   Klyuchevskoy **front map**, 51, 55
   Kronotsky **front map**, 20
   Kurile Islands 82
   Pauzhetka **front map**, 55
   Ploskiy Tolbachik **front map**, 225
   Sheveluch **front map**, 55
   Siberia 2, 55, 57
   Tao-Rusyr, Kurile Islands **front map**, 55
rust 147
Rylstone, NSW 175
Ryukyu Island, Japan 82
   Kikai **front map**, 55

sag structures 107, 108
SAGE *see* Stratospheric Aerosol and Gas Experiment
Saint *see* St.
Sakurajima, Japan **front map**, 50
samarium 147. 35
samarium-neodymium dating 35
San Andreas fault, USA 180
San Sebastiano, Italy 67
sandstone 70, **114**, 118, **176**, 200, 205, 209
Sandy Bay, Tas 199, **map 218**
sanidine **155**, 226
Santa Maria, Guatemala **front map**, 224
Santorini, Greece **front map**, 3, 55, 224
Sao Miguel Island, Azores (Portugal) 49
Sao Tome, Gulf of Guinea 25
sapphires 47–8, 117, **156**, 157, 160–1, 206
sapphirine 161
Sarcopetalum 139
Sasha Seamount, East Pacific 27
Savai'i Island, Western Samoa 16
Schank, Mt, SA **front map**, **16**, 17, 32, 33, 100, 103, 113, **184**, 186, 198, **map 217**
schist 107, 206, 209
Scholfield, Mt, Qld 117
scoria 146, 198, 210, 236
scoria cones 102–3, 186, 196, 197, 198, 210
scoria deposits 43, 45
Scottsdale, Tas 153
scree 128
sea floor mapping 26, **26**
SeaMARC 236
seamount 236
seamount 6c, East Pacific **24**
seamount chains 95
   Indian Ocean 176
   Tasman Sea **map 173**
Searles Point, Lord Howe Island (NSW) 201, **map 220**

Sebastopol, Mt, Qld **154**
secondary minerals 145, 146–7
sedimentary rocks 70
seismic 236
seismic activity 62
seismic zones 188, **map 215**
Selfs Point, Tas 199, **map 218**
Selwyn's fault, Vic 191, **map 192**
serpentine mud 96
Seven Sisters, Qld 102, 206, **map 221**
Shadwell, Mt, Vic 197
Shaft, The, Mt Eccles, Vic 197
Shag Rock, Sth Is, NZ 207, **map 220**
shale 70, **114**, 118, **176**, 205, 236
Shaw Islands, Qld 179
sheet lava **75**
sheets 115, 116
Sheffield, Tas 135
shell fossils 142, **142**
Sheveluch, Russia **front map**, 55
shield volcanoes **13**, 18, 200, 203, 207, 236
Ships Stern, Binna Burra, Qld 203, **map 220**
Shishaldin, Alaska, USA **front map**, 20
shoshonite 155, **155**, 236
SHRIMP *see* Special High Resolution Ion Micro Probe
Siberia, Russia 2, 55, 57
siderite 227
Siding Springs, NSW 203, **map 219**
Sif Mons, Venus 7
silica 72, 122, **137**, 146, 151, 236
silica-rich lavas 84
silica-rich rocks 147
silica-satisfied basalts 150
silicate lavas 148
silicates 157, 161
silica-unsatisfied basalts 150
siliceous 236
silicified plants 138
silicon 70, 84
silicon dioxide 72
silks 160, 236
sills 115, 116, 118, 174, **176**, 205, 209, 236
Silver Plains, Qld 188
Skipton Cave, Vic 126
Skittleball Plains, Tas 136
skylights 125
slab 236
slate 206
slips 128
Smeaton Hill, Vic 196, **map 217**
Snake River basalt 94
Socompa, Chile 21
sodium 72
sodium oxide 72
Soela seamount, Tasman Sea 172
soils, volcanic 42–3, **42**
Solander Island, Sth Is, NZ 88
Solomon Islands **map 216**
   Billy Mitchell **front map**, 55
   Cook (submarine volcano) **front map**, 22
Solomon Sea plate 79
somma 236
Somma, Mt, Italy **front map**, **14**, 16, 66, 67
somma stage **123**, 16, **19**
somma volcanoes 18
South Australia
   basalt fields 101
   Blue Lake **front map**, **16**, 17, 109, 198, **map 217**
   Bluff, The 113
   Bootlace Cave, Mt Gambier 198
   Browne's Lake 17, 109, 198, **map 217**
   Burr, Mt, Range volcanic field 198
   Devil's Punchbowl 109, 198, **map 217**
   future volcanism 186
   Gambier, Mt **front map**, **167**, 17, 32, 100, 109, 113, **184**, 186, 197–8, **map 216**, **map 217**

South Australia *continued*
   Gawler Range 157
   Kangaroo Island 174
   Kingston 186
   Lake Terrace 197
   Leg of Mutton Lake 17, 109, 198, **map 217**
   Muirhead, Mt 100, 113
   Orroroo 176
   Orroroo kimberlite dykes 158
   Port Augusta 176
   Razorback, Mt 198
   Robe 186
   Schank, Mt **front map**, **16**, 17, 32, 33, 100, 103, 113, **184**, 186, 198, **map 217**
   Terowie 176
   Valley Lake 17, 109, 198, **map 217**
   volcanic tour 197–8, **map 217**
   xenoliths 165
South Esk River, Tas 134–5
South Sandwich Islands
   pumice raft dispersal **27**, 29
Southern Alps, Sth Is, NZ 37
Southern Climatic Oscillation 50
Southern Crater, Nth Is, NZ 212, **map 222**
Southern Highlands, NSW 133, 136
Southern Ocean magnetic pattern 76
Southern Ocean ridge 27, **29**, 70
Special High Resolution Ion Micro Probe (SHRIMP) **36**
sphene 38
spherulites 127, 177, 206, 236
spheruloids 127, 236
spindle-tailed volcanic bombs 199
spinel 157, 161, 163, 226, 236
   ulvospinel 226
spinel peridotite 72, 74
spinel peridotite mantle 163
Spit Beach, Otago Harbour, Sth Is, NZ **208**, 209, **map 222**
spreading 236
spreading ridges 75
spreading sea floors 72–8, 172
Springsure, Qld 161, **168**, 236
St Augustine, Alaska, USA 55
St Helena (UK) 94
St Helens, Mt, Washington, USA **front map**, 6, 10, 15, 16, 21, 52, 55, 59, 60, 61, 94, 111, 225
St Kilda, Vic 193
St Martin, Mt, Qld 113, 164, 188
St Michaels Cave, NSW 119
St Peter's Dome, Qld **168**
St Pierre, Martinique (France) 10, 59, 115, 224
St Vincent, La Soufrière **front map**, 224
Stabiae, Italy 3
Staircase Range, Qld 134
Stanley, Tas **112**, 113, 199, **199**, **map 218**
starting plumes 96
steam **12**
steam explosion vent 198
Steaming Cliffs, Nth Is, NZ 212, **map 222**
Steep Island, Tas 110
stellerite 152, 227, 236
Stevens Island, Qld 110
Stevens Point, Lord Howe Island (NSW) 201, **map 220**
stilbite 152, 227, 236
stocks 85, 237
Stone River, Qld 113
Stony Range, Qld 113
stony rises 43
stratosphere 50, 237
Stratospheric Aerosol and Gas Experiment instrument (SAGE) 51, 52, 236
stratovolcanoes **13**, 18, 86, 87, 174, 177, 237
stretching 186
Stromboli, Italy **front map**, 55, 102
Strombolian deposits 102, 103, 108, 110,

113, 237
Strombolian eruption **12**
Stromboli-style explosions 210
strontium 63, 171
strontium 87. 35, 37 *also*Sr87: Sr86.35
structures of volcanoes **10–11**
Stuart, Mt, Qld **179**, 180
Sturgeon field, Qld 187
Styx River, Tas 136
subduction 71, **73**, 180, 237
subduction zone volcanoes 78–88, **79**, 180
subductive andesite rocks 154–5
subglacial volcanoes 23
submarine eruptions 64
submarine ore deposits 47
submarine slides 21–2
submarine vents 26–9
submerged volcano 12
submersible craft **74**
subsidence formula for sea floor 76
sulphate acid aerosols 54
sulphide minerals 25, 148
sulphur 50, 54, 145, 226
 mining 45–6
Sulphur Banks, Hawaii, USA **46**
sulphur deposits
 Io 7
sulphur dioxide 52, 54, 61
 Io 7
sulphuric acid 50, 51, 52, 54
Sumatra *see* Indonesia
Sumbawa *see* Indonesia
summit crater 12, 237
superplumes 95
Surat Basin, Qld 122
surge 12, 237
Surprise, Lake, Vic 103, **194**, 197
Surprise, Mt, Qld 34, 36, 126
Surprise Rock, Binna Burra, Qld 203, **map 220**
Surtsey, Iceland **front map**, 22, 109
Surtseyian deposits 109
Surtseyian eruption **12**, 237
Sydney, NSW, future eruption scenario 190–1, **map 192**
Sydney Basin, NSW 55, 116, 118, 119–22, 172, 175, 187
syenite 70, 85, 115, 118, 154, 209, 237
syncline 122–3, 237
synclinorium 123, 237
synthetic rocks 162

Taal, Philippines **front map**, 17, 60, 224
Table Cape, Tas 113, 164, 198, 199, **map 218**
tacharanite 153, 227, 237
tachylyte 146
tadpole tectonics 96
Takahe, Mt, Antarctica **front map**, 23
Talasea region, Papua Niugini 3
talus 128, 237
Tamar River, Tas 138
Tamar Valley, Tas 128
Tambora, Indonesia **front map**, 20, 54, 55, 224
Tamborine, Mt, Qld 127
Tanzania 2
 Kilimanjaro, Mt **front map**, 8
 Laetoli 2
Tanzanian rift 89
Tao-Rusyr, Kurile Islands, Russia **front map**, 55
Tara Creek, Qld 143
Taranaki, Mt, Nth Is, NZ **front map**, 87, 88, 181
Tararan, Qld 113, **149**
Tarawera, Mt, Nth Is, NZ **front map**, 12, **14**, 15, 19, 60, 61, 88, 210, 211–12, 224
Tarawera Dome, Nth Is, NZ 212, **map 222**
TAS (Total Alkali and Silica) 237
Tasman fold belt, Tas 123, 156
Tasman Line of Fire 172–4, **map 173**

Tasman Sea **map 169**, **map 170**, 78, 94–5, 160, 164, 169, 172, 174
 future volcanism 185–6
Tasmania, Australia **121**
 Badger Corner, Flinders Island 138
 Bell Bay 128
 Blinking Billy Point 113, 199, **map 218**
 Blow, The **122**
 Blue Tier 38
 Boat Harbour 160
 Borrodaile Plains 135
 Bow Hill 163, 164, **164**
 Bradys Lookout **129**
 Breises tin mine **129**
 Brittons Swamp 142
 Burnie 127, 128, 198, **map 218**
 Bushy Park **130**
 Cam River 135
 Cameron West, Mt 128
 Cape Barren Island **184**
 Circular Head 199, **199**, **map 218**
 Claremont 199, **map 218**
 Cocked Hat Hill 135, 138
 Cygnet, Port 118, **155**
 Dee River 136
 Derby **129**
 Derwent River system **132**, 135–6, 138, 199, **map 218**
 Devonport 198, **map 216**, **map 218**
 Don Heads 198, **map 218**
 Duck Point Bay **134**
 emergent volcanoes 110
 Flat Topped Bluff 110
 Flinders Island 153, **184**, 185
 Petrifaction Bay 138
 Forth River 134
 fossil beds, Wynyard 198
 Fossil Bluff **141**, 141–2, **142**, 198, **map 218**
 Frankford Highway **176**
 future volcanism 185–6
 Gads Hill 134, 135
 Geilston Bay 141
 Glenora 138
 Great Lake 128, **134**, 136
 Grim, Cape **109**, 110, **110**
 Guildford courses 135
 Hellyer River 123, 138
 Hercules 123
 Hicks, Mt **map 218**, **132**
 Hobart **114**, 115, 199, **map 216**, **map 218**
 Lady Barron, Flinders Island **184**
 Launceston 134–5, **184**, **map 218**
 Liawenee 136
 Lorinna 134
 Lune River 115, **137**, 138
 Lyell, Mt **122**, 123
 Macquarie Plains 138
 Maggs Mountain 134
 Margate 199, **map 218**
 Marrawah 142
 Marrawah-Redpah district 110
 Mersey River 134
 Mersey-Forth Rivers 138
 Moina 135
 Nive River 128
 Nive-Derwent River confluence 136
 Nut, The **112**, 113, 199, **199**, **map 218**
 Ouse River 128, 136
 Paloona 135
 Portland, Cape 118, **118**
 Que River 123
 Read, Mt 123
 Redpa 142, 153
 Reynolds Island 136
 Ringarooma River 129, **129**, 138
 Robbins Island 110
 Rokeby 199, **map 218**
 Rose Rivulet 134
 Rosebery 123
 Rosevears 128
 Sandy Bay 199, **map 218**
 Scottsdale 153

Tasmania *continued*
 Selfs Point 199, **map 218**
 Sheffield 135
 Skittleball Plains 136
 South Esk River 134–5
 Stanley **112**, 113, 199, **199**, **map 218**
 Steep Island 110
 Styx River 136
 Table Cape 113, 164, 198, 199, **map 218**
 Tamar River 138
 Tamar Valley 128
 Tasman fold belt 156
 Trefoil Island 110
 volcanic tours 198–9, **map 218**
 Weldborough 38, 129
 Wellington, Mt **114**, 115, **map 218**
 Wynyard 141–2, **142**, 198, **map 218**
 zeolites 153
Tasmantid seamounts, Tasman Sea 169, **169**, 170
Taupo, Lake, Nth Is, NZ **front map**, 59–60, 189, 224
Taupo, Nth Is, NZ 32, 38, 88, 181, 189, 212
Tavurvur, New Britain, Papua Niugini **front map**, 80, **80**, 225
Taylor, G.A.M. 86
Taylor Point, Sth Is, NZ 209
Taylors Cave, Qld 126, **126**
Tazieff, Haroun 6, 16
Te Tokaru Reef, Nth Is, NZ 210, **map 223**
Te Wairoa buried village, Nth Is, NZ 212, **map 222**
tectonic plate 237
tectonism 69
Tengger, Indonesia **front map**, 19
Terowie, SA 176
terraces 128, **212**
Terror, Mt, Antarctica **175**
Tertiary period 128, 138, 139, 140–3, 152, 237
Tharsis Tholus, Mars 7
thermo-luminescent dating 33, 237
tholeiite 151, 174
tholeiitic 237
tholeiitic basalts 72, 75, 79, 82, 88, 95, 106, 150, **151**, 169, 170
tholeiitic olivine basalt 199
thomsonite 153, 227
thorite 237
thorium 73
thorium 230. 37
thorium 232. 35
thorium-lead dating 35
Three Kings, Nth Is, NZ 210, **map 222**
thunder eggs 48, 127, 177, 206, 237
Thursday Islands, Qld 180
Thurstons Lava Tube, Hawaii, USA 125
Thylacoleo 143
Tibberowuccum, Mt, Qld 118, **map 220**
Tibrogargan, Mt, Qld 118, **202**, 204, **map 220**
tilt meters 61
Timaru, Sth Is, NZ 207
Timor, NSW 134, 200
Timor trough 82
tin 129
Tinbeerwah, Mt, Qld 118
Tingammara, Qld 140
titanium oxides 148
titanium silicates 148
TM *see* Landsat Thematic Mapper per
Toba, Lake, Indonesia 20–1, 21, 54, 180
Tolai people 80
Tomah, Mt, NSW 191, **map 192**
tomography, seismic 62
TOMS *see* Total Ozone Mapping Spectrometer
Tonga
 Falcon Island **front map**, 22
 Lau Basin 47, 82
Tonga-Kermadec arc 82
Tongariro, Mt, Nth Is, NZ **front map**, 87

Tongariro region, Nth Is, NZ 210, 212
Toomba, Qld 32, 103
Toombullup, Vic 186
Tooraweenah, NSW 200, **map 219**
Tootie, Mt, NSW 191, **map 192**
Torres Strait 187, 188, **189**
Total Alkali and Silica (TAS) classification 146
Total Ozone Mapping Spectrometer (TOMS) 51, 52, 237
tourism in geothermal areas 48–9
tours, volcanic field 195–213
Towac, Mt, NSW 111
Towada, Japan **front map**, 55
Tower Hill, Vic 32, 103, **106**, 108, 197, **map 217**
'Tower Hill' (Guérard) 104
Townsville, Qld **179**, 180
Trachodon 56
trachyandesite 115, 158
trachyte 43, 72, 84, 85, 89, 92, 95, 111, 115, 116, 118, **149**, 154, **154**, 156, 175, 177, **179**, 188, **200**, 200, **202**, 202, 203, 207, 208, 209, 237
tracking eruptions 50–1, 52–3
Transantarctic Mountains, Antarctica 95, 115
transform faults 27, **76**, 237
Transit Hill, Lord Howe Island (NSW) 201, **map 220**
transitional basalts 150
travertine spring deposit 141
tree-fern fossil **137**, 138
Trefoil Island, Tas 110
trenches **79**, 237
 sea 78
Trentham Falls, Vic 128
Triassic period **114**, 174, **176**, **178**, 179, 180, 205, 237
Triceratops 56
Trichosurus 142
triple rift junctions 78
Tristan da Cunha (UK) 94
troposphere 50, 237
trout cod fossil **140**, 142
Tryggvason, H. **64**
tsunamis 64, 237
tuff rings 12, **13**, 18, 106–8, **107**, 196, 197, 210
tuffs 44, 106, **106**, 119, **119**, 177, 198, 199, **199**, 200, 201, 203, 207, 209, 210, 237
Tuhua, Nth Is, NZ *see* Mayor Island
Tuluman, Admiralty Islands, Papua Niugini **front map**, 22
Tumbarumba, NSW 161
tumuli 125, 237
Tunbubudla, Mt, Qld 118, 203, **map 220**
Tunnel Cave, Mt Eccles, Vic 197
tunnels 125–6
Turkana, Lake, Kenya 111
Turtle Rock, Binna Burra, Qld 203
tuya 23, 237
Tweed, Qld 16, 136, **166**, 167, 172, 203
Tweed Shield, Qld 115, **166**, 167
twin crystals 152–3
twinned drainage 134
twinning 237
Tyrannosaurus rex **56**
Tyrendarra lava flow, Vic 197

Uluwan, New Britain, Papua Niugini **front map**, 19, 55, 79
Una Una, Indonesia 55 synonym of Colo **front map**, 225
Unanderra, NSW 152
Undara, Qld **34**, 36, 103, 126, **126**
undercover crust 162–3
underwater vents 109
United States Geological Survey 59
Unzen, Mt, Japan **front map**, 6, **8**, 21, 60, 61, 63, 224–5
Uplands, Vic 186
upper colonnade 127

upper mantle 79, 163, **165**
upwelling 237
uranium 73, 148, 160, 161
uranium 235, 35, 37
uranium 238, 35, 37
uranium-lead dating 35, 37
uranopyrochlore 237
Uruguay, amethyst mining 48
USA
  Akutan, Alaska **front map**, 55
  Aleutian Islands, Alaska
  Aniakchak, Alaska **front map**, 55
  Augustine, Alaska **front map ii–iv**, 55
  Black Peak, Alaska **front map**, 55
  Cascades volcano observatory, Washington 60
  Craters of the Moon, Idaho 15
  Devil's Post Pile, California 127
  Geysers, The, California 49
  Grand Canyon, Arizona 129
  Great Rift Zone 15
  Haleakala, Hawaii **front map**, **15**
  Halemaumau, Hawaii 17, **60**, 126
  Hawaii 1, 8, 18–19, 21–2, 88, 92, 101
  Hualalai, Hawaii **front map**, 18
  Katmai, Mt, Alaska **front map**, 55
  Kilauea, Hawaii **front map**, 4–5, 18, **19**, **46**, **60**, 125
  Kohala, Hawaii 18
  Kuai, Hawaii 92
  Kupaianaha, Hawaii **front map**, 4, **5**, **124**
  Loihi (submerged), Hawaii **front map**, 18, 60, 92
  Maui, Hawaii **15**, 92
  Mauna Kea, Hawaii **front map**, 18
  Mauna Loa, Hawaii **front map**, 18, **19**
  Mazama, Oregon **front map**, 55
  Molokai, Hawaii 92
  Ninole, Hawaii 18
  Oahu, Hawaii 92
  Okmok, Alaska **front map**, 55
  Old Faithful geyser, Yellowstone, Wyoming 48
  Patoa, Hawaii 49
  Pavlov, Alaska **front map**, 55
  Pu'u 'O'o, Hawaii **front map, vii, viii**, 4, **4**, 225
  Redoubt, Alaska **front map**, 50, 63
  Rio Grande rift, New Mexico 89
  San Andreas fault 180
  Shishaldin, Alaska **front map**, 20
  St Augustine, Alaska 55
  St Helens, Mt, Washington **front map**, 6, **10**, 15, 16, 21, 52, 55, 59, 60, 61, 94, 111
  Sulphur Banks, Hawaii **46**
  Thurstons Lava Tube, Hawaii 125
  Yantarni, Alaska **front map**, 55
  Yellowstone, Wyoming **front map**, 94

Valle De Bove, Italy 21
Valley Lake, SA 17, 109, 198, **map 217**
Vanuatu
  Kuwae **front map**, 55
Vega Island, Antarctica 23
Vegetable Creek, NSW 139
vents 101–12, 210, 237
Venus 7
  Alpha Regio highlands 7
  Cleopatra crater 7
  Gula Mons 7
  Maat Mons 7
  Sif Mons 7
vermiculite 147, 227
vesicles 127, 237
vesicular 237
Vesuvius, Mt, Italy **front map**, 3, 12, **14**, 16, 55, 66–7, **67**, 101, 111, 224
Vexuvio space observatory 60
Victoria, Australia
  Alcove, The, Mt Eccles 197
  Alvie, Mt 108
  Anakie, Mt 161

Victoria continued
  Bacchus Marsh 142
  Ballarat 132, 139, 196, **map 217**
  basalt fields 101
  Bellarine Peninsula **map 192**, 191
  Berwick 131, 139
  Blowhard, Mt 106
  Bo'ok, Mortlake 100
  Bridgewater, Cape 113
  Brocks Monument 196
  Bullenmerri, Lake 108, 164, 197, **map 217**
  Bullenmerri Crater 37, 108
  Buninyong, Mt 196, **map 217**
  Byduk Caves 126
  Camels Hump 196, **map 217**
  Campaspe River 127
  Camperdown 197
  Colac 196–7, **map 217**
  Colac, Lake 108, 196
  Coode Island 193
  Corangamite, Lake 108, 196, 197
  Corio Bay **map 192**, 193
  Coragulal, Lake 108
  Cosgrove 169
  Daylesford 139
  Dog Rock 143
  Eccles, Mt **front map**, 32, 103, 125, 126, **194**, 197, **map 217**
  Elephant, Mt 8, **10**, 19, 102–3, 197, **map 217**
  future volcanism 186
  Geelong 143, **map 192**, 193, **map 217**
  Gisborne, Mt 106
  Gnotuk, Lake 108, 164, 197, **map 217**
  Gnotuk Crater 108
  Gothic Cave 126
  Grant, Cape 113
  Hamilton, Mt 132
  Hamilton fauna 142
  Hampden Shire Quarry **101**, **102**
  Hanging Rock (Mt Diogenes) **30**, 37, 196, **map 217**
  Harman Valley Flow 32
  Harman Valley flow 32, 125
  Hobsons Bay **map 192**, 193
  Hollowback, Mt 106
  Keilambete, Lake 108
  Kooroocheang, Mt 125, **125**, **map 217**
  Lady Julia Percy Island 22, 110, 184, **map 217**
  lava volcanoes 106
  Lawaluk 106
  Leura, Mt 100, **106**, 108, 117, **163**, 197, **map 217**
  Macarthur 197
  Macedon 168, **168**
  Macedon, Mt **map 217**
  Mondilibi 106
  Mornington Peninsula 193
  Mortlake 197, **map 217**
  Mt Eccles 32, **map 217**
  Mt Eccles National Park **194**, 195, 197
  Mt Elephant 8, **10**, 19, **map 217**
  Mt Emu Creek 132
  Napier, Mt 103, 126
  Natural Bridge, Mt Eccles 197
  Nelson, Cape 113
  Noorat, Mt **102**, 103, 197, **map 217**
  Organ Pipes National Park 127, 196, **map 217**
  Phillip Island 153
  Pisgah, Mt 106
  Pit, The, Mt Eccles 197
  Plenty flow 132
  Porndon, Mt 106, 111–12, 125, 197, **map 217**
  Port Fairy 193, 197
  Port Phillip Bay 191, **map 192**, 193
  Portarlington 191, **map 192**, 193
  Portland 113, **map 217**
  Purrumbete, Lake 107, 108, 197, **map 217**

Victoria continued
  Red Rock 108, 196, 197, **map 217**
  Redruth 139
  Rouse **133**
  Rouse, Mt **133**, **map 217**
  Rowsley fault **map 192**, 193
  Selwyn's fault 191, **map 192**
  Shadwell, Mt 197
  Shaft, The, Mt Eccles 197
  Skipton Cave 126
  Smeaton Hill 196, **map 217**
  St Kilda 193
  Surprise, Lake 103, **194**, 195, 197
  Toombullup 186
  Tower Hill 32, 103, **106**, 108, 197, **map 217**
  Trentham Falls 128
  Tunnel Cave, Mt Eccles 197
  Tyrendarra lava flow 197
  Uplands 186
  volcanic tour 196–7, **map 216**, **map 217**
  Wannon River 133
  Warnambool 197, **map 217**
  Werribee plains 106
  Western Plains 43
  Woodend 196
  zeolites 153
Victory, Mt, Papua Niugini **front map**, 86
Villarrica, Chile **front map**, **84**, **85**
viscous 237
Voight, B. 65
Volcan de las Tres Virgines, Mexico **front map**, 84–5
volcanic bombs **86**, **101**, 102, 103, 163, **163**, 197, 199, 206, 209, 237
volcanic boomerangs 169–72, **maps 170**, 184
volcanic features, Australasian region **map 216**
volcanic trends, Eastern Australia **map 215**
'Volcano, Town and Western Plains' (Docking) **105**
volcanoes defined 237
volcano distribution **front map**, 25–9
volcano types 18–19
volcanologists 3
volcanology 237
Vulcan, New Britain, Papua Niugini **front map**, 80, **80**, 225
vulcanian 237
Vulcanian deposits 103, 110
vulcanian eruption **123**
Vulcano, Italy **front map**, 55, 102

Wahanga Dome, Nth Is, NZ 211, **map 222**
Waiheke Island, Nth Is, NZ 62
Waikarohini Geyser, Nth Is, NZ 211
Waikato River, Nth Is, NZ 132
Waikiti Geyser, Nth Is, NZ 211
Waimangu, Nth Is, NZ **front map**, **58**, 61, 88, 212, **map 222**
Waiowa, Papua Niugini **front map**, 86
Wairakei, Nth Is, NZ 49, 88, 153
wairakite 153
Waitakere Range, Nth Is, NZ 62, **211**
Walcha-Nundle district, NSW 134
Walleroo, Qld 187
Wannon River, Vic 133
Warning, Mt, NSW 16, **114**, 115, **166**, 167, 203
Warnambool, Vic 197, **map 217**
Warrumbungles, NSW **98**, 115, 133, 136, 142, 170, 171, **map 171**, 200, 202–3, **map 216**, **map 219**
Washington see USA
water vapour 51
weathering of volcanic rocks 42–3
Wedge Island, Qld **116**, 141
Weijermers, R. 97
Weldborough, Tas 38, 129
Wellington, Mt, Nth Is, NZ **front map**, 32, **map 223**

Wellington, Mt, Tas **114**, 115, **map 218**
Werribee plains, Vic 106
Werris Creek, NSW 48
Western Australia
  Archaean greenstones **156**, 157
  Argyle diamond mine 47, **47**
  Argyle pipe 47, 157, 158, **159**, 163, 165
  Bunbury basalt 176
  Carnarvon Basin 176
  Fitzroy trough 189
  Kambalda **156**, 157
  Kimberley 189
Western Samoa
  Matavanu **front map**, 16
  Savai'i Island 16
wetspots 92
Weymouth, Cape, Qld 180
Whakamaru ignimbrites 38
Whakarewarewa, Nth Is, NZ **front map**, **40**, 49, 211, **212**, **map 222**
Whangaehu River, Nth Is, NZ 17
Wheeler, Mt, Qld 116, 206, **map 221**
Wheeo Creek, NSW 133
Wheeo flow, NSW 133
White Cave, Binna Burra, Qld **202**, 203, **map 220**
White Cliffs, NSW 165
White Island, Nth Is, NZ **front map**, 19, **20**, 46, 63, 87, 88, 189, 210, 212
Whitsunday Islands, Qld 177, 179
Wild River, Qld 136
Wildhorse Mountain, Qld 118
Wilson, Mt, NSW 191, **map 192**
Windera, Qld 148
windows 125
Windsor, NSW 191, **map 192**
Wingen, NSW 100, **101**
within-plate 237
within-plate volcanoes 88–95, 180
Witori Caldera, New Britain, Papua Niugini **front map**, 19
Wollondilly River, NSW 133
Wollongong, NSW 187, 188
Woodend, Vic 196
Woodlark Basin, Papua Niugini 28, 47
Wyangapini, Mt, Qld 161
Wyer Island, Australia 108, **189**
Wynyard, Tas 141–2, **142**, 198, **map 218**
Wynyardia bassiana 141–2
Wyoming see USA

xenocrysts 145, 157, 161, 237
xenoliths 145, 162–3, **163**, 164–5, 237

Yantarni, Alaska, USA **front map**, 55
Yellowstone, Wyoming, USA **front map**, 94
Yeppoon, Qld 205–6, **map 221**
Yerranderie, NSW 123
Yetholme, NSW **179**
Yungaburra, Qld 206, **map 221**
York, Cape, Qld 180, 188
Yucatan, Mexico 57

Zaire
  Nyamuragira **front map**, 90, **90**
  Nyiragongo 16, 90
Zeehan seamount, Tasman Sea 172
zeolites 122, 127, 147, **147**, 152–3, 227, 237
  and gemstones 47–8
zircon 35, **36**, 37, 38, 148, **148**, 157, **157**, 160, 161, 170, **175**, 186, 226, 237
zirconium silicate 148
Zygomaturus 143